Energy Systems Engineering Handbook

Energy Systems Engineering Handbook

Henry Oman

ENGR
621.4
Om15e

Prentice-Hall, Inc.
Englewood Cliffs, New Jersey 07632

Prentice-Hall International, Inc., *London*
Prentice-Hall of Australia, Pty. Ltd., *Sydney*
Prentice-Hall Canada, Inc., *Toronto*
Prentice-Hall of India Private Ltd., *New Delhi*
Prentice-Hall of Japan, Inc., *Tokyo*
Prentice-Hall of Southeast Asia Pte. Ltd., *Singapore*
Whitehall Books, Ltd., Wellington, *New Zealand*
Editora Prentice-Hall do Brasil Ltda., *Rio de Janeiro*
Prentice-Hall Hispanoamericana, S.A., *Mexico*

© 1986 by

PRENTICE-HALL, INC.

Englewood Cliffs, N.J.

*All rights reserved. No part of this
book may be reproduced in any form or
by any means, without permission in
writing from the publisher.*

Library of Congress Cataloging-in-Publication Data

Oman, Henry, 1918–
   Energy systems engineering handbook.

   Includes index.
   1. Power (Mechanics)   2. Power resources.   I. Title.
TJ163.9.O44   1986   621.4   85-28233

ISBN 0-13-277294-9

Printed in the United States of America

Introduction

A comparison of energy and power sources is made complicated by many factors, the key ones being differing units of measure, efficiency of conversion, and costs that change continuously. Units of measure vary from cubic feet for natural gas to langleys of received solar radiation. Efficiency of conversion by heat engines depends on temperature, with higher temperatures bringing better efficiency as well as problems. Causes of changing costs range from technical breakthroughs such as copper-indium-selenide solar cells to varying rates of interest. The objective of this book is to give the professional engineer the understanding, techniques, and basic data for making valid comparisons of diverse energy and power projects, ranging from engines that propel ships to the extraction of energy from garbage landfills.

It is interesting how seemingly good comparisons lose their validity as time passes. For example, in nuclear versus coal power generation comparisons in the early 1970s, nuclear power generation was a clear winner. Public utilities that built plants promptly have for the next thirty years benefitted from gold mines of low-cost power. Those that were slow in building plants had to respond to endless changes in government regulations and a doubling of interest cost. Going even further into history, Grand Coulee Dam, completed in 1943, is one of the lowest-cost producers of electricity in the world. However, if Grand Coulee Dam had been designed in the 1980s and was to be built in the 1990s, it couldn't compete with coal or nuclear power plants.

As you read these words, many of the contemporary published energy comparisons have already become obsolete! For example, if copper-indium-selenide solar cells are introduced at one-tenth the price of silicon solar cells, all previous solar-power cost predictions become invalid. Thus, when you, as an engineer, are assigned to make a power-energy comparison, you would do well to make your own comparison with up-to-date data, rather than relying on someone else's published work.

I have made hundreds of such comparisons—or "trade studies"—ranging from deep earth to Mars. The deep-earth study was for a military installation that had to survive for over a year with no outside resources for cooling, combustion air, or exhaust disposition. The Mars-headed spacecraft had a choice of solar-cell arrays or a radioisotope-heated thermoelectric generator.

To make energy and power comparisons, you need three things—technology, up-to-the-minute costs, and a technique. By technology, we mean the development status and performance of an energy process. For example, to convert sunlight to electricity you can use silicon solar cells, made by many firms, with efficiency ranging from 10 to 16%, depending on what price you want to pay. This technology is developmentally mature. In contrast, a solar coal-gasification experiment deposited some 40 percent of the incoming energy into the product gas, and 70 percent efficiency can be predicted. However, no factory is producing the apparatus for this process, and cost estimates are speculative. Thus, the technology for this solar gas process is immature. To help you on technology, this book will provide the basic characteristics in the important energy sources and conversion equipment, along with references to further data.

Some costs are easy to find. The *Wall Street Journal* publishes the prices of petroleum products, some chemical reactants, and metals. Manufacturers happily quote prices on their catalog products. Estimates in this book can be used if they are adjusted for subsequent inflation. On the difficult-to-get end of the scale are nuclear power plant costs. Estimating the cost of such a plant is a task for an experienced consultant who charges a fee. I can only provide the costs of representative completed power plants.

The technique for making energy and power comparisons is the basic theme of this book. The alternatives being compared must be as equivalent as possible, particularly in terms of output and reliability for meeting common requirements. This is not a trivial task when oil is sold in barrels, coal in tons, and uranium enrichment in separation work units. Achieving equal reliability also is complex. For example, a few extra solar-cell strings and batteries added to a solar power plant can give 0.9999 power availability. For a remote mountain-top repeater station, the alternative diesel engine plant, to be equivalent in reliability, would consist of several engines having 1,000-hour mean-time-to-failure ratings, plus automatic switchover. If continuous power is required, several uninterruptible power units would be added, plus scheduled service and provision for prompt repair of failures, all of which enter into life-cycle cost comparisons.

Introduction

In this book, energy and power sources are described in sufficient quantitative detail so that professional readers will understand their design and operation. Design data for conventional equipment and plants is not provided—it is available in handbooks. Chapters are grouped into the areas of energy systems engineering, heat engines, energy sources, water power, nuclear power, solar power, energy storage, electrochemical energy conversion, economics, moving energy, transporting people and freight, and heating buildings. Related sources and conversion are treated together—for example, water turbines that relate only to hydro power sources.

Henry Oman

Acknowledgments

This book was developed from lecture notes for a "Modern Energy Systems" class at The Boeing Company. Our objective was to enable a professional engineer to qualify for power and energy assignments at Boeing. Such assignments varied from energy management systems in public utilities to photovoltaic-versus-Brayton-cycle trades in solar power satellites that would generate 10 gigawatts. Two other instructors, Charles J. Bishop and Joseph W. Gelzer, participated in teaching this course at different times. I appreciate the contributions they made to this work.

Many colleagues at Boeing and in technical societies have generously contributed time, ideas, data, and critiques to this book. I am grateful to them and I have acknowledged their contributions in the text. The following have been especially helpful.

E. Richard Brown	Boeing Aerospace Company
Clark Dodge	Washington State Ferries
Sid Gross	Boeing Aerospace Company
Ronald D. Hamrick	Alaska State Ferries
William Hesse	Consultant
Paul Juhaz	Washington State Energy Office
Roy Holub	University of Wisconsin
Frank Minteer	Boeing Aerospace Company

Steven E. Moloney	Consultant
Jerry Morey	Morbark Industries
Len J. Nuttall	United Technologies
William B. Oaks	Boeing Aerospace Company
Richard Shaw	Boeing Aerospace Company
Howard Steele	Boeing Aerospace Company
Ron Stepien	General Electric Company
W. J. R. Thomas	Rolls Royce Ltd.
Joseph Szymborski	GNB Batteries, Inc.

I am especially appreciative of the patience, help, and encouragement given by my wife, Earlene M. Oman, during the two years that went into writing this book.

Contents

INTRODUCTION iii

ACKNOWLEDGMENTS ix

1 ENERGY SYSTEMS ENGINEERING: A TECHNIQUE FOR EVALUATING ALTERNATIVES 1

How Deep Do We Go? 2
Setting Up the Energy Systems Engineering Evaluation 4
Requirements are Real, Derived, and Assumed 4
Postulating Alternatives, Even Strange Ones 6
How Much Preliminary Design Do We Need? 7
Life-Cycle Cost—The Key to Energy Systems Engineering Computations 9
Ranking of Alternatives 12
Evaluation of Elements That Defy Quantification 13
Reliability Predictions and Calculations 15
The Strange Units That Measure Energy 17

2 ENGINES THAT MAKE POWER FROM HEAT 19

Can You Beat Carnot-Cycle Efficiency? 21
The Perfect Engine, Invented by a Nineteenth-Century Minister 23

Rankine Cycle, the Steam Engine, and Turbine 28
Diesel Cycle That Propels Trucks, Buses, Trains, and Ships 36
Otto Cycle Engines for Cars, Mowers, Boats, and Chain Saws 44
The Gas Turbines—New Techniques for More Power, Higher Efficiency 47
Peaking Power Plants 58
Combined Cycles: The Route to 50 Percent Efficiency 60

3 WORLD ENERGY SOURCES THAT WILL DISAPPEAR IN THE 2000's 66

Coal and How to Enrich It 67
Petroleum, a Dwindling Resource 75
Natural Gas: From Wells and Artificial Sources 88
Natural Gas Economics 94
Synthetic Fuels from Coal 95

4 RENEWABLE POWER FROM WATER 97

Hydro Power Cost Is Interest 106
Medium Size Hydro Plant—Life Cycle Cost 107
Pumped Hydro: A Big Rechargeable Battery 111
Power from Tides and Ocean Thermal Gradients 117
Geothermal: Free Heat from the Earth 119
Geothermal Resources 120

5 NUCLEAR FISSION—PRODUCES NO ACID RAIN, CO_2, OR NOX 130

Nuclear Power Plant Configurations 134
High Temperature Gas Cooled Reactor for High Efficiency 138
Breeder Reactors and Fusion Power 149

6 SOLAR POWER: RENEWABLE BUT WATCH THE COST 151

Quantity and Quality of Sunshine 152
Cost of Storing Solar Energy 158
Solar Cells, for Milliwatts to Megawatts 159
Solar Cells for Spacecraft 167

Design of Solar Arrays for Spacecraft *169*
Terrestrial Solar Array Design *170*
Solar versus Coal for Power Generation *174*
Solar Heat Collection for Buildings and Water Heating *176*
Solar Heated Engines *178*
The Future of Solar Energy *180*
Wind Power: Derived from Sunlight *183*
Windmills on the Farm *184*
Wind Turbine Power Output *185*
Cost of Wind Power *188*
Small Wind Turbines *188*
Gas Turbines: The Alternative to Wind Power *188*

7 STORING ENERGY FOR NIGHTS AND DAYS WITH NO SUN OR WIND 192

Storage Batteries *193*
Lead Acid Batteries *196*
Megajoules in Superconducting Magnetic Storage *211*
Flywheels for Short- and Long-Term Storage *212*

8 ELECTROCHEMICAL CONVERSION AVOIDS CARNOT LIMITATIONS 215

Primary Batteries: Old and New *216*
Cells that Make Power from Fuels *223*

9 ECONOMICS, THE FINAL DECISION IN COMPARISONS 229

What Determines Price? *231*
Supply and Demand Economics *232*
Elasticity Relates Price and Quantity *235*
Interest and Future Expenses *237*
Learning Curves for Multiple Units *239*
Economic Opportunities from Power Grids *240*

10 MOVING ENERGY IN WIRES, PIPES, AND SHIPS 242

Electric Power Transmission 242
Example of High-Availability Transmission Network 247
Hauling Energy in Pipes and With Vehicles 252
Sending Oil through Pipelines 253
Oil Tankers for Efficient Transportation of Energy 256
Tanker Cost 257
Natural Gas Pipelines and Ships 258

11 MOVING PEOPLE WITH LESS ENERGY 260

Passenger Ships 264
Ferries for Less Luxurious Travel 267
Ferry Power Plants: Mostly Diesel Engines 268
Hydrofoils for Passengers 272
Buses Carry People Efficiently Wherever Roads Go 275
Airplanes that Fly Passengers and Freight 277
Hauling Freight—By Airplane or Ship? 285
Bicycles, Sailplanes, and Sunpower 286
Sailplanes for Fast Transportation without Fuel 288
Solar Powered Airplanes and Cars Work 289
Routes to Energy Efficient Cars 289
Reducing the Energy Required by the Car 293
Electric Automobiles 295
Conserving Energy in Transportation 298

12 MOVING FREIGHT WITH SHIPS, TRAINS, AND TRUCKS 300

Hauling Cargo with Ships 300
Economics of Ship Operation 305
Trucks for Highway Freight 310
Costs of Trucks and Trucking 312

13 HEATING BUILDINGS WITH LESS ENERGY 314

Measurement of Heat for Buildings 316
Conserving Heating Energy 316
Heating Strategies 317

Calculating Heat Loss from a Building *320*
Optimizing Insulation Level *322*
Energy Heat Sources for Buildings *327*
Heat from High-Temperature Sources *327*
Firewood *328*
Low-Temperature Heat Sources for Buildings *329*
Gas Turbine Cogeneration *335*
Pumps That Move Heat *335*

14 NON-FOSSIL FUELS: HYDROGEN, ALCOHOL, AND WOOD 341

Hydrogen: A Good Fuel but Hard to Store *341*
Methanol and Ethanol for Fuels *346*
Alcohol from Agriculture Products *347*
Burning Wood for Power Generation *349*
Fuel from Forest Waste *350*

INDEX 357

Energy Systems Engineering Handbook

1

Energy Systems Engineering: A Technique for Evaluating Alternatives

When we wander through the nation's junkyard of energy projects, we find three causes for failure—equipment that didn't perform the way it was expected to perform, changes in economic climate, and energy system engineering breakdowns.

Upon entering the Centralia, Washington, steam plant on a tour, I was surprised to learn that our guide was the superintendent of the plant. On previous visits, the guide was a foreman. We soon learned why. Spread over the generator room floor were parts of two 700-megawatt (MW) turbine generators. One turbine had a blade failure, and the inspector had found an imminent failure in the second turbine. In contemplating a six-month shutdown of a power plant that normally generates almost a million dollars worth of power a day, the superintendent had concerns about his job security. Reports on design failures of energy-processing equipment have been too numerous, but they are outside the scope of this book.

A changing economic climate trapped the Washington Public Power Supply System (WPPSS), which started its plants four and five at a time of growing power demand, low interest rates, high employment, moderate inflation, and a ready market for tax-exempt bonds. Every one of these factors changed, and the two plants had to be cancelled after $2.2 billion had been spent on engineering, purchasing machinery to beat inflation, and some construction.

Systems-engineering failures often are rationalized to other causes. Power generation from low-temperature geothermal wells is an example. Assigned causes of power generation failure include failure of equipment to perform with needed efficiency or scale-forming dissolved minerals in the geothermal fluid. Not mentioned is the simple fact that on a life-cycle cost basis a coal-burning power plant would have produced electricity at lower cost.

Energy systems engineering is a discipline for evaluating alternatives in energy sources, processes, and uses. A simple example is comparing the cost of alternative fuels. Heat from oil sold by the gallon can be compared with heat from natural gas sold by the thermal unit by converting both costs to cents per 100,000 Btu. On the other end of the complexity scale is evaluation of a public utility's options for supplying higher loads ten years in the future.

HOW DEEP DO WE GO?

Some problems are worth zero depth. An example is the energy used in entertainment. The promoters of the Memorial Day auto race at Indianapolis are not much interested in a serious analysis showing how energy could be used more effectively in their race cars. Similarly, serious energy systems engineering probably is not warranted in activity having goals measured in advertising or publicity success.

The possible depths of energy systems engineering analyses covers a wide range. One extreme was a study of batteries on the customer's side of the electric power meter, done by Walter J. Stolte and Stevem W. Eckroad of Bechtel for the Electric Power Research Institute.[1] A man-year of engineering plus computer support went into this sophisticated analysis that evaluated all costs and sensitivities for batteries that reduce the power consumer's demand charge, which can be around $4 to $15 per kW. Cost of batteries, peak-load duration, and demand charge were important factors. Battery life, balance-of-plant cost, and inventory cost turned out to be much less important. The discounted cost of replacing batteries was so small that the best scenario was to discharge the batteries deeply and sacrifice life. Figure 1-1 shows some of the many sensitivity plots that they produced.

Someone who has a load-leveling opportunity might find that the most practical route to valid results is to commission Bechtel to enter new conditions into the computer model and generate a new set of sensitivity plots.

At the other end of the depth-of-study is this problem: Does it pay to generate electric power by stoking the boiler with wood costing $90 a cord? A simple calculation will show that a 4,000-lb. cord of dry wood will deliver around 5,500 Btus of heat per pound, which corresponds to 6,448 kWh of heat per cord.

---

[1]Stolte, W. J., and Eckroad, S. W. "Feasibility Assessment of Customer-Side-of-the-Meter Applications for Battery Energy Storage." Report EM-2769, Electric Power Research Institute (Palo Alto, CA), pp. S4–S6.

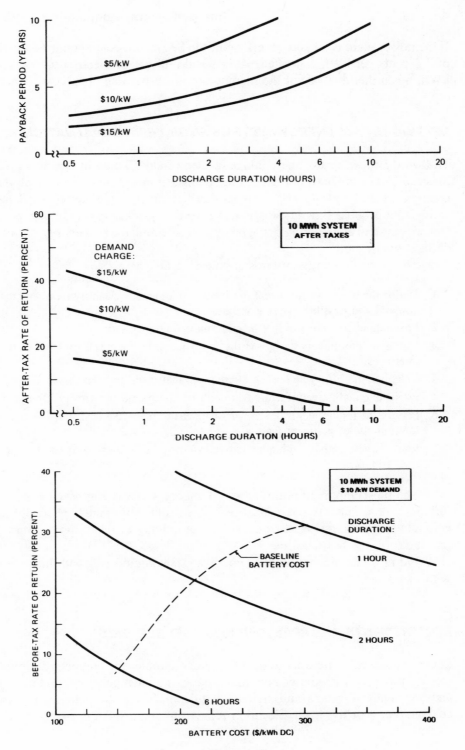

Figure 1-1

If the power plant is 30 percent efficient, the $90 a cord corresponds to a fuel cost of 4.6 cents per kWh generated. If the local utility sells electricity for 4 cents a kWh, no further analysis of the many other wood-burning plant costs is needed.

SETTING UP THE ENERGY SYSTEMS ENGINEERING EVALUATION

The word *system* means many things in many fields. To a physician, "system" means a part of the body. To an electrochemist, it means an electricity-producing reaction. In public-utility work, it means an electrically integrated power-handling entity. In this book, I avoid the use of the word "system" except in one meaning, "energy system engineering," the process of selecting the best energy source for a defined application.

The steps in energy systems engineering are these:

1. Define the requirements—real, derived, and assumed. Quantify requirements with nominal and possible extreme values.
2. Postulate alternative ways of meeting the requirements.
3. Configure preliminary designs to the depth necessary for cost estimating, and define interfaces.
4. Calculate the life-cycle cost of alternatives that meet the requirements.
5. Evaluate quantitatively factors that can't be interpreted in terms of life-cycle cost.
6. Perturb the requirements to see what happens to life-cycle cost. Explore upper and lower limits of pertinent factors.
7. Rank the alternatives with quantitative evaluation of life-cycle cost and other pertinent factors.

The most important contribution of energy systems engineering is the perturbing of requirements and evaluation of the results. With the support of modern computers, the energy systems engineer can quickly evaluate many alternatives and focus on the best choices.

We now examine the steps in energy system engineering and illustrate them with examples.

REQUIREMENTS ARE REAL, DERIVED, AND ASSUMED

Some requirements are easy to quantify. For example, the engineer designing an electric power source knows how much power he must generate, and its voltage and frequency. Voltage and frequency tolerances can be estimated from the characteristics of the load. Elevation relative to sea level, the range of outdoor

Requirements are Real, Derived, and Assumed

temperature, the maximum wind velocity, and the earthquake zone are readily obtained.

Not so easy to define is the requirement for power availability or reliability. Power availability is what the user wants—how many hours per year is he going to be without power? Reliability, the common measure of the performance of equipment, is usually specified in terms of the mean-time-between-failure (MTBF). For example, the U.S. Army Mobility Engineering Research and Development Command specifies the minimum MTBF of its standard 400-kW diesel generator set as 1,000 hours.

Also important is the mean time to repair (MTTR). If it takes four hours to get the generator set running again, then for every 1,000 hours (h) of operation, four hours are lost.

Thus, each year the users are without power (Po) for:

$$Po = 8,766 \text{ h}/1,000 \text{ h} = \text{about 9 times}$$

At four hours for repairs each time, the user's total no-power time is thirty-six hours.

Power availability is a function of MTBF and MTTR. The classical formula for availability (A) is:

$$A = \frac{MTBF}{MTBF + MTTR}$$

Another way of looking at it is to compare on-time versus total time. In the example, availability would be $(8,766 \text{ h} - 36 \text{ h})/8,766 \text{ h/year} = 0.996$, the same as obtained with the classical formula.

If a 0.9999 power availability is required, redundant diesel generator sets must be provided. A public utility normally can supply power with high availability, but it won't guarantee availability. One of the utility's coal or nuclear power plants might have an availability between 60 and 80 percent. However, the utility is part of a regional grid that has many generators and appropriate spinning reserve that isolate the user from the effect of a single generator failure. On the other hand, an isolated power plant does not have such a grid backing it, so it needs redundancy if it must have the power availability normally expected from a public utility.

Another key requirement is allowable power-off period. If the power-off period is going to last for more than a few cycles or sixtieth's of a second, an installation with computers on the line needs an uninterruptible power supply with a battery backup. If a power interruption of several minutes is permissible, the starting of a standby diesel generator is feasible. Sometimes in a new installation it pays to design the computers so that they store their active computations in a nonvolatile memory as the power decays. This avoids the need for an uninterruptible power supply. Such are the kinds of perturbations of requirements that are possible for the energy systems engineer.

Derived requirements. From the basic requirements, derived requirements can be allocated. For example, consider a steam power plant in which the overall plant efficiency is the output in terms of power flowing out in the transmission lines divided by the input energy content of the coal entering the boilers. Some of the generated power must be allocated among the plant auxiliaries. Among the biggest auxiliaries in a steam power plant are the pumps that force feedwater into the boiler. When energy was cheap, these multistage pumps were driven by constant-speed induction motors rated to supply required feedwater at a time when the main turbine was at full power. At part loads, reduced flow was achieved with feedwater regulators that throttled the pump output. Throttling a fluid flow wastes energy because the pump still generates a pressure corresponding to its speed. The alternative is to drive the pumps with variable-speed motors so that the water flow can be reduced by slowing down the pump, reducing consumed power.

Eugene Kempers has developed the technique of analyzing the power consumption of variable-speed feedwater pump motors. He showed that a $540,000 per year saving was possible in Iowa Public Service Company's 321 MW George Neal 2 plant.[2] Here, the feedwater pumps deliver up to 16,000 horsepower (hp). A key input to his analysis is the hours of operation per year at less than full load. Kempers determined the hours per year that his plant operated at each 25-MW power output increment between 0 and 321 MW. This became the derived requirement for his optimization of the boiler feedpump configuration.

Assumptions. Requirements that cannot be determined precisely become assumptions. For example, the number of successive days of cloud-obscured sky cannot be predicted in advance, yet the assumed value may size the energy storage in a solar power plant. All important assumptions need to be identified with their limits and flagged for perturbation in the energy system engineering process.

POSTULATING ALTERNATIVES, EVEN STRANGE ONES

A key to energy systems engineering is the consideration of all reasonable alternatives. Some can be dismissed quickly on the basis of data provided later in this

---

[2]Eugene Kempers, "Avoid Pitfalls in Economics of Boiler-Feed-Pump Variable Speed," *Power,* March 1984, pp. 53–57.

book. For example, geothermal heat is practical only in places where hot water can be pumped from the ground at reasonable depth. Otherwise, the pumping power is better invested in a heat pump. As for the generation of geothermal power from hot steam, Geysers, California, is the only place in the United States where the steam has been plentiful and hot enough.

New ideas for energy equipment are presented in papers at the annual Intersociety Energy Conversion Engineering Conference, held in even years on the west coast and in odd years on the east coast of the United States. Proceedings of past conferences are available in libraries or can be purchased from the sponsoring societies: AIAA, IEEE, AIChE, ANS, ASME, SAE, and ACS.

HOW MUCH PRELIMINARY DESIGN DO WE NEED?

The next step is to carry preliminary designs to the point where component costs and life-cycle costs can be estimated. For some power plants, this is not difficult. For example, the preliminary design of a diesel power plant can be a block diagram, a plan view, and a list of engines, tanks, valves, pipes, controls, and switchgear (Figure 1-2). The plant might be copied from a handbook or derived from a previously built plant. Component costs are readily available from suppliers.

In contrast, consider a fuel cell. Its simple electrochemical process is implemented with an array of pressurized reservoirs, controls, regulators, valves, pipes, and pipe joints, all of which eat up allocated reliability. Many of these components do not have published failure rates because field experience is lacking, and no one has performed the costly tests that establish reliability. The energy systems engineer then has to estimate the reliability of components, using failure rate sources such as MIL-HDBK-217, Rome Air Development Center's (RADC) "Non-Electronic Reliability Notebook," and IEEE's "Reliability Survey of Industrial Plants" (March/April 1974). References (3) and (4) provide data for power transmission network analyses.

The important point is that the key to power availability may lie in the accessories, and the achievement of a reliability requirement may be possible only with layers of redundancy, which must be considered in making life-cycle cost estimates.

---

[3]Stephen A. Mallard and Virginia C. Thomas, "A Method for Calculating Transmission System Reliability," *IEEE Transactions on Power Apparatus and Systems,* March 1968, pp. 824–34.

[4]E. C. Yerks, Jr., "A Practical Approach to System Reliability Analysis," *IEEE Transmission and Distribution,* January, 1974.

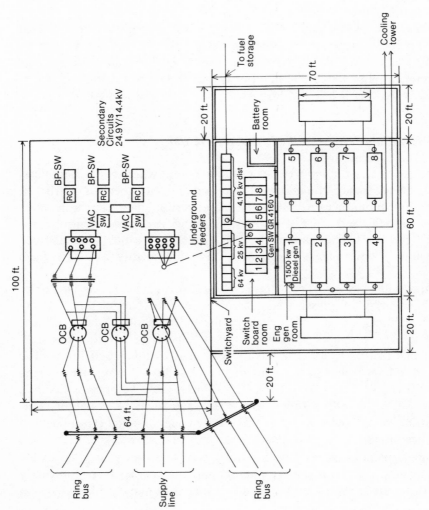

Figure 1-2

LIFE-CYCLE COST: THE KEY TO ENERGY SYSTEMS ENGINEERING COMPUTATIONS

Life-cycle cost is the most important criterion in evaluating energy handling equipment and power plants. Life-cycle cost is the dollars that the owner has to expend during the life of the apparatus. It starts with procurement of components, labor of installation, test and initial operation. After that come fuel, lubricants, purchased power, operators' salary and fringe benefits, maintenance, repair parts, overhauls, interest, and decommissioning. Estimating life-cycle cost is complicated by inflation, which makes future expenditures more costly than present ones. On the other hand, money spent today could alternatively earn interest until it is needed in the future.

Life-cycle cost is the total cost of a process or project pertinent to the owner, but owners have differing interests. Some examples are:

- Projects and equipment for the Department of Defense generally have no interest expense. Warships, guns, and warplanes are bought with funds appropriated by the United States Congress, and interest cost is not pertinent. The remaining life-cycle cost elements, except perhaps the salaries of military personnel, are evaluated.
- A developer building houses is mostly interested in the size of the customer's mortgage payments, which are related to construction cost and interest rate. Fuel and energy savings are useful if they help sell houses.
- A builder of office buildings measure the value of investments on their return for a period like five years. He compares his return from energy efficient construction and equipment with his alternative return from constructing more buildings.
- Public utility managers convert life-cycle costs into terms of cost per kilowatt hour of electric power delivered to customers.
- Energy payback has been used in evaluating solar cell manufacturing. For example, silicon solar cells in a solar power satellite in geosynchronous orbit would within the first year generate as much energy as it took to build the cells.

Power plants being compared in life-cycle cost must be equivalent in output and power availability. Figure 1-3 shows how we compared a small nuclear plant with diesel generators, gas turbines, and fuel cells for a power plant that generates power for a year. We sought a power availability of 0.999. Nuclear plants in public-utility service have an availability of around 0.8, so we paralleled three reactors, each having two closed-cycle gas-turbine driven generators to satisfy the reliability requirement. Two reactor plants were needed during peak load.

Nine 1,000-kW diesel generator sets were needed because of their assumed 1,000-h mean time before failure. Two of these nine generator sets are redundant when the load is at its peak. The Brayton-cycle gas turbine has a better reliability, so only one extra one was needed.

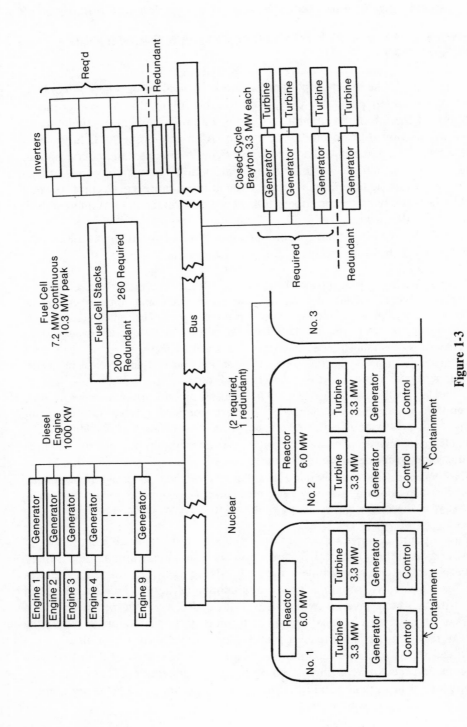

Figure 1-3

Life-Cycle Cost: The Key to Energy Systems Engineering Computations

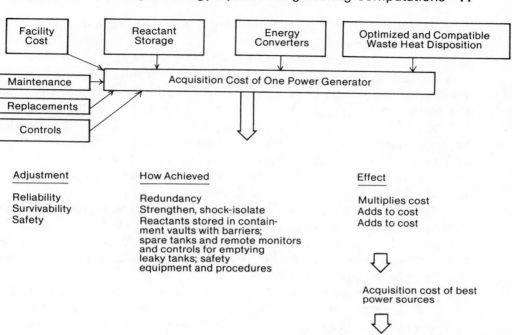

Figure 1-4

Each of the fuel cells will have a failure rate of 0.18 per million hours—a very small number. However, the cells are assembled ninety in series in a stack, the smallest replaceable unit. Twenty-five stacks are required for replacing failures during the year's operating life. Ways of calculating reliability of redundant units are described later.

Evaluation is simplified by quantifying as many factors as possible into life-cycle cost. Figure 1-4 shows how we collected all elements of acquisition cost and adjusted this cost for other factors. The acquisition cost of our power generating plant included its heat sinks for absorbing conversion-cycle losses, energy converters, storage of fuels, maintenance, overhaul services, replacement parts, and installation expense. The cost of a single power plant is then increased by its share of any development expense required to achieve appropriately low risk, redundancy to achieve required reliability, and provisions to meet safety standards.

Development expense and safety need to be examined carefully when new apparatus is being considered. This added cost can range from a simple event such as a delay in start-up because components failing during a test had to be replaced, to the scrapping of a 1,000-MW steam turbogenerator, because after years of trying, reliable operation could not be achieved. The best example of the cost of safety is the installation of the first 4.8-MW fuel-cell power plant in New York City. Its fuel, naptha stored in two 25,000 gallon (gal) tanks, was new to the fire authorities. The process of getting approval to operate was so time-consuming that the fuel cells degraded before they could generate power.

Estimates of the cost of fuel, labor, parts, overhead, and overhaul services required during the operating life of a plant need to be adjusted for inflation accumulated by the time these items are required. One specialist on life-cycle cost studies says that the prediction of future economic events is not possible, so he uses a 6 percent per year inflation adjustment.

Operating personnel is a big element in life-cycle cost if continuous attendance is required at a plant. A twenty-four-hour-per-day watch needs at least five persons for the three shifts, weekends, vacations, and sickness. Overhead, taxes, vacations, sick benefits, and retirement plans are expenses that can make the labor cost be twice the pay of the workers.

Interest on investment is another important element in life-cycle cost. Interest rate depends on the quality of the investment as well as the prime rate at the banks. While current interest rates are published in the *Wall Street Journal* and other media, predicting future rates is not possible.

The computation of life-cycle cost is a straightforward procedure. However, effective energy systems engineering requires computation of life-cycle cost of alternative approaches as well as perturbations of inputs to the process. Using a computer to produce an electronic spread-sheet program speeds the process.

RANKING OF ALTERNATIVES

Ranking the life-cycle costs is a first step in energy systems engineering evaulation. Some alternatives will be obvious rejects, and others will be so nearly equal that further optimization and perturbations of inputs to establish sensitivities are needed.

For example, once we evaluated power sources for a one-year survival of an underground military base. In the first screening, we found that it was clearly inappropriate for power plants to use geothermal heat, cyclohexane fuel, and lithium. Other power sources looked good. Our next step was to engage experienced subcontractors to make trades and optimize designs. For example, a team at General Electric evaluated twelve different kinds of fuel cells, concluding that the best was one using as an oxidizer chlorine, and as a fuel, hydrogen generated by reacting iron balls with hydrochloric acid. Another team at Rockwell International evaluated four kinds of nuclear reactor power plants and developed a preliminary design of a super-safe liquid-metal cooled fast-neutron reactor. We

Evaluation of Elements that Defy Quantification

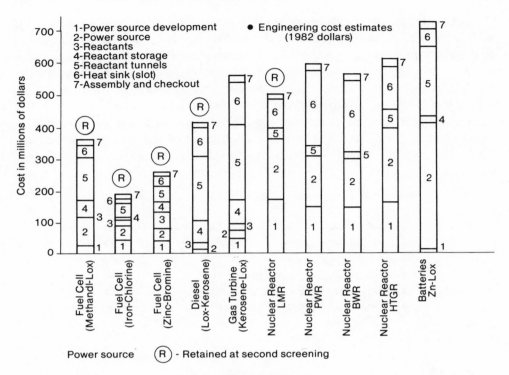

Figure 1-5

calculated the costs of tunnels for housing these power plants, their reactants, waste products, and cooling water. In Figure 1-5 are plotted the results of the study.[5]

We concluded that the iron-chlorine fuel cell was best, but development risks indicated that a back-up power source should be simultaneously developed until all potential problems had solutions.

EVALUATION OF ELEMENTS THAT DEFY QUANTIFICATION

Experienced decision makers don't immediately accept the results of polished numerical analyses. Valid analytical conclusions have been upset by factors that could not be analyzed numerically.

[5]W. G. Harris and H. Oman, "Chlorine as an Oxidizer for Fuel Cells and Its Containment," *Proceedings of the 19th Intersociety Energy Conversion Engineering Conference,* ANS (1984), pp. 810–14.

Peter D. Hindley and associates developed a useful approach for dealing with decision elements that cannot be quantified directly.[6] Their basic analysis for evaluating eighty possible capital improvements to Pacific Gas and Electric's utility plant produced an investment efficiency ratio. This ratio indicates how much a resource alternative reduces revenue requirements for each dollar of capital spent. "More specifically, the investment efficiency ratio is defined as the ratio of the present value of revenue requirement savings to the present value of capital outlays," explains Hindley. The most attractive option is the one that has the highest positive value. The following ratios, from his paper, were the most attractive:

Project	Average score
Conservation programs	5.03
Load management programs	−2.02
Improvements to existing plants	1.10
Development of the Geysers	1.81
Development of preferred resources	−0.04
Power purchases	5.16
Conventional resources	−0.05

Notice that further development of the Geysers geothermal resource doesn't look good yet. Conservation came out well because it reduces the energy that has to be generated in oil-burning plants, in effect shifting power generation to lower cost hydro and nuclear resources. Power purchases came out high because with relatively low-cost transmission lines, surplus power could be imported from the Pacific Northwest.

The key step was to perturb this analysis by introducing fractional weightings for different scenarios. For example, a low-risk scenario injected units of 0.03 value into the calculation with ten of them going into the lead-time element. The environmentalist scenario had multiples of 0.04, with fourteen of them applied to the regulation constraints factor. The ratepayer scenario had 0.05 value units, with ten units applied to levelized cost of power, mills/kWh.

Each scenario produced different investment efficiency scores. Installing a new generator at the Geysers geothermal field looked good in all scenarios, so Pacific Gas and Electric applied for a permit to build it.

[6]Peter D. Hindley et al., "Economic Assessment of Alternative Generating Sources," *Proceedings of the 18th Intersociety Energy Conversion Engineering Conference,* ANS (1984), pp. 1039–44.

Reliability Predictions and Calculations

RELIABILITY PREDICTIONS AND CALCULATIONS

Failure rates are specified in mean time between failures (MTBF). For example, a catalog of U.S. Army standard diesel generator sets lists MTBFs of around 1,000 hours (h). This failure rate can be derived by combining the failure rate of components, or by analyzing the operating history of engines. Collecting reliable failure rates from operating history is possible in a disciplined industry such as the domestic airlines. Less statistically significant would be failure rates of diesel engine components from the logging or construction industry.

The probability of failure of parallel redundant units is the product of the failure probability of the individual units. For example, Lister makes a special diesel engine that has a well-documented failure history of 16,700 h MTBF. The probability of no failure (P_s) during one year for this engine is:

$$P_s = e^{-t} = e^{-(8,766 \text{ h}/16,700 \text{ hours})}$$
$$= 0.5916$$

where $t =$ the operating time (1 year)/MTBF

For two engines in parallel redundancy the success probability (Ps) is:

$$P_s = 2P_s - P_s^2 = 0.833$$

A reasonable operating scenario is that the first failed engine will be fixed within twenty-four hours after it has failed. The repair period is called mean-time-to-repair (MTTR). The power availability (A_s) is the portion of time that power can be expected to be available from the power source. For a single Lister engine, the power availability is:

$$A_s = \frac{MTBF}{MTBF + MTTR} = \frac{16,700 \text{ h}}{16,700 \text{ h} + 24 \text{ h}} = 0.9986$$

With two engines in a redundant configuration, a power outage occurs only if the second engine fails while the first is being repaired. The availability of the redundant engines (A_t) becomes:

$$A_t = 2A_s - A_s^2 = 2(0.9986) - (0.9986)^2 = 0.999998$$

For example, a hospital might have public utility power with an outage rate of 4 h per year and a diesel engine with a MTBF of 1,000 h. Availability of utility power (Au) then is:

$$Au = \frac{8766 \text{ h} - 4 \text{ h}}{8766 \text{ h}} = 0.99954$$

The availability of the standby generators (Ad) would be:

$$Ad = \frac{MTBF}{MTTF + MTTR} = \frac{1{,}000 \text{ h}}{(1{,}000 + 24) \text{ h}} = 0.9766$$

The availability of power from the utility and standby generator can then be combined to give the availability of power in the surgery room (Ao):

$$Ao = Au + Ad - (Au \times Ad) = 0.99954 + 0.9766 - (0.99954 \times 0.9766)$$
$$= 0.99999$$

There also is a finite probability that an engine won't start. Assuming that it starts 95 percent of the time, the availability of power in the surgery room becomes 0.999966.

A common problem is the failure rate of components in a series chain. For example, the failure mechanisms of a starved-electrolyte 8-cell storage battery and their probability of occurring during a 12.5-year period are:

Element	Quantity	Reliability
Cells (positive grids)	8	0.99928
Pressure relief valve seal	8	0.99206
Jar seam	8	0.99977
Jar-to-cover seal	8	0.99977
Cover-to-post seal	8	0.99977
Intercell connections	14	0.99900
Terminal connections of battery	2	0.99986
Battery assembly		0.9895

The reliability of the battery assembly is the product of the reliabilities of the individual elements. From the rate of failures per year, the mean time between failures of a battery (MBTBF) can be calculated:

$$MBTBF = \frac{8766 \text{ h} \times 12.5 \text{ yr}}{(1 - 0.9895)} = 10{,}000{,}000 \text{h}$$

Neither customer nor manufacturer is likely to be around to see if the battery did fail in 10,000,000 h. Actually, the life of the battery is limited to some twenty years by corrosion of the positive grid. During that active life, an installation of batteries will experience a failure at the rate of 10,000,000 h. divided by the number of batteries.

THE STRANGE UNITS THAT MEASURE ENERGY

Energy system engineering comparisons would be simpler if the International System of Units (SI) units prevailed throughout the world. In the real world, crude oil is measured in barrels, coal is sold by the ton, and the electric power meter reads in kilowatt hours. Fortunately, hand calculators simplify energy systems engineering calculations. For example, Texas Instruments' TI Converter changes entered metric values to English values and vice versa with two touches of buttons. Programmable hand calculators with continuous memory can remember dozens of conversions pertinent to any particular problem. These calculators cost less than $75.

In this book, alternative units are printed in parentheses only when pertinent. For example, the forty-two-gallon petroleum barrel is used throughout the world in oil transactions and statistics. There is no point to printing the metric equivalent of 0.159 cubic meters per barrel with every mention of barrels of oil.

With a hand calculator capable of chain multiplication and division, many of the energy-related conversions can be derived. For example, from the relationship that one cubic centimeter of water weighs one gram and one inch is 2.54 cm, an engineer can easily calculate the weight of water in a cubic meter (1,000 kg or one metric ton), and a cubic foot (28.32 kg). From the SI equivalent of a pound (0.4536 kg), one can calculate the weight of water in English units. The amount of water in an acre foot can be found by dividing the cubic feet in a one-square-mile lake one foot deep by 640, the number of acres per square mile.

The relationship, 1 British thermal unit (Btu) = 1055.06 joules (J), is worth remembering because it relates English heat to SI heat. For example, a joule is a watt-second (W-s), so a kWh is:

$$1 \text{ kWh} = \frac{1,000 \text{ watthour} \times 60 \text{ s} \times 60 \text{ min} \times J \times \text{Btu}}{\min \times h \times 1 \text{ W-s} \times 1055.06 \text{ J}} = 3,412 \text{ Btu}$$

In SI units, electrical and mechanical energy are measured in joules, a joule being also one newton meter (N-m). In English units, the equivalence of mechanical to electrical power is that one horsepower-hour = 0.746 kWh. One horsepower represents work done at the rate of 550 foot-pounds (ft·lb) per second.

Other useful conversions are:

1 joule = 6.24146×10^{18} electron volts
= 0.239 calories
= 0.10197 kilopond meter
= 1.1126×10^{-17} kilograms of mass converted to energy

Some useful approximations are:

- One pound of coal releases from 10,000 to 14,000 Btu when burned, depending on the quality of the coal. The best steam power plants can deliver one kWh for every 9,000 Btu delivered to the plant in coal.
- Many diesel engine performance calculations are based on a heating value of 137,750 Btu/gallon for fuel weighing 7.1 lb/gallon.
- Gasoline, when burned in air, releases 18,000 Btu/lb.
- Sunlight on a bright day has on the surface of the Earth an intensity of about 1 kW per square meter.

The letter standing for a quantity is capitalized if it stands for someone's name. For example, A is capitalized when it is an abbreviation for ampere. When e represents an electron, as in electron volts (eV), it is not capitalized.

Equivalents pertinent to particular energy processes are defined in later chapters.

2

Engines that Make Power from Heat

All power sources in which fuel is burned are heat engines. Part of the heat released from the burning fuel appears as power in the output shaft of the engine. The rest of this heat is lost in the exhaust stack or engine cooling fluid. Efficient engines convert all possible heat of the fuel into mechanical energy. For example, of the heat energy available in gasoline, 20 percent goes into propelling a typical car and 80 percent heats the atmosphere with exhaust gases and cooling air. In a 50 percent efficient turbo compound marine diesel engine, only one-half of the fuel energy goes into heating the ocean and atmosphere.

Heat engines use turbines or pistons to extract mechanical energy from an expanding gas, called working fluid. In a Rankine cycle, the working fluid during expansion is gas, but in part of the cycle it is water or another liquid. When the combustion is external, as with Stirling and Brayton cycles, the working fluid need not be air but could be a more advantageous gas such as hydrogen or helium. Useful heat engines range in size from a small Stirling cycle unit that fits into an artificial heart to a steam turbine that drives an 1,100-megawatt (MW) electric power generator.

Heat engines are important because they generate nearly all of the electrical and mechanical power that is used in the world. The exceptions are hydro, wind, and solar power, but even these can be considered to be parts of a total heat engine.

Features of heat engines are:

- The commonly used heat engine thermodynamic cycles are the Rankine, Brayton, Diesel, and Otto cycles. Some ten basic cycles have been invented, each having variations.
- Heat input not converted to mechanical power must be dissipated in some way.
- The heat power delivered in a liquid or gas is proportional to the mass rate of fluid flow and its temperature above that of a heat sink. This makes low-temperature engines and turbines larger than high-temperature ones.
- Diesel engines usually are used to drive electric generators when less than 1,000 kW is needed. An engine and its auxiliaries cost around $200 to $300 per kW, and the fuel consumption is around 75 gallons/megawatt h.
- Most power plants generating over 100 MW use steam turbines. For new plants, the use of gas turbines and combined cycles needs to be considered.

The limits attainable in heat engine performance are these:

- With a given source and sink temperature, no engine or combination of engines can have an efficiency higher than that of a Carnot cycle for those temperatures. The Stirling and Ericcson cycles theoretically could attain Carnot-cycle efficiency. Other cycles have lower efficiency limits.
- In external-combustion engines such as steam turbines and Stirling cycle engines, the heat must go through a membrane that has high pressure on one side. In steam turbines, this wall has been the tubes in the boiler, and 1200°F (650°C) seems to be a practical limit. This in turn limits the efficiency of these plants to under 40 percent.
- In diesel engines, the source temperature of the cycle is that in the cylinder where the fuel burns. Over 50 percent engine efficiency can be obtained in small high-speed engines if the surfaces of valves, pistons, and cylinders are a ceramic that can withstand higher temperature than can metal parts.
- Gas turbine efficiency is limited by the temperature that the first row of turbine blades can withstand. Internal cooling and ceramic surfaces are the route to over 40 percent efficiency.
- Power plant efficiencies of over 50 percent are being achieved in combined cycles where a steam turbine runs from steam generated in an unfired boiler heated by gas turbine exhaust.
- Engines receiving heat at temperatures below the boiling point of water have not been practical because they are so large.
- Water-moderated nuclear reactors are limited to around 706°F (375°C), the temperature at which water ceases to be a defined liquid. The efficiencies of these plants have been under 30 percent.

Many terms are used in characterizing heat engines. Turbogenerators consume pounds of steam per hour and deliver kilowatts of electricity. Steam power plant efficiency is expressed as heat rate, which is the energy in the fuel consumed when generating one kilowatt hour (kWh) of delivered power. Efficiency of American diesel engines is measured in pounds of fuel consumed per horsepower hour. European engines are rated in grams of fuel per kWh. Airplane engine output is measured in pounds of thrust. Some of the conversion factors useful in heat-engine calculations are:

$$
\begin{aligned}
1 \text{ kWh} &= 3412 \text{ Btu} = 1.341 \text{ horsepower hours} \\
&= 3.6 \times 10^6 = \text{joules (J)} \\
1 \text{ gallon diesel oil} &= 137{,}750 \text{ Btu} = 145{,}334 \text{ kilojoules (kJ)} \\
1 \text{ Btu} &= 1055.06 \text{ J}
\end{aligned}
$$

Heat engine performance is related to the source and sink temperatures, expressed in degrees Rankine (R) or Kelvin (K). The definition of *kelvin* includes the word *degrees*. These are absolute temperatures that are related to common units of Celsius (C) and Fahrenheit (F) as follows:

$$
\begin{aligned}
R &= {}^\circ F + 459.67 & K &= {}^\circ C + 273.15 \\
&= 1.8 \text{ K} & &= (1/1.8) \text{ R}
\end{aligned}
$$

In some analyses, engine efficiency is not important. An example is comparing diesel and gas-turbine standby generators. The generators run so little that the fuel consumed is not significant. On the other hand, the cost of power from a continuously running plant relates directly to the plant's efficiency.

CAN YOU BEAT CARNOT-CYCLE EFFICIENCY?

Conversions from one energy form to another where heat is not involved can have efficiencies that approach 100 percent. For example, large generators convert mechanical power to electricity with over 95 percent efficiency. Chemical-to-mechanical conversion in muscles can exceed 50 percent efficiency.

The efficiency with which energy in the form of heat can be converted to mechanical or electrical energy is limited to the efficiency of the Carnot cycle. The thermodynamic processes in energy conversion are treated rigorously in textbooks on thermodynamics.[1] The Carnot-cycle efficiency limit is important because so often higher-than-Carnot efficiency is claimed for heat-to-mechanical energy processes. One expression for Carnot efficiency (η_c) is:

$$\eta_c = \frac{T_h - T_c}{T_c}$$

[1] For example, J. P. Holman, *Thermodynamics* (New York: McGraw-Hill Book Company, 1974).

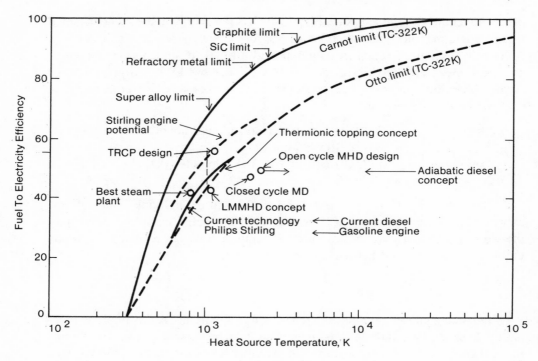

Figure 2-1

where T_h = The absolute temperature of the fluid entering the conversion process, either K or R

T_c = The temperature at which heat is rejected by the process, in K or R

Heat pumps and refrigerating machines have corresponding Carnot limits. An engine with higher than Carnot efficiency could theoretically drive a heat pump that could supply the heat input to the engine, achieving perpetual motion. Most engines are well below Carnot cycle in efficiency for their source and sink temperatures as shown in Figure 2-1 which was developed by W. R. Martini.[2] Non-engineers still make mistakes because they do not understand the efficiency limits of heat engines. For example, in the 1970s the California State Patrol ordered a fleet of steam engine powered patrol cars.

[2]W. R. Martini, "Whither Stirling Engines," *Proceedings of the 17th Intersociety Energy Conversion Engineering Conference,* IEEE, 1982, pp. 1675–80.

THE PERFECT ENGINE, INVENTED BY A NINETEENTH-CENTURY MINISTER

In 1816, twenty-six-year-old Reverend Robert Stirling, one year out of divinity school and just ordained into his first parish, invented the Stirling engine, which works on one of the few thermodynamic cycles that is limited only by Carnot efficiency. Watt had already invented the steam engine, but thirty years were to elapse before Joule identified the mechanical equivalence of heat and Sadi Carnot had yet to define the cycle efficiency of engines.

The Stirling cycle consists of four reversible processes with which pistons in a cylinder can convert heat to mechanical power in a rotating shaft. The Stirling cycle is important because its theoretical efficiency is the best that is possible, and Stirling engines have been built for some 160 years. Features of the Stirling cycle engines are:

- Stirling cycle engines, other than models, are not in volume production (in 1985), so cost estimates are only predictions.
- With refractory-metal parts, the potential efficiency of Stirling cycle engines is over 60 percent. The best diesel engines in 1985 were around 50 percent, and coal-burning steam plants were under 40 percent.
- The Stirling cycle engine, being a closed cycle, can use any heat source. Engines have operated on concentrated sunlight and on burning rice hulls.
- For a Stirling engine powering a 3,000-pound automobile, a 60 miles-per-gallon fuel consumption has been predicted.
- A Stirling heat pump need not be electrically powered.
- The problems in Stirling cycle development have been the leakage and friction in seals, and the durability of the high-temperature heat exchangers.

Some of the limits with Stirling-cycle engines are:

- With a 1500°F (815°C) heat source and a 100°F (38°C) heat sink, the Carnot efficiency would be 71.4 percent.
- One designer of a Stirling automobile engine, extrapolating from performance of an actual United Stirling engine, predicts an efficiency of 76.9 percent of Carnot.

The pure Stirling cycle consists of an isothermal compression, a constant-volume heating, an isothermal expansion, and a constant-volume cooling (Figure 2-2). Practical engines do not achieve the pure cycle, but they are close enough to get respectable efficiencies. A basic Stirling engine has a hot space, heater, regenerator, and cold space (Figure 2-3). The gas flow passages have no valves. The required motion of the pistons can be approximated with a rhombic drive. Engines such as ones made by United Stirling in Sweden have double-acting

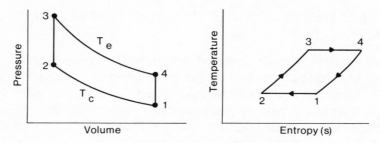

Figure 2-2

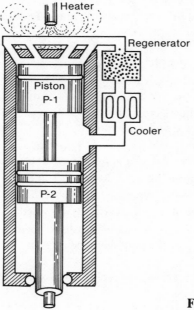

Figure 2-3

cylinders arranged with cross-coupled fluid flows so that only one piston per cylinder is required and the pistons drive simple gear-coupled crankshafts.

W. R. Martini published a good description of the steps in the Stirling cycle.[2] The simplest analysis for predicting the performance of a Stirling engine was published by J. R. Senft,[3] who elaborated on a performance constant developed by W. T. Beale. W. R. Martini developed a computer analysis that has

---

[3] J. R. Senft, "A Simple Derivation of the Beale Number," *Proceedings of the 17th Intersociety Energy Conversion Engineering Conference*, IEEE, 1982, pp. 1652–55.

predicted the actual performance of a General Motors 4L23 engine within 10 percent. With the computer, he can optimize parameters such as pressure ratio and temperatures for the highest efficiency or other desired characteristic.[4] More sophisticated analyses are described by Israel Urieli and M. Kushnir.[5] The tools are available for optimizing and designing Stirling engines.

Status of Stirling engines. During the first 100 years after the invention of the stirling engine, dozens of firms built engines, mostly for pumps and fans. One amusing use was for a popcorn wagon, which needed mechanical power as well as heat for popping corn. A Rankine-cycle engine would have required a troublesome boiler. Duplex Vacuum Company of Chicago built the engine for these wagons.[6] Convenient availability of electric power and low-cost motors made stirling engines obsolete for driving fans and pumps. Otto-cycle gasoline engines replaced the hot-air engines for other uses.

Where Stirling engines may be used. A Stirling engine is unique in that it can convert any heat power into mechanical power, and the engine need not be big. For example, in remote regions where rice is raised, the hulls must be thrashed from the grain with a 4-kW machine called a "rice huller." Delivered diesel fuel is costly. With funding from the Asia Foundation, Sunpower developed a 4-kW engine and a heater that burns rice hulls to produce gas at a temperature up to 1,000°C for the engine. The engine efficiency was only about 10 percent, but now the plentiful rice hulls could be substituted for diesel fuel. An objective was to design an engine that could be manufactured in the region where it would be used. Before starting on their design, Sunpower engineers spent three weeks in Bangledesh evaluating the capabilities of local foundries and machine shops, and checking on the availability of materials.[7]

W. R. Martini, a Stirling engine consultant, suggests that Stirling engines would be good in heat pumps, remote power plants, solar heat-power plants, heavy-duty motive power, and wherever quiet power is needed. The quietness of

---

[4]W. R. Martini, "A Revised Isothermal Analysis Program for Stirling Engines," *Proceedings of the 18th Intersociety Energy Conversion Engineering Conference,* AIChE, 1983, pp. 743–48.

[5]Israel Urieli and M. Kushnir, "The Ideal Adiabatic Cycle—A Rational Basis for Stirling Engine Analysis," *Proceedings of the 17th Intersociety Energy Conversion Engineering Conference,* IEEE, 1982, pp. 1662–68.

[6]A. Ross, "Stirling Cycle Engines," (Solar Engines, Phoenix, AZ, 1977) p. 38.

[7]J. Gary Wood, Bruce J. Chagnot, and Lawrence B. Peaswick, "Design of a Low Pressure Air Engine for Third World Use," *Proceedings of the 17th Intersociety Energy Conversion Engineering Conference,* IEEE, 1982, pp. 1744–48.

Stirling engines is impressive, since the burner is the main source of noise. The high efficiency, plus ability to use heat as an energy source, makes the Stirling heat pump attractive. A doubling of the price of crude oil from $30 to $60 a barrel would certainly rekindle interest in Stirling engines. Martini places the Stirling engine efficiency potential between the best of the topping cycles and the limits imposed by materials and the Carnot cycle (see Figure 2-1, Reference (2)].

Martini observed that eight large organizations, including Ford and General Motors, have developed Stirling engines, but none has chosen to build a factory for manufacturing engines at a competitive cost. Presumably the risks in trying to compete with conventional diesel and otto-cycle engines were great enough to discourage serious interest.

The free-piston Stirling engine has all moving components in a sealed tube, so that escape of the working fluid need not be a problem. The power can be extracted with a linear alternator when electric power is to be generated, or the unit can be developed as a pump for heating or cooling. The analysis has progressed well, but development has been plagued with component problems, as explained by William T. Beale.[8] The main difficulties that he identifies are in seals, loading, heater heads, and control.

Future of Stirling cycle. Stirling cycles have been successful in pumping heat. For example, Norelco makes a small laboratory air liquifier that is based on the Philips Stirling engine. R. J. Vincent calculates that with components that he has tested, a Stirling-cycle kinematic heat pump can be built to deliver 1030 watts at 90°F (32°C). The input consists of 190 W of power and 890 W of heat at 20°F (−6.7°C). The corresponding coefficient of performance (COP) is 6.3, which is 92 percent of Carnot.[9] The COP of the best of the small air-to-air electric heat pumps is around three. The working fluid in the Stirling pump is helium at a mean pressure of 150 psi (1034 kPa). The pressure ratio is 2 to 1.

Development problems in Stirling engines. Sliding seals and heaters have been the life-limiting components of Stirling engines. The displacer piston needs a modest-pressure seal to keep the working fluid in its cycle. The piston-rod

---

[8]William T. Beale, "The Free Piston Stirling Engine: 20 Years of Development," *18th Intersociety Energy Conversion Engineering Conference,* AIChE, 1983, pp. 689–93.

[9]Ronald J. Vincent, William D. Rifkin, and Glendon M. Benson, "Test Results of High Efficiency Stirling Machine Components," *Proceedings of the 17th Intersociety Energy Conversion Engineering Conference,* IEEE, 1982, pp. 1867–74.

seal not only retains the high-pressure working fluid, but it also keeps working fluid from being polluted with oil.

An external-combustion Stirling engine is unique in that it requires a high-temperature heat exchanger. External combustion is also needed in the Rankine cycle, but steam temperatures are generally kept below 1100°F (593°C), whereas for the Stirling cycle to be more efficient than a diesel engine, the source temperature must be above 1500°F (815°C). In the internal combustion cycles, Diesel, Otto, and Brayton, the heat is released directly into the working fluid. The heat does not have to be conducted through a diaphragm with high pressure on one side.

A successful heater consisting of a nest of tubes is used by United Stirling of Sweden for its automobile engine. An important requirement of this heat exchanger is that it doesn't leak after repeated startups and shutdowns, which could number up to one thousand a year in a car.

An interesting example of a Stirling engine for an automobile was described by D. G. Beremand, who summarized work sponsored by NASA Lewis Research Laboratory.[10] Engines were developed by Mechanical Technology Inc. with United Sterling AB of Sweden as a subcontractor. Their eighth Mod I engine with hydrogen at 15 MPA (2,175 psia) as a working fluid achieved 33 percent efficiency in a full-up automotive configuration, with all auxiliary and control losses included (Figure 2-4 from Ref. 11). The engine recovers exhaust heat in a heat exchanger that pre-heats the combustion air. The Mod I engine had accumulated over 9,000 h of test time in February 1985, with the longest mean time between failures in any engine being 1,170 h. Engine Number 7 had accumulated 2,340 running h. Beremand noted that in United Stirling tests, rod seals had run 3,000 h without failure.

Pioneer Engineering and Manufacturing Co., a NASA Lewis contractor, estimated that production of the engine would cost $808. Pioneer predicted the following "installed in vehicle" costs for an annual production of 300,000 engines:

Spark Ignition, Otto	Diesel	Stirling
$1802	$2198	$2017

[10]D. G. Beremand, "DOE/NASA Automotive Stirling Engine Project Overview '83," *Proceedings of the 18th Intersociety Energy Conversion Engineering Conference*, AIChE, 1983, pp. 681–88.

[11]Noel P. Nightingale, "Automotive Stirling Engine Development Program, Semi-annual Technical Progress Report for Period: January 1–June 30, 1984," *Mechanical Technology Inc.*, NASA CR–174749, October 1984, pp. 5–13.

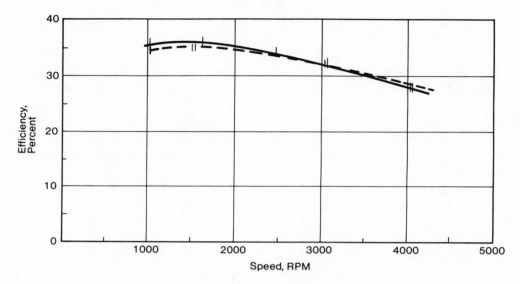

Figure 2-4

RANKINE CYCLE, THE STEAM ENGINE, AND TURBINE

In a Rankine thermodynamic cycle a pressurized gas, generated by boiling a liquid with heat, is expanded through a piston engine or a turbine to cause motion and generate power. Steam power plants, operating on the Rankine cycle, generate most of the electric power being consumed in the U.S. The input heat is obtained from nuclear reactors, or from burning coal, oil, natural gas, wood, or garbage.

Rankine-cycle power plants are important because there is no foreseeable practical substitute for building more of them as demand for electrical power grows. Brayton-cycle gas turbine plants can generate power as efficiently, but they cannot run directly on cheap coal. Likewise diesel engines can't run on coal. Solar power plants require ten times the area, construction materials, and money that is needed for steam power plants with the same output.

Energy features of Rankine cycle power plants include:

- Efficient operation is achieved by making the power plants large, up to 1,100 megawatts electrical (MWe) per unit. They can then have energy-conserving refinements such as multi-stage feedwater heating with bleed steam, superheating and reheating of the steam, and economizers that extract heat from the gases being discharged into the stack.

Rankine Cycle, the Steam Engine, and Turbine

- A good location for a steam plant has access to cold river, lake, or ocean water for circulating in the condenser. Such locations are generally unavailable, so various forms of cooling towers are used.
- In most U.S. steam power plants, the fuel energy not delivered to the power transformers appears as heat, mainly in the condenser cooling water. Bottoming cycles that would recover energy from this cooling water have not been practical. In Europe steam-plant losses are often used to heat buildings, a process called "district heating."
- Rankine-cycle steam engines cannot compete with diesel and gas-turbine engines for propelling trucks, buses, airplanes, and small boats. For propelling automobiles, steam engines are a novelty that sometimes achieves sales in spite of historic boiler-maintenance difficulties.
- Limiting the emission of sulfur, nitrogen-oxygen compounds, and flyash are costly processes in coal-burning steam power plants.

A boiler producing steam for a reciprocating steam engine is the simplest mechanization of the Rankine-cycle engine. These steam engines reached their climax in the 700,000 lb Mallet locomotives that hauled freight trains from Cheyenne to Laramie, and the triple-expansion engines that propelled the World War II Liberty ships.

Some of the limits of Rankine cycle heat engines include:

- Being a heat engine, the efficiency of a Rankine cycle power plant is limited by the heat source temperature, which is the inlet temperature of the high-pressure turbine. The limiting Carnot-cycle efficiency is around 64 percent for a power plant with a 1100°F (593°C) maximum temperature and a 100°F (38°C) sink temperature. The best achieved efficiency is around 40 percent.
- A team of engineers from Westinghouse and Combustion Engineering created specifications for an advanced-development plant that has a net efficiency of 41 percent, which corresponds to a heat rate of 8,330 Btu/kWh. The turbine throttle pressure would be 4,500 psig (31 MPa).
- A turbine inlet temperature of 1100°F (593°C) is a traditional limiting temperature for steam power plants. A higher temperature unit was built, operated briefly, and ultimately scrapped because of troubles. Steam pressures as high as 3,300 psig (22.7 MPa) are used.
- For various reasons nuclear power plants are limited to temperatures under 706°F (374.5°C). Above this temperature, water ceases to act as a pure liquid.
- Most U.S. power plants have turbines rated under 1,100 MW power output. Kraftwerk Union A.G., in the Federal Republic of Germany, is building at 1,300 MW nuclear plant. New Soviet nuclear plants are rated 1,500 MW.
- A successful 200–3000 watt Rankine cycle power plant made by Ormat uses an organic working fluid and propane as fuel. Approximately 3,000 units have been built for low-power remote radio repeaters.

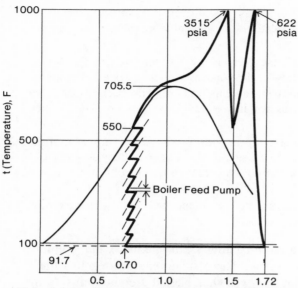

Figure 2-5

The power output of steam turbines is commonly expressed in horsepower (hp), and the output of a turbo-generator set is given in kilowatts (kW). A perfect plant would have a heat rate of 3412 Btu/kWh. The best steam plants have a heat rate of around 9,000 Btu/kWh.

Factors affecting efficiency of a steam plant. The heat path of the Rankine cycle in a modern steam turbine power plant is shown in Figure 2-5, which comes from Babcock and Wilcox's steam design manual.[12] The temperature of the fluid, water or steam, in °F is plotted on the vertical axis and the entropy of the fluid in terms of Btu/lb./degree °F on the horizontal axis. At any point on the heavy curve the entropy multiplied by the absolute temperature gives the energy content of a pound of water or steam in Btu. Within the dome-shaped curve lie the energy contents of mixtures of water and steam. Water at 32°F contains zero energy by convention. For example, at 100°F, which is 559.6 R, the entropy of the water is 0.1295 and its energy content (Ew) is:

$$Ew = 0.1295 \times 559.6 = 72.5 \text{ Btu/lb}$$

On the right side of the dome the fluid is steam, with the edge of the dome representing a saturated condition in which any energy extraction would allow

[12] "Steam, Its Generation and Use," Babcock & Wilcox, New York, p. 2–16.

Rankine Cycle, the Steam Engine, and Turbine

water droplets to form. For example, the energy in a pound of saturated steam at 212°F, corresponding to 671.67 R, would be 1,150 Btu/lb.

In raising water from 32°F to 212°F, its average temperature is 581.62 R. Its entropy at 212°F is 0.3119. The energy content at 212°F is then

$$Ew = 0.3119 \text{ Btu/lb.-R} \times 581.67 \text{ R} = 181.4 \text{ Btu/lb.}$$

The difference between the energy content of steam and water at 212°F, 969 Btu/lb., represents the latent heat of vaporization.

The temperatures and pressures in Figure 2-5 correspond to those of Southern California Edison's Alamitos steam station where 3.326 million lb. of steam/h is generated by a boiler that can be fired with gas or oil. The superheated steam is delivered to the high-pressure turbine by the boiler at 3,515 psia and 1000°F. This steam contains 1516.1 Btu/lb. The output of the high-pressure turbine, containing 1,262 Btu/lb. at 622 psia, is reheated back to 1000°F in the boiler, and then delivered to the intermediate-pressure turbine. The low-pressure turbine expands the steam to a condenser operating at 91.7 F°. The steam entering the condenser still has 993.4 Btu/lb of heat.

A simplified diagram of this steam plant in Figure 2-6 shows the flow of heat. The plant actually has two generators, one being a 3,600 rpm unit driven by

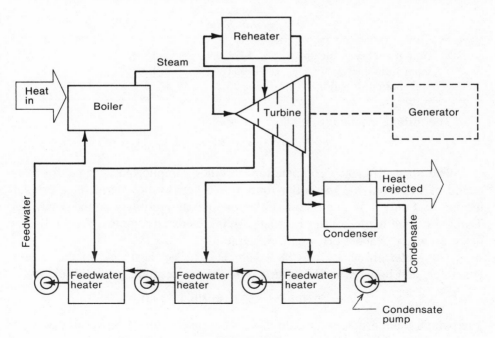

Figure 2-6

the high-pressure and intermediate-pressure turbines. The other, an 1,800 rpm unit, is driven by the low-pressure turbine. All three turbines contribute steam to eight feedwater heaters that raise the temperature of the water to 530°F before it enters the boiler. The feedwater pressure is raised by a condensate pump at the bottom of the condenser, a booster pump after the first three feedwater heaters, and finally by two half-capacity turbine driven boiler feed pumps.

The efficiency of the plant is its power output divided by its heat input. Available to the turbine are the two temperature drops from Figure 2-5:

	Enthalpy of Steam in Btu/lb	
	High-pressure turbine	Intermediate and low pressure turbine
Input	1516.1	1516.1
Output	1262.8	993.4
Total	253.3 +	522.7 = 776.0

The boiler delivers the following heat energy:

	Energy in Btu/lb of steam	
	Main boiler	Reheater
Output energy of steam	1424.0	1516.1
Input fluid energy	531.8	1262.9
Boiler contribution =	892.2 +	253.2 = 1145.4

The turbine efficiency (η_t) then becomes:

$$\eta_t = 776.0/1145.4 = 67.7 \text{ percent}$$

The plant efficiency computation becomes complicated because for every pound of steam leaving the boiler, only 0.826 gets to the intermediate-pressure turbine, and only 0.58 lb goes out of the low-pressure turbine into the condenser. The rest is bled for feedwater heating, driving feedwater pumps, the air ejector, make-up water evaporators, and a de-aerator.

The boiler efficiency was 88.8 percent. The net heat rate for the turbine, generator, and boiler was 8523 Btu/kWh. This corresponds to an efficiency:

$$3412 \text{ Btu/lb}/ 8523 \text{ Btu/lb} = 0.40$$

The overall plant efficiency would be less because of losses in transformers and power consumed in fuel handling equipment, condenser circulating pumps, cooling tower blowers, and other auxiliaries. The Carnot efficiency of the plant would

Rankine Cycle, the Steam Engine, and Turbine

be the difference in heat source and sink temperatures, divided by the source temperature:

$$c = \frac{(Th - Tc)}{Th} = \frac{(1000°F + 459.67°F) - (91.7°F + 456.67°F)}{(1000°F + 459.67°F)} = 0.622$$

Duke Power's Bellews Creek Steam Station, one of the best in the U.S., achieved in 1979 a heat rate of 8987 Btu/kWh, which corresponds to an efficiency (η_b) of:

$$\eta_b = \frac{3412 \text{ Btu/kWh}}{8987 \text{ Btu/kWh}} = 0.38$$

A Rankine cycle power plant is limited in efficiency by inevitable losses in the condensation of the working fluid. If a use can be found for the cycle losses—for example, district heating—then the energy recovered from the fuel burned can approach 80 percent. Some of the other losses in the Rankine cycle power plant are:

- Throttling losses when the turbine is operated at partial load.
- Boiler feed pump losses, particularly when throttled for partial loads.
- Practical heat exchangers used for feedwater heating have a temperature drop between input and output fluids.
- Friction in pipes and passages.
- Steam consumed by the air ejectors that maintain vacuum in the condenser.

Boiler feed pumps are generally the biggest single consumer of power in a generating station. For example, the pump motors for Iowa Public Service Company's George Neal II generator, rated 321 MWe, deliver 11,000 hp when the station is producing full power. The contribution by the pumps to the heat rate of generated power is 150 Btu/kWh at full load and 200 Btu/kWh at half load.[13]

Cost of operating crew. Operation of a power plant requires personnel to monitor feedwater quality, boiler conditions, coal pulverizers, coal stockpile management, turbine conditions, cooling-tower performance, and other elements. Sudden changes in boiler water level and coal-pulverizer fires require immediate action.

The Centralia, Washington, coal-burning steam station has to pay the salaries of a crew of about 200 who work 40-hour weeks. The plant normally

---

[13]Eugene Kempers, "Avoid Pitfalls in Economics of Boiler-Feed-Pump Variable Speed," *Power*, 128, No. 3 (March, 1984), 53–57.

generates 1240 MWe from two turbine-generator units. The power production per man hour (Pm) at normal load is:

$$Pm = \frac{1240 \text{ MWe} \times 8766 \text{ h/year}}{200 \text{ men} \times 8 \text{ h/day} \times 5 \text{ days/week} \times 52 \text{ weeks/year}}$$
$$= 26.1 \text{ MWh/man hour}$$

If the power at the bus is worth 2 cents per kWh, then each man hour produces electricity worth $523.

The crew and plant are most productive when the plant is generating rated load. Reducing the power output of the plant by throttling the turbines to half power would make the plant the equivalent of a 620 MWe plant, but the operating manpower would not be significantly affected.

Boilers for generating steam. The boiler, where the steam is generated, is a chamber lined with firebricks and water pipes in which the water is boiled. Hot gases leaving the chamber pass through superheater and reheater pipes before entering an economizer where the combustion air is preheated (Figure 2-7). The flue gases then pass through electrostatic precipitators which, sometimes in combination with a bag house, remove fly ash. Plants burning high-sulfur coal may have flue-gas desulfurizers.

Coal for most American boilers is conveyed from coal storage to hoppers that feed pulverizers. The pulverizers grind the coal into a fine powder before it is blown into the boiler. Important parasitic loads in the power plant include the coal pulverizers, and forced-draft and induced-draft blowers. Approximately 80 to 90 percent of the energy of the coal appears as heat in the steam. The rest is lost in the form of condensable steam and sensible heat in the stack gases and in convection and radiation loss from the external surfaces of the boiler.

Each coal-burning boiler is a custom design, configured for the characteristics of the coal that will be used. Changing the type of coal can require the rebuilding of the boiler. The boiler is erected in the field, making it a costly element in the power plant.

Cost of steam plants. Many steam plants in the U.S. are fired with oil and natural gas. New oil and gas burning plants are not likely to be built because these fuels can be used more efficiently in combined-cycle power generating plants. El Paso Electric, in 1984, paid $7.14/million Btu for oil fuel, $3.62 for natural gas, and $0.83 for coal.

Important variables that affect the cost of coal-burning steam plants are the type of coal, flue-gas clean-up requirements, availability of cooling water, and the cost and productivity of construction labor. Most large steam plants completed in the middle 1980s cost $1,000 to $1,500/kWh. Colstrip 3, rated 700,000 kW,

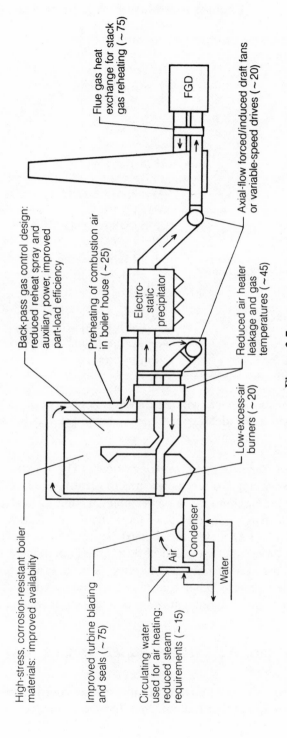

Figure 2-7

TABLE 2-1. General Specifications for an Advanced Plant

	General Electric, Babcock & Wilcox team	Westinghouse, combustion engineering team
Gross rating	725 MW	824 MW
Net rating	674 MW	773 MW
Net heat rate	8875 Btu/kWh	8330 Btu/kWh
Net thermal efficiency	38.5%	41%
Steam throttle pressure	4500 psig (31 MPa)	4500 psig (31 MPa)
Steam throttle temperature	1050°F (566°C)	1100°F (593°C)
Steam reheat temperatures	1075°F (579°C) and 1100°F (593°C)	1050°F (566°C) for both
Final feedwater temperature	580°F (304°C)	604°F (318°C)
No. of feedwater heaters (including deaerator)	9	9
Condenser back pressure	1.5 in Hg	2.5 in Hg
Excess combustion air	17%	15%
Air preheater exit gas temperature	300°F (149°C)	270°F (132°C)
FGD inlet gas temperature	235°F (113°C)	220°F (104°C)
FGD exit gas temperature	120°F (49°C)	120°F (49°C)
Stack gas temperature	170°F (77°C)	170°F (77°C)

cost $1,400/kW. Minnesota Power, which generates 89 percent of its power with coal and lignite, charged its residential customers an average of 5.6 cents/kWh in 1984.

Improvements in Rankine cycle performance. The Electric Power Research Institute commissioned two research teams to identify the limits of current fossil fuel technology for Rankine cycle power plants (Table 2-1). Note that the throttle pressure (4,500 psig) is higher than most plants use today. The steam temperatures (1050° and 1100°F) are the highest being used at the time of the study. The teams concluded that with careful design and component testing, the new high-efficiency plants could have availability as high as is attained in the 1,000 existing coal-fired plants. Other routes to high availability are redundant components, design for ease of maintenance, good spare parts inventory, and on-line diagnostics.

DIESEL CYCLE THAT PROPELS TRUCKS, BUSES, TRAINS, AND SHIPS

In the Diesel cycle, air in the cylinder is compressed during the upstroke of a piston. The air becomes so hot that fuel injected into the compressed air immediately burns. Fuel injection continues during part of the downstroke. The fuel flow is then cut off and the combustion products and nitrogen continue to expand and do work during the rest of the downstroke.

The Diesel cycle was invented by Rudolph Diesel in 1893. He hoped to run this internal combustion engine with coal as a fuel, but lacking success, he switched to oil. Diesel engines are important because they can convert the energy of fuel oil to mechanical power with efficiencies of over 50 percent.

Engine manufacturers, in competing for the truck-engine market, have developed mechanical components of their engines to the point where mechanical breakdowns are rare and a 500,000 mile service life between overhauls is being achieved.

Energy features of diesel engines include:

- A representative range of fuel consumption is 0.275 lb/horsepower (hp) h for a MAN-B&W 9800 hp 110 rpm marine propulsion engine to 0.35 lb/hp hour for a Cummins 82 hp engine in a Kohler 50 kW generator set.
- Large engines can be manufactured for $50 to $100/kW. Kohler Co. suggests $250 to $300/kW for installed generator sets in the 100 to 1000 kW range.
- One of the highest predicted mean-time-between-failures is 16,700 hours, for a 15 kW Lister generator set. The cost of this set is about three times that of an ordinary diesel generator set.

Diesel engine output is measured in horsepower or kilowatts. The kilowatt rating can be ambiguous for engines coupled to generators because it can refer to either the engine itself or the generator output. The generator output is around 10 percent lower than the engine output because of generator losses. Engine efficiency is calculated by dividing the engine-shaft power output by the input in terms of rate of fuel energy consumption. Cummins Engine Co. bases its engine ratings on fuel that weighs 7.1 lb/gallon (851 grams/liter) and contains 137,750 Btu/gallon (38.4 megajoules/liter).

The power output of an engine at its output shaft may not be the true output power. For example, an engine with a radiator mounted on the same bedplate as the engine, generally has a cooling fan that is belt-driven from the engine crankshaft, and all of the power in the output shaft is available for driving loads. Sometimes the engine is in a building and the radiator is mounted outdoors where the air is circulated through it with an electric-motor-driven fan. This motor power must be recognized in efficiency comparisons. The fan power for a Cummins 82 hp engine is 2 hp (1.5 kw) at 1800 rpm.

Thermodynamic cycle. The thermodynamic cycle of a diesel engine begins at point 1 in Figure 2-8 with an isentropic compression of the gaseous

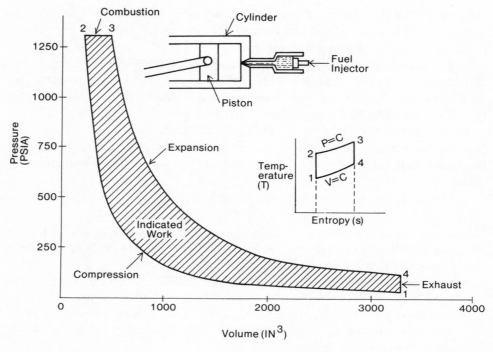

Figure 2-8

working fluid in the cylinder. At the top of the stroke, point 2, where the gas temperature can be around 2000°C, fuel under pressure, up to 14,000 pounds per square inch gage (psig) [96.5 megapascals (Mpa)], is injected into the cylinder. The burning fuel adds heat to keep the gas at approximately constant pressure as the gas expands during the first part of the power stroke of the piston. This expansion is represented by line 2–3 in the pressure-volume diagram. Then fuel is cut off at Point 3, and the working fluid expands and cools to around 1100°F (593°C) at Point 4, where the working fluid containing combustion products is exhausted.

In a two-stroke engine, the cylinder volume is then scavenged with fresh air from a blower, and the compression starts again. In a four-stroke engine the next upward stroke clears the cylinder of combustion products, and a downward stroke in a naturally aspirated engine sucks in fresh air for the next cycle. A supercharged engine has compressed air waiting in the manifold for filling the cylinder when the intake valve opens.

The area in the enclosure 1-2-3-4 in Figure 2–8 is a measure of the energy released by one cycle of the engine operation. The energy/cycle multiplied by the number of cycles/minute is a measure of the power produced by the cylinder. Real pressure-volume diagrams only approximate the shape of the theoretical diagram shown. Corresponding points of a diesel cycle are plotted on a temperature-entropy diagram in Figure 2–8.

Features of diesel engines. Diesel engine manufacturers have developed standard cylinder sizes and crankshaft configurations that they assemble in combinations appropriate for the rating being sold. The wearout and failure mechanisms have been identified and corrected.

The three major families of diesel engines are truck diesel engines and their derivatives, railroad engines, and low-speed marine engines. The mass-produced components for truck engines can be used in engines having up to 16 cylinders and producing up to 1,500 hp (1,120 kW). The larger railroad engines, with ratings up to 6,000 hp (4,480 kW), are produced in smaller quantities and hence cost more. Shaft speeds between 900 rpm and 1,800 rpm are common for the big units derived from truck engines, as well as the railroad engines. The large marine engines, often custom designed and built for a specific ship, operate at propeller speed. For example, the MAN-B&W "ultra long stroke" 12,480 hp engine has a speed of 110 rpm.

The exhaust gas from a diesel engine, having a temperature of around 1000°F (538°C) contains valuable heat energy. The Cummins 16 cylinder KTA 3067 engine, when running at its 1,500 hp rating, releases some 55,500 Btu/minute (58.5 MJ/minute) into its exhaust from the fuel fed into the engine. The turbocharger is a practical way to recover some of this energy. It consists of a single stage gas turbine that drives a rotary compressor. It absorbs heat energy, cooling the exhaust gas to around 700°F (371°C). The compressor pressurizes the air entering the cylinders, contributing power that finds its way into the output shaft.

Compressing the air also heats it, reducing the efficiency of the compression stroke in the engine. Both efficiency and power output of the engine can be improved by cooling the compressed air with the engine coolant returning from the radiator. For example, a Cummins 1,500 hp engine has an efficiency of 36.9 percent with turbocharging and aftercooling.

Performance of diesel engines. The performance of a diesel engine varies with engine speed, load, and atmospheric pressure and temperature. Engine manufacturers provide performance estimates for different operating conditions. Figure 2–9 maps the performance of a large engine, showing how efficiency varies with load and shaft speed.

A diesel engine loses power when elevation and air temperature are raised. For example, a Cummins engine equipped with a turbocharger and aftercooler

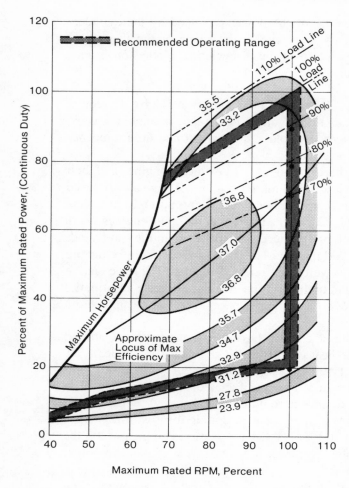

Figure 2-9

will deliver rated power as long as the elevation is less than 5,000 ft (1524 m) and the temperature is under 110°F (43°C). Then the engine must be derated 4 percent/1,000 ft (305 m) of higher elevation and 1 percent for every 10°F (5.6°C) of higher temperature. Derating starts at a lower elevation for engines that are not turbocharged.

Designing diesel power plants. The designer of a standby or prime power plant need not be concerned about the thermodynamics optimization of the engine. He will normally select a developed engine that satisfies his requirements, and design an installation around that engine. The selection and installation of an engine involves these factors:

- Part-load performance, if required. Some engines build up carbon in their cylinders if operated at idle or part-load for long periods.
- Response to load changes, and in generating sets, to electrical faults.
- De-rating for long life.
- Fuel supply tanks and pumps.
- Cooling and exhaust apparatus.
- Mounts for isolating engine vibration and for earthquake shock.
- For power generation, transfer switches and grounding.
- Protection from weather, and provisions for starting when the air temperature is low.

Building codes complicate an otherwise simple installation. For example, NFPA-110T, *Emergency and Standby Power Systems,* requires in a Level 1 installation, remote visual alarms for 17 engine parameters, ranging from low fuel in the main tank to the position of the shutdown damper if one is used. Codes pertaining to transfer switches and grounding in standby power plants are particularly complicated.

A useful book that deals with the design of diesel engine installations is published by Kohler.[14] It describes all the elements that must be incorporated in an engine installation and suggests design approaches. For example, it notes that pipes and tank-surfaces exposed to diesel fuel should not be galvanized because the fuel causes the zinc to peel off and flow into fuel injectors. Terms used in diesel power generation technology are defined, and 17 pages of tables in the appendix provide data for calculating component sizes. Sample specifications for procuring engines and generators are also provided.

Slow speed marine engines. Truck, railroad-locomotive, and standby power engines generally operate at speeds between 1,000 and 2,600 rpm, with piston speeds around 28 to 35 ft/second (8 to 12 meters/second). They generally have a piston stroke that is about the same as the piston diameter. For example, the big Cummins engines are assembled from cylinders 6.25 inches (159 mm) in diameter and have a stroke of 6.25 inches. Twelve cylinders provide 2,300 cubic inches (37.7 liters) of displacement. At its highest rating the engine produces 100 hp (74.6 kW) per cylinder.

Marine propulsion engines in contrast are designed for direct coupling to a ship's propeller shaft. An example is the three-story tall MAN-B&W 6L60, which has a 600 mm (23.6 inch) bore and an 1,800 mm (70.9 inch) stroke. The crankshaft speed is only 110 rpm. At full load, the engine delivers 12,480 brake

[14] "Engineer's Guidebook to Power Systems," Kohler Co., Kohler, Wis., 53044.

hp (9,310 kW), and the piston travels at 3.18 m (10.4 ft) per second. Some of the engine characteristics are plotted in Figure 2–10 for 15°C (59°F) at sea-level air pressure. The fuel consumption at 75 percent load is 125 grams/brake horsepower (bhp) h. The corresponding efficiency (ηd) is:

$$\eta_d = \frac{bhp - hr \times 453.69g \times 2545 \text{ Btu} \times 7.1 \text{ lb} \times \text{gallon}}{125 \text{ g} \times \text{lb} \times \text{bhp-hr} \times \text{gallon } 137{,}750 \text{ Btu}} = 0.476$$

This efficiency corresponds to 0.2755 lb of fuel/horsepower h.

This efficiency is remarkable, and is being achieved in only the best combined-cycle plants. Some of the engine parameters of the 6L60 at maximum power (12,480 bhp) are:

Maximum pressure in cylinder	125 bar	1812 psi
Compression pressure	105 bar	1522 psi
Scavenging air pressure	2.8 bar	41 psi
Turbo inlet	385° C	725° F
Turbo outlet	253° C	487° F

Assuming a 2,256 K maximum combustion temperature in the MAN B&W 6L60 engine, its Carnot-cycle efficiency (η_c) with a 253°C or 526°K turbine outlet temperature would be

$$\eta_c = \frac{2256K - 526 \text{ K}}{2256 \text{ K}} = 76.7\%$$

The key to the high efficiency is the long stroke that extracts energy from the hot gases by expanding and cooling them. Note the turbocharger outlet temperature of 487°F. The corresponding temperature in a Cummins 36-percent efficient 1,600 hp engine is 990°F (532°C). The Cummins engine rotates at 1,800 rpm and has a friction corresponding to 277.5 hp (207 kW). Its piston speed at 1,800 rpm is 1875 ft/minute (9.5 m/sec). The low piston speed of the MAN B&W marine diesel may also contribute to low friction losses.

Future of diesel engines. The conventional diesel engine has been developed to the point where efficiency improvements are hard to achieve. The long stroke marine engine achieves high efficiency, but along with that comes the cost of weight and volume. Increasing compression ratio in high-speed diesel engines introduces combustion problems. Increasing combustion temperature makes exhaust valves degrade faster, jeopardizing the trouble-free engine life that is important to the engine users.

The exiting new development is the adiabatic engine with ceramic parts. In a technical sense the word adiabatic means that no heat is transferred to or from the working fluid during a compression or expansion. The term "adiabatic" applied

Diesel Cycle that Propels Trucks, Buses, Trains, and Ships

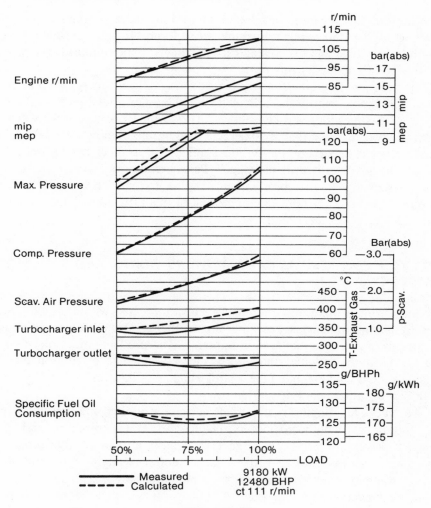

Figure 2-10

to an engine suggests that the engine heat is not released to the environment, which is not quite true. However, this engine is operated at such a high temperatures that its losses are radiated directly into the surroundings, eliminating the need for power-consuming cooling-water pumps and fans. Ceramic surfaces in the cylinder, piston top, and valve faces are designed to withstand the high-combustion temperatures. The high efficiency comes from having cylinder walls at exhaust-gas temperature, so that the combustion stroke is truly adiabatic in that no heat is lost to the cylinder wall during expansion.

American and Japanese developers of ceramic diesel engines predict efficiencies as high as 50 percent, making possible an 80 miles-per-gallon diesel automobile. R. Sekar and his colleagues describe the opportunities for improving diesel-engine efficiency by this route.[15]

Another route to high efficiency is recovering exhaust energy with a turbine that is coupled to the output shaft. Brown Boveri has developed such a turbine for Sulzer's RTA super-long stroke engines. The turbine operates only when the engine is delivering more than a 50 percent load because at lower loads all of the exhaust energy is needed by the turbocharger. At 85 percent load a fuel rate of 117 grams/horsepower hour is claimed.[16] This corresponds to a remarkable 50.8 percent efficiency. A large turbo compound engine, the MAN B&W L90MCE, has a 115 grams/horsepower-hour rate. The cost of this efficiency is the size of the engine, 13.6 m (44.6 feet) high.

OTTO CYCLE ENGINES FOR CARS, MOWERS, BOATS, AND CHAIN SAWS

In the Otto cycle a mixture of fuel and air is compressed within a cylinder and ignited by a spark. The subsequent burning of fuel creates a hot gas which does work by pushing on the piston as it expands. The products of combustion are exhausted from the cylinder, and the cycle repeats. The cycle is named for Nikolaus A. Otto who exhibited the 4-stroke engine in Paris in 1867 and patented it in 1877 with a partner, Eugene Langen.

The Otto cycle engine is important because it is used in most cars, small trucks, motorcycles, outboard motors, chain saws, hedge trimmers, and snowmobiles. It drives the propellers of general aviation airplanes. High production volumes have justified tooling with which the complex engines are built at low cost. For example, a complete set of brand new precisely-machined pistons and detachable cylinders for a Volkswagen 4-cylinder air-cooled horizontally-opposed engine could be bought by an owner for around $50 in the mid-80s.

Energy features of the Otto cycle engine are:

- The Otto cycle engine is the lowest cost machine for converting the energy of gasoline fuel into mechanical power.

[15]R. Sekar, L. Tozzi, R. Kano, "New Perspectives for Advanced Automobile Diesel Engines" *Proceedings of the 18th Intersociety Energy Conversion Engineering Conference,* (AIChE, 1983), pp. 600–608.

[16]"Booster Will Cut RTA Fuel Consumption to 117 g/bhp h," Marine Engineering Log, July 1984, p. 69.

Otto Cycle Engines for Cars, Mowers, Boats, and Chain Saws

- For powering small general aviation airplanes it is the most practical combination of high power-to-weight ratio and reasonable cost.
- For standby power Otto cycle engines are risky and not permitted by many codes because gasoline contains molecules that eventually polymerize into gum that clogs carburetor jets. Generator sets rated above 100 kW are usually powered by diesel engines.

At one time diesel engines appeared to be the route to high miles per gallon in automobiles. However, the Otto cycle engine, refined with computer-controlled fuel injection and spark advance, and coupled to a carefully matched transmission, has been able to equal and even exceed the diesel-engine fuel economy for automobiles. The limits of Otto cycle engines are these:

- Otto cycle engines have no lower limit in power rating. Model-airplane engines have outputs of a fraction of a horsepower.
- Large stationary engines burning gaseous fuel often use spark ignition because gases, in contrast to diesel fuel, are hard to ignite by compression.
- The efficiency of an Otto cycle engine is limited to about 30 percent because higher efficiency could be obtained only with compression ratios that are above detonation limits.

Units used in measuring diesel engine performance also apply to Otto cycle engines. The term ''spark-ignition'' engine is often used to distinguish Otto cycle engines from diesel engines.

A basic requirement of gasoline engines is that the fuel vaporize before it is admitted into the cylinder. This means that the fuel must not contain heavy hydrocarbons. The light ones have less energy, so the range of energy content of gasoline is 112,000 to 119,000 Btu per gallon. Diesel fuel has around 137,750 Btu per gallon. The hydrocarbon content of gasoline is also constrained by the need to avoid pre-ignition or ''knock'' in high-compression engines. Anti-knock qualities, at one time achieved with lead additives, are generally achieved by tailoring the hydrocarbon content of the gasoline.

Otto cycle diagram. In Figure 2–11 we start at Point 1 with the cylinder filled with an air-fuel mixture which is compressed in a polytropic process as the piston rises. No heat is added or subtracted intentionally between 1 and 2, but the temperature rises as the mixture absorbs compression energy. At some instant near the top dead center of the piston travel, the mixture is ignited and it burns quickly, releasing heat which further raises the pressure and temperature of the working fluid. Then the expanding gases push the piston downward with a force which is carried to the crankshaft by the connecting rod, producing output power.

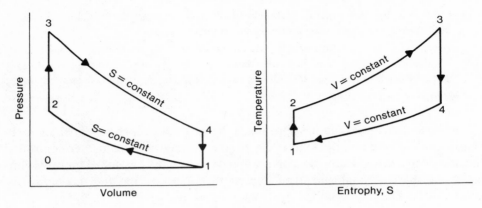

Figure 2-11

At Point 4 the exhaust valve opens, releasing the products of combustion and the hot working fluid.

In a two-stroke engine the gas in the cylinders is next scavenged and replaced with a new mixture before the piston starts upward again in its next cycle. In a four-stroke engine the piston rises with the exhaust valve open, clearing the cylinder. Then with the exhaust valve closed and the intake valve open the descending piston sucks in a new charge of fuel and air to return to Point 1. The cycle then repeats.

The area in the pressure-volume diagram within 1,2,3,4 represents the work done by the piston in one cycle. This work per power stroke times the strokes per minute is proportional to the power generated by the engine. As with the diesel engine, the theoretical pressure-volume and temperature-entropy diagrams are not the ones actually observed.

Design of Otto cycle engines. The engine as designed must be producible in an automatic factory. Some of the key requirements of a successful automobile engine are:

- Emission limits should be met with minimum unburned fuel discarded into the exhaust for burning in the catalytic converter.
- The required engine output torque can be anywhere between zero and maximum available at any vehicle speed.
- Whether cold or hot, the engine must start promptly and run smoothly.
- Intake air temperature can vary from around $-40°F$ ($-40°C$) to $+130°F$ ($°54$ C), and pressure altitude may vary from below-sea-level to above 12,000 ft (3,660 m).
- Fuel quality and battery voltage can vary.

The Gas Turbines—New Techniques for More Power, Higher Efficiency

- The running speed of the engine can vary from idle, around 800 rpm, to perhaps 4,500 rpm at maximum power. Some engines operate at even higher maximum speeds.

In Otto cycle engines the routes to high efficiency are higher combustion temperature and higher compression ratio. The combustion temperature can be no higher than that obtainable with a stoichiometric mixture of fuel and air. Higher compression ratios without detonation can be had with aviation gasoline, which is costly and limited in supply. Injection of water and alcohol into the intake helps to reduce detonation, but with the complication of extra tanks and pumps. An approximate expression for the efficiency of a gasoline engine (η_g) is:

$$\eta_g = 1 - \frac{1}{(R_k)^{k-1}}$$

where k = The gas constant of the working fluid
$R_k = \frac{V1}{V2}$ = compression ratio
$V1$ = Cylinder volume at the start of the cycle
$V2$ = Cylinder minimum volume

This relationship limits the efficiency of the Otto cycle engine to about 30 percent because high compression ratios cause the gasoline-air mixture to detonate rather than burn. Lead and other additives can make compression ratios of over 10 feasible, but most automobile engines operate with compression ratios of around 7 to 9. Among the highest was a 1971–74 American Motors V-8 engine with a ratio of 13.6. One of the lowest was a 4-cylinder engine built in 1971–73 with a ratio of 4.8. In contrast, a Cummins KTA 306 sixteen-cylinder diesel engine has a compression ratio of 15.5, and the Volkswagen Rabbit diesel uses a compression ratio of 23. The General Electric LM 5000 gas turbine has a compression ratio of 30.

High miles-per-gallon records are generally set by diesel-powered vehicles. However, by using a knocking-resistant iso-octane fuel, a University of Saskatchewan team won a 2.58 kilometer competition with 702.9 miles per gallon.

THE GAS TURBINES—NEW TECHNIQUES FOR MORE POWER, HIGHER EFFICIENCY

Three essential elements of a modern Brayton cycle gas turbine are a rotating air compressor, a chamber in which burning fuel heats the compressed air, and a turbine which extracts power by expanding the heated compressed air. The output power can come from either the reaction from the high-velocity exhaust stream which gives thrust for a jet airplane, or a rotating shaft that drives generators or other equipment.

Brayton cycle gas turbines are important because they are the best propulsion engines for jet passenger and cargo airplanes. The alternative Otto cycle gasoline engines have been superseded for all but small airplanes. A by-product of the intense development of ever better aircraft gas turbines has been a series of industrial gas turbines which compete with diesel engines and Rankine-cycle steam turbines in efficiency as well as reliability and cost.

Energy features of gas turbines are:

- Gas turbine power plants installed in the middle 1980s cost around $200/kW.
- Most gas turbines require refined fuel oil or natural gas. Some are running with treated crude oil and gas made from coal. Solid fuels have not been successful.
- Part-load efficiency is poorer than can be obtained from diesel engines and steam turbines.

Some of the limits of Brayton cycle gas turbines are:

- Units derived from Boeing 747 airplane propulsion engines can drive a 20 MW generator. A more powerful propulsion engine for airplanes is not likely to be developed because there is little interest in airplanes larger than the 747. However, larger gas turbines are being built for industrial and utility use.
- The use of ceramic first-stage turbine blades is only beginning. Fuel efficiencies of over 50 percent are possible.
- Efficiency of simple-cycle turbines will also rise as a result of higher pressure ratios, which are achievable with careful computer-aided compressor design, and tricks such as shroud cooling.
- As long as gas turbines cannot burn coal, their fuel will cost approximately four times that of the coal for steam power plants, except in areas of the world where petroleum is not heavily taxed, or natural gas has no better market.

Gas turbine cycles. A basic gas turbine has two sets of blades, mounted on disks. One set is arranged to compress the air and the other to extract power from the compressed air after it has been heated by burning fuel (Figure 2-12a). Air flowing through the blades experiences friction losses, and so the compression is not reversible. On the other hand, the air is heated by the friction as well as the compression, and this heat is partially recovered in the turbine. Thus, the friction heating partly reduces the fuel flow needed in the burner to maintain required turbine inlet temperature. The process, called "polytropic," is described

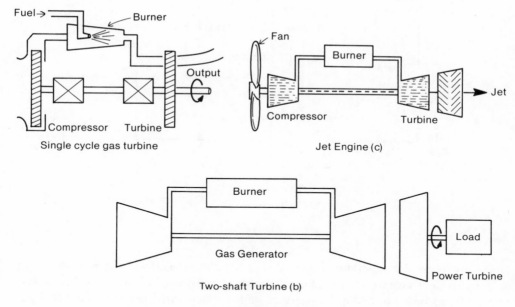

Figure 2-12

by Kolin.[17] He observes that an arrangement of spinning paddles among vanes is similar to the apparatus used by Joule to measure the mechanical equivalent of heat.

The simple gas turbine consists of a compressor, burner, and turbine, with exhaust air from the turbine being discharged into the atmosphere. The turbine is on the same shaft as the compressor, as is the gearing that extracts power from the shaft. This type of turbine is useful for driving electric generators because full air flow is going through at all times, and a sudden load increase requires only an increase in fuel flow.

For propulsion of ships and land vehicles, a better arrangement has two sets of gas turbine blading, one to drive the compressor and the other to drive the load (Fig. 2–12b). This allows the compressor to be slowed when the load is low thereby reducing fuel consumption. Some aircraft propulsion turbines have two spools of turbine blades, one to drive the compressor and the other to drive a turbofan (Figure 2–12c). Exhaust gas from the engine also contributes thrust.

If a gas turbine is 30 percent efficient, then 70 percent of the energy in the burned fuel is discharged in the gas turbine exhaust. A way to recover this lost

---

[17]Ivo Kolin, "The Evolution of the Heat Engine," (London: Longman Group Limited, 1972) p. 15.

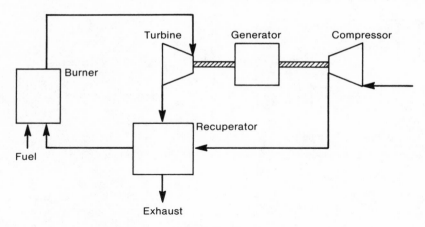

Figure 2-13

energy to use a recuperator (Figure 2–13). Common to all such approaches is the need to duct large volumes of air through heat exchangers that *must* have low pressure drops. For example, we once built a regenerated Boeing model 502 gas turbine. As a result of the pressure drops in the regenerator and the back pressure added to the turbine exhaust, the regenerated turbine was less efficient than the simple engine. Heat exchangers with high efficiency can be built and they will not load the turbine with backpressure (Figure 2–14).

Garrett AiResearch makes metal heat exchangers with 85–90 percent effectiveness. However, the effectiveness falls to 50–60 percent at 2000°F (1090°C). Above 2000°F the heat exchangers must be made from ceramics. Kongsberg's KG 3-R gas turbine, rated 1.47 MW, has a regenerator with 88 percent effectiveness. The turbine achieves 36 percent efficiency with a 9 to 1 pressure ratio and 1100°C (2012°F) turbine inlet temperature. The unit is 6.1 m (20 ft) long, 1.9 m (6.2 ft) wide, and 2 m (6.6 ft) high. It weighs 9,800 kg (21,600 lb).

A 27 percent reduction in fuel rate is being achieved with a regenerator in Soyland Power Cooperative's compressed air storage plant in Illinois. Here, previously compressed air stored in underground tunnels is heated and delivered to a turbine that generates peak-load power. The regeneration pays off because the air from the underground storage is cold and high-recuperator effectiveness is not needed to produce a useful temperature rise.

Possible combinations of gas turbines and equations for calculating their performance are provided by Kolin.[18]

[18] Ivo Kolin, *The Evolution of the Heat Engine* (London: Longman Group Limited, 1972), pp. 12–13.

The Gas Turbines—New Techniques for More Power, Higher Efficiency 51

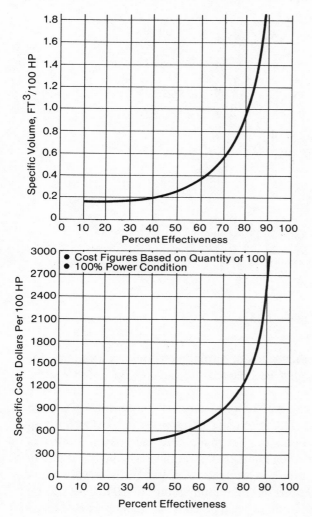

Figure 2-14

Recent gas turbine development. The gas turbine illustrates how billion-dollar development programs, well-motivated and properly directed, can produce heat engines that outperform traditional ones.

The motivation for turbine engine development was the high productivity of the commercial jet airplane. For example, a fully loaded 747 can generate around $25,000 per hour from passenger fares while flying. The cost of engines for these airplanes was less important than their reliability and fuel efficiency. The result-

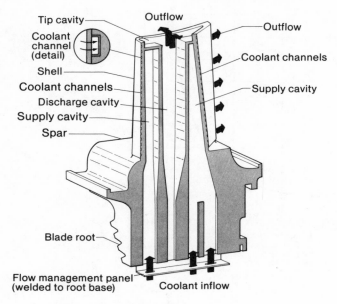

Figure 2-15

ing tools and plants for building thousands of airplane engines are available for building power units that could also be adapted for power generation on the ground.

The achieved performance and reliability of aircraft gas turbines has been remarkable. The Rolls Royce RB211-535 engine is rated for 33,000 h between shop visits. Careful design of compressors has made practical pressure ratios as high as 30 to 1. This is significant when compared to ratios of around 8–10 to 1 in Otto cycle engines and 14–20 to 1 in diesel engines. Cooling of the blades in the first stage of the turbine has made turbine inlet temperature of 2012°F (1100°C) practical. The blades are cooled with air routed from the compressor through passages in the shaft and through the turbine disk to the interiors of the blades. This cooling air is then discharged from the blades in a direction that contributes to the output of the turbine (Fig. 2–15 from Ref. 19). For example, General Electric in its uprated LM 2500 engine, is using a 2214°F turbine inlet temperature. By cooling the metal in the turbine blades in the first row to 1400°F, the firm was able to show 25,000 operating hours between hot-section repair.

[19]"Advanced Cooling of Utility Gas Turbine Engine Components," *EPRI Journal*, October, 1982, p. 36.

The Gas Turbines—New Techniques for More Power, Higher Efficiency

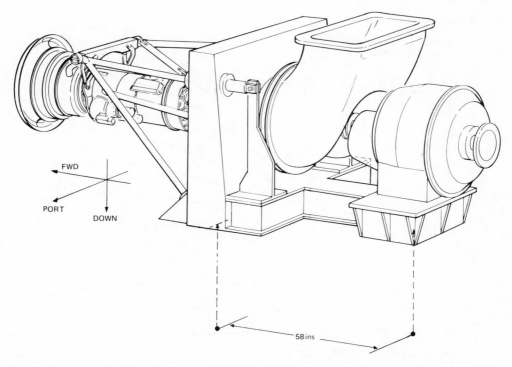

Figure 2-16

Engines with over 2000°F (1093°C) turbine inlet temperatures are getting over 34 percent efficiency. Installed beside Tokyo Electric's Sudegaura station is a 124 MW AGTJ100A gas turbine designed to operate at 1300°C (2372°F) turbine inlet temperature. Its simple-cycle efficiency of over 39.4 percent surpasses most industrial diesel engines and coal-burning power plants.[20]

Practical gas turbines for power generation and mechanical drive are often combinations of jet engines and additional stages. The jet engine, consisting of the compressor, burner, and high-temperature turbine blades and disks, brings the benefit of costly development that produced reliability and efficiency. The additional stages and mechanical adaptation are then furnished by licensees who also assemble and market the power plants.

Gas turbine characteristics. A Rolls Royce 6000 hp Marine Tyne engine illustrates gas turbine characteristics. This unit is 168.25 in (4.273 m) long, 52 in

[20] "Energy News and Trends," *Gas Turbine World,* May–June, 1984, p. 60.

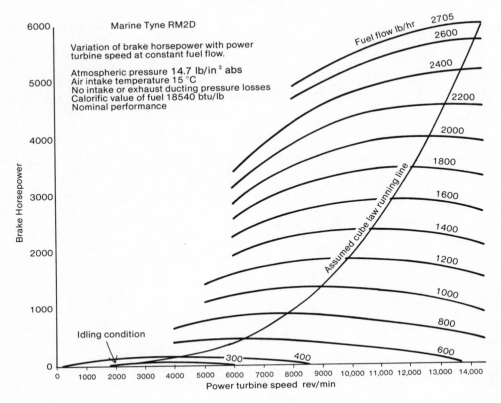

Figure 2-17

(1.32 m) wide, and 67 in (1.7 m) high (Figure 2–16). The primary reduction gearing of the Tyne has a ratio of about 4 to 1, to reduce the maximum speed at output to a value suitable as an input to the ship's main reduction gearbox. The maximum speed of the propeller of a major warship is, typically, around 250 rpm. At all other shaft speeds, less than full power is required from the turbine, so the hot gas-producing section of the turbine can produce less gas and run slower during cruise. The gas-producing unit can continue running at its idling speed even when the propeller shaft is stopped. The idling fuel flow is 300 lb (136 kg)/h. The limiting fuel flow at full power is 2,800 lb (1,270 kg)/h, a value that relates to the maximum allowable turbine inlet temperature for this unit.

The gas generating unit can be operated at less than full speed and full fuel flow when the ship is cruising at less than maximum speed (Figure 2–17). At lower speeds, the power required to drive the propeller falls off along a speed-cubed curve. However, the specific fuel consumption of the gas turbine goes up as the load is reduced (Figure 2–18).

An important feature of the engine is that, unlike the typical diesel engine, it

The Gas Turbines—New Techniques for More Power, Higher Efficiency 55

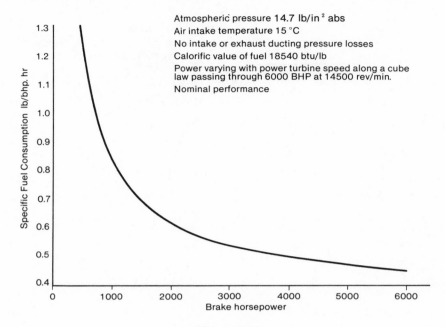

Figure 2-18

is not torque-limited. If the ship's power-speed characteristic alters, because of hull fouling for example, the gas turbine can readily apply more torque at lower speed.

The decreasing fuel efficiency of a gas turbine at part load may not be important in an ocean-going ship that travels at cruise speed most of the time. It is very important in trains, trucks, and automobiles that rarely need maximum power for propulsion. For this reason, simple gas turbines are unsuitable for these land vehicles. However, the addition of intercoolers between compressor sections, and a recuperator (also called "regenerator") to recover waste heat in the power turbine exhaust, can improve part-load efficiency significantly. The Avco LYcoming AGT 1500 is an example of a recuperated engine that is used by the US Army for Battle Tank propulsion.

Intake air temperature affects permissible gas turbine output. The International Standards Organization (ISO) suggests rating gas turbines at 15°C temperature and sea-level pressure. The output of the Rolls Royce Tyne RM2D engine as a function of air temperature is plotted in Figure 2–19. Hot air is less dense than cold air. The air compressor is a constant-volume device that pumps more lb/minute of cold air than hot air. However, higher compressor speeds are

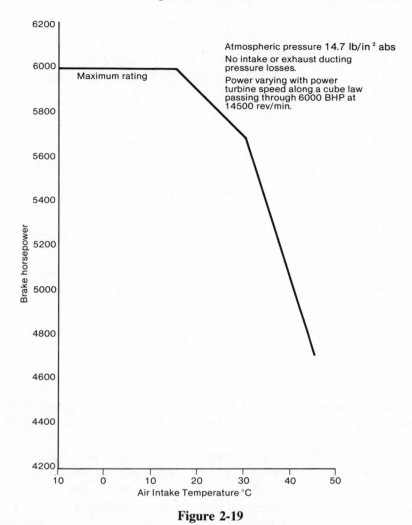

Figure 2-19

needed to pump this additional air. In the Tyne, stress considerations impose a limit on low-pressure compressor speed between 15 and 30°C.

The limiting Carnot efficiency of a turbine depends on the temperature difference between the heat source and sink temperatures. The heat source temperature is limited by the allowable temperature at the turbine inlet. The sink temperature is that at the compressor inlet. The higher the sink temperature, the smaller the temperature difference, and the lower the efficiency. Hence, from 30°C and upward, power turbine entry temperature limits the achievable power.

The rated output is quoted for a turbine on a test stand with no inlet or

The Gas Turbines—New Techniques for More Power, Higher Efficiency

exhaust ducts. When installed, the turbine usually has ducts that direct the inlet and exhaust noise upward, and the ducts often contain silencing material. For example, Rolls Royce suggests that a possible duct pressure drop for a marine installation of their Tyne turbine would be 4 in (10.16 cm) of water in the intake ducts, and 6 in (15.24 cm) of water in the exhaust ducts. These pressure drops reduce the 30°C rating of the unit from 5,700 bhp (4,252 kW) to 5,500 bhp (4,103 kW), and increase the specific fuel consumption from 0.462 lb/bhp-h to 0.475 lb/bhp-h.

Another example of derating is a Brown Boveri installation of two Type 11 units for SCECO, South Shadig, in Saudi Arabia. The units had an ISO rating of 72,500 kW. The rating, when installed at the site and burning treated crude oil, is 48,720 kW; this is 33 percent lower than the ISO rating.

In another installation at Rusail, Oman, John Brown "Frame 9E" units rated 107.85 MW under ISO conditions deliver 82.056 MW when run at 50°C ambient temperature. At 40°C they were able to produce 94 MW.

Single shaft gas turbine. A two-shaft gas turbine, such as the Rolls Royce RM2D, works well for marine power. It also works well for driving alternators that generate power for utility grids. The grid normally has so many big generators supplying its power that it looks like an infinite bus to a gas turbine driving a generator. Any change in the alternator shaft torque will affect neither frequency nor alternator speed. Thus, the output power of the turbine generator plant can be controlled by merely adjusting the fuel flow to the burners.

Most gas turbine plants on utility grids are used to help carry peak load and shut down when they not required. This is because they can be started and brought to full load in about two minutes. Steam plants, in contrast, require several hours to be brought on load from a shutdown condition. For example, Puget Power with 247 MW of gas turbines at Fredonia, Washington, expects to run them no more than 500 h/year burning oil or natural gas. At other times, the utility's coal-burning and hydro plants supply the load.

A two-shaft turbine, such as in Figure 2–12b would not work well as an isolated power plant. At low load the gas-producing section would be operating at less than full speed; a sudden increase in load would tend to stall the generator-driving turbine until the gas-producing section gains speed and output. This change in speed takes time, perhaps seconds.

We once adapted the Boeing Model 502 two-shaft turbine to drive a generator. The only way we could control speed with quickly-varying loads was to install a wastegate at the compressor outlet. When load was dropped the wastegate would open. Fortunately, fuel was not very expensive at that time.

A single-shaft turbine arranged as shown in Figure 2–12a is common for driving alternators that are not connected to a utility grid.

Installation of gas turbines. The output of a gas turbine plant is less than the turbine ISO rating because of temperature, altitude, and duct losses. The generators and reduction gears also have losses, around 5 percent for a big generator and 1–2 percent in the gears. Power is consumed in the plant by fuel pumps, lube-oil pumps, and the blowers that send air about the gas turbine to cool the turbine surfaces and mounted accessories. Turbine and gearbox oil must be cooled, either with pumped water or blown air. For example, the Rolls Royce RM2D gear-box oil releases 405,000 Btu/h and the turbine oil releases 221,000 Btu/h. The lube oil pump circulates 1,400 imperial gal/h to carry away this heat, and the turbine consumes about one pint of oil/h.

Acoustic silencing is normally needed. The peak unsilenced intake sound level for the RM2D turbine is 128 dB at the octave band centered around 4000 Hz. A common form of silencing is a right-angle bend in ducts lined with acoustical absorber to direct the exhaust flow upward.

Periodic washing of deposits from compressor blades may be required in marine and other salt-air or contaminated-air installations. The washing apparatus sprays distilled or demineralized water into the compressor blades while the engine is being cranked.

Cadmium and zinc can be undesired contaminants in fuel and oil. Rolls Royce prohibits the use of cadmium plating and zinc galvanizing in lines and tanks that contact fuel.

Gas turbines are provided with ports that can be opened for boroscope inspection of critical parts such as turbine blades and burner cans.

Starting a gas turbine. To start a gas turbine, the compressor has to be brought up to a speed where it pumps enough air through the burners to avoid melting turbine blades. Small turbines can be started with battery power. Aircraft propulsion gas turbines usually have air starters. In the tail section of most modern jet airplanes is a battery-started auxiliary gas turbine that supplies electricity for the airplane while it is on the ground, and compressed air for starting the main engines.

A single-shaft turbine for generating electric power can be started from utility power, using the generator as a motor, provided that the generator has damper windings and appropriate reduced-voltage starting control. One make of gas turbine equipped for "black start" has on its bedplate a diesel starting engine driving a pump that powers a hydraulic starting motor.

PEAKING POWER PLANTS

Low first cost, quick startup, and unmanned operation with remote control are all important requirements of peaking power plants. Peaking power is needed by electric utilities for a few hundred h/year; for example, to carry air conditioning load during hot summer days. An alternative to buying peaking-power generators is to build more base-load plants, which cost interest when not being used. Another alternative is buying power from other utilities if they can spare it and if transmission paths are availabile. For those who have it, hydro power, if backed by water storage, is ideal for peaking power.

The Brayton cycle gas turbine is the common peaking-power generator, costing only $200–$500/kW. The high cost of fuel oil or natural gas for the peak load plant is not as important as it would be for a base load plant. Furthermore, the efficiency of gas turbines almost matches that of large, well-designed steam plants. Thus, if oil or natural gas is to be burned, it is better burned in a gas turbine.

Compressed-air energy storage for peaking power. In compressed-air energy storage for utilities, the air is pumped into underground caverns and heated before being expanded in a turbine. Thus, the thermodynamic cycle is essentially a Brayton cycle, with loss of heat following compression of the working fluid. The performance of a compressed air storage installation is measured by its heat rate, which is the Btu of heat added to the compressed air to generate power, not considering the energy going into the compression of the air. For example, the Huntorf plant in Germany has a heat rate of 5,500 Btu/kWh. This compares with 9,000 Btu/kWh for the best coal-burning power plants.

The cost of the energy required to compress the air might not be significant. For example, a hydro plant that is releasing water over its spillway can provide free power when there is no other market for its power. Similarly, a nuclear plant at night can produce power for only the small cost of the uranium fuel.

An example of compressed-air storage is a 220 MW plant built by Soyland Power Cooperative in Decatur, Illinois. Suitably impervious rock was found 1,800–2,000 ft underground, below which a 275,000 cubic yard (210,000 cubic meter) excavation in the form tunnels will store the compressed air at 57 atmospheres (5.8 Mpa). A water column with a surface reservoir floods the underground storage space as air is released, essentially maintaining constant air pressure. As a result, the plant can run for 11 hours at full power with a heat rate of 4,100 Btu/kWh. Using turbine exhaust to preheat the air rising from storage contributes to the low heat rate.

An older 290 MW plant at Huntorf, West Germany, stores compressed air in a solution-mined salt dome. Without water pressurization, the air pressure diminishes as the stored air adiabatically expands during extraction. For each kWh stored, the variable-pressure plant requires six times the underground volume required by a plant with water-column pressurization.

The estimated cost of the Soyland power plant in 1982 dollars was 156 million, or $709/kW. Charging power will come from coal and nuclear power plants.

COMBINED CYCLES: THE ROUTE TO 50 PERCENT EFFICIENCY

A combined cycle is a set of heat engines in which the heat rejected by one or more engines is the energy input for other heat engines. For example, the exhaust of a gas turbine can be ducted into an unfired boiler to generate steam for a Rankine cycle turbine.

Energy features of combined cycles include:

- Combined cycles offer efficiencies of 45–50 percent in modules with rated output of around 100 megawatts electrical (MWe). The Rankine cycle power plants that approach 40 percent efficiency must be much larger to justify the efficiency enhancing auxiliaries needed.
- Combined cycle plants are generally assembled from factory-built units, shortening construction time. Building a conventional Rankine-cycle steam plant takes five to eight years. A combined cycle plant can be erected in two to three years, saving interest paid on funds during construction. For example, Vulcan Materials Company's 100 MW plant at Giesmat, Louisiana, was in operation one year after site preparation started.
- The cost of a combined cycle plant is 30–50 percent of that for a Rankine cycle plant having the same rating. For example, the cost of the 100 MW Vulcan plant was $48 million.[21]
- The gas turbine in a combined cycle plant requires a liquid or gaseous fuel, which costs more than coal, the common fuel in Rankine cycle steam plants. In the mid-1980s natural gas cost $2–$3/million Btu and coal cost $1–$2/million Btu.
- A combined cycle plant can respond to load changes three times as fast as a steam plant can.
- A gas turbine with a Rankine cycle bottoming turbine seems to be the most practical combined cycle power plant.

The limits of combined cycle power plants include:

- The efficiency of combined cycle power plants can only approach the Carnot cycle efficiency for the source and sink temperatures used. A bottoming cycle on a Rankine cycle steam plant cannot alter the Carnot cycle efficiency of the plant.
- Gas turbine inlet temperature is a key to combined cycle efficiency. Practical temperatures improve as new and more efficient aviation gas turbines are devel-

[21] "Combined Cycle for Vulcan Plant," *Cogeneration*, September-October, 1984, p. 4.

oped. A high compression ratio, which contributes to high simple-cycle turbine efficiency, is not important in combined cycle plants. A high compression ratio produces a cooler turbine exhaust that reduces the efficiency of the steam bottoming cycle.
- The exhaust gases of a diesel engine are at a temperature of 500°–900°F (260°–480°C), which is suitable for generating steam in a bottoming cycle boiler. However, almost 45 percent of the engine losses appear at around 212°F (100°C) in the engine cooling water. The energy available at this low temperature far exceeds that needed to heat the feedwater for the boiler, and hence all of it is not usable for producing power.

The term *specific work* is a measure of the performance of gas turbines used in combined cycles. This is the ratio of power output in kilowatts to the air flow in pounds per second.

Mechanization of combined cycles. In a typical combined cycle the exhaust of a gas turbine is ducted through an unfired boiler into the atmosphere. The boiler generates steam for a turbine that supplements the gas turbines in producing power (Figure 2–20). In the boiler, the exhaust normally flows first through a steam superheater, then through an evaporator, and finally through economizer heat exchangers. In this manner all feasible heat is extracted out of the gas, with the discharge being around 250°–350°F (121°–176°C). Design objectives include operating the turbines at their highest practical efficiencies and avoiding wet-steam problems.

The power output of a combined cycle is controlled by regulating the fuel flow into the turbine. Output is increased by admitting fuel at a faster rate. This raises exhaust gas temperature, causing the boiler to produce more steam, which increases the steam turbine output. The fuel is usually petroleum, or natural gas which is delivered at a pipeline pressure of 250–350 lb/in^2. Gas at this pressure requires no further boosting before admission into the burners in the engine. In contrast, the control of a coal burning steam plant is more complex. A change in power output requires changes in fuel firing rate, feedwater flow, and air flow in the boiler.

A gas turbine power plant can be carrying one half of its rated load within 10 minutes after being started. A hot steam turbine in a combined cycle can be carrying load one hour after being started. Steam turbines for combined cycle plants are usually small units that are designed for fast response so that they can follow varying load. The load can be carried by a cold steam turbine three hours after a startup order is received; the gas turbine can be started and stopped as load varies. If the utility has multiple turbines, then one or more can be operated at full or part load as required.

The use of an unfired boiler introduces inherent safety and reliability. Safety

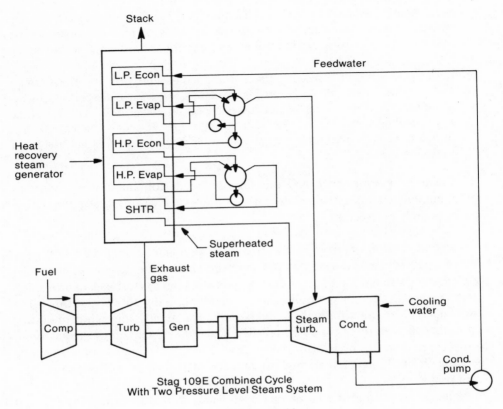

Figure 2-20

codes require that fuel-burning steam boilers be watched, because a low water level or overpressure can result in an explosion. The flames in a high-performance steam boiler can melt boiler tubes in which no water or steam is flowing. In an unfired boiler the temperature of the boiler tubes can become no hotter than the exhaust temperature of the gas turbine, which is below the melting temperature of boiler materials.

R. L. Messerlie and A. O. Tischler propose that the same turbine accept both hot gas and steam and convert the heat to electric power efficiently.[22] The steam would be generated in an unfired boiler that is heated by the gas turbine exhaust. In a test an Allison Model 501 KB gas turbine, which is a power-

[22]R. L. Messerlie and A. O. Tischler, "Test Results of a Steam Injected Gas Turbine to Increase Power and Thermal Efficiency," *Proceedings of the 18th Intersociety Energy Conversion Engineering Conference,* AIChE (1983), pp. 615–625.

Combined Cycles: The Route to 50 Percent Efficiency

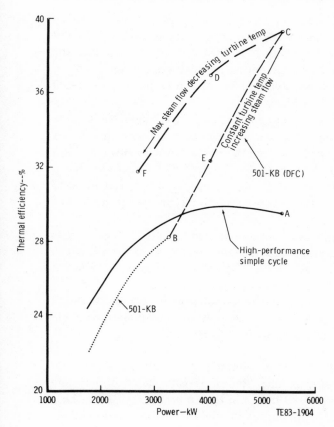

Figure 2-21

producing adaptation of an aircraft gas turbine engine, was modified with the installation of two steam manifolds around the outer combustion case. The results of this concept are plotted in Figure 2–21. "Cheng/DFC" is the steam injection condition. The power output of the turbine was increased from 3,260 kW to 5,370 kW by the admission of steam. The turbine efficiency increased from 29.1 to 39.5 percent. Thus, power output and efficiency were improved with only a bigger generator and an unfired boiler being required.

A disadvantage of the steam injection is that makeup water of high purity must be pumped into the boiler. The spent steam is lost in the gas turbine exhaust. In ordinary combined cycles, the steam is condensed and pumped back into the boiler; this avoids buildup of boiler scale.

Fuel for combined cycles. Combined cycle power plants had been discouraged in the U.S. by the Power Plants and Industrial Use Act of 1978, which prohibited the use of oil as a fuel in new power plants that operate over 1,500 h/year. Combustion gas turbine plants can perform their peaking service within the 1,500 hour limit. On the other hand, the owner of a higher efficiency plant would like to operate his equipment more than 1,500 hours a year to reduce the cost of capital per kWh generated. An exception from the act is required for burning natural gas or oil in a combined cycle plant. The trend has been to grant such exceptions for natural-gas fueled plants.

One alternative is to convert the coal to gas, clean the gas, and burn it in the combustion turbines. This process has losses, discussed in another chapter. A coal-gasification combined-cycle plant at Southern California Edison's Cool Water station generates 100 MWe. Illinois Power's Wood River operation has a 600 ton/day kiln that supplies 145–165 Btu/cubic ft gas for a boiler, with a 37 MWe combustion turbine planned. Note that natural gas has an energy content of around 1,000 Btu/cubic ft.

In Japan, where oil and natural gas are not restricted for turbine fuel, Tokyo Electric Power Co. is installing several combined cycle plants. General Electric is supplying two 1,000 MWe plants. Each of GE's units consists of a 108.8 MWe gas turbine that supplies heat to an unfired boiler that generates steam for a 34 MWe steam turboalternator. The arrangement is shown in Figure 2–20. With natural gas fuel, a 47 percent efficiency is expected. Additional combined-cycle plants have been ordered from other firms by Tokyo Electric.

Cost of combined cycle plants. In a combined cycle much of the power is generated by low cost gas turbines, rather than by costly steam boilers and steam turbines. Furthermore, the unfired boiler in the combined cycle can be a factory assembled unit, rather than being built in the field with construction labor.

In 1983, Pakistan requested bids for two 300 MW power plants. The received prices ranged from $206–$291/KW for the gas turbine part of the plant. The bids did not include the heat recovery boiler or steam turbine generator set. Almost two-thirds of the plant capacity was to come from the gas turbines.

At the same time Al Taweelah, in the United Arab Emirates, opened bids for a nominal 300 MW ISO rated gas turbine plant with heat recovery steam generators for seawater desalinization. The 55°C rating at sea level of the plant was to be 200 MW. The scope of the contract was "turnkey" in that it included the machinery packaged for outdoor installation, foundations, distillate storage tanks, cranes, and exhaust stacks. Some suppliers quoted a five-year supply of spare parts. Quoted prices ranged from $238–$605/kW without spares. Spares added from $5–$170/kW. Note that boilers but no steam turbines were included in these prices. Deliveries ranged from 22–24 months.

At a session sponsored by Alsthom Atlantique for representatives from utilities, General Electric showed that operating h/year is an important parameter

Combined Cycles: The Route to 50 Percent Efficiency

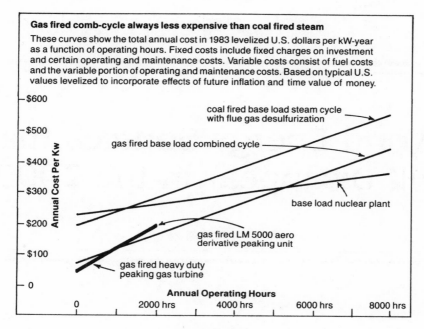

Figure 2-22

in comparing the cost of power generation with combined cycles and other plants (Figure 2–22).

General Electric observed that a strategy called "progressive generation" usually wins in comparison of power generation options. This scenario starts with a site having enough area for gas turbines, boiler, steam turbines, coal gasifier, and coal storage. The initial installation is natural gas fired gas turbines for carrying the utility's peak load. A five-year exception is arranged to allow the turbines to run on natural gas fuel. Extensions of the exception are arranged later. Boilers and steam turbines are added to improve efficiency as the utility's load grows and operating time per year increases to where the new plant carries base load. Coal gasification is added only when its use can show lower operating cost.

3

World Energy Sources that Will Disappear in the 2000s

Natural gas, petroleum, and coal are fossil fuels that will ultimately be exhausted if not saved in the very near future. Uranium is not a fossil fuel, but it too can become exhausted. However, the scope of most energy systems engineering problems is less than 30 years, so the important task is estimating fuel costs when depletion becomes a contributor to the price of fuels. Other factors such as inflation, cartels, and taxation also affect the price of fuels.

Coal resources are so great that during a 30-year period, depletion of coal resources is not likely to affect cost. These resources are so scattered that cartels are unlikely. At one time coal use was constrained in industrial cities because of smoke pollution. Smoke pollution need not be a problem with properly designed boilers. Fly ash can be controlled with electrostatic precipitators and bag houses. Sulfur emission can likewise be controlled. However, nothing can be done about the carbon dioxide released when coal is burned.

The price of natural gas tends to relate to the price of oil because industrial users can convert their boilers to burn oil if the price difference between gas and oil is too great. Rising gas prices can motivate such switches and can also motivate marginal producers to seek and drill new wells.

The cost of oil sold by major producers has little relation to the cost of production. Taxes charged by governments of countries in which the oil is produced and used usually exceed the cost of producing the oil. Cartels, at times, control oil production to maintain its price.

Coal and How to Enrich It

Measuring energy from oil, gas, and coal. Measurements of fossil energy quantities involve big numbers. For example a quarter-page advertisement in the *Wall Street Journal* showed the chairman of Diamond Shamrock emerging from a meeting with the company's president to present an announcement about Oregon. He said that Oregon's hydrocarbon search had previously featured 200 dry holes. Diamond Shamrock, with high-tech science, had hit 6 million cubic ft of natural gas per day!

The oil equivalent of this gas (Oe) is:

$$Oe = \frac{6 \times 10^6 \text{ cu ft} \times 1000 \text{ Btu} \times \text{gallon oil} \times \text{barrel}}{\text{day} \times \text{cu ft} \times 138{,}500 \text{ Btu} \times 42 \text{ gallons}}$$
$$= 1030 \text{ barrels/day or } 43 \text{ barrels/h}$$

In comparison, the Trans Alaska pipeline delivers 1.6 million barrels per day. Just one of the pumping stations on the Trans Alaska pipeline burns 3,300 barrels a day to power the pumps.

The British thermal unit, which is equivalent to 1055.06 joules, is a useful unit for comparing coal, gas, and oil energy. Approximations useful in comparing energies are:

Fuel	Quantity	Typical Btu Released When Burned
Petroleum	1 gallon (7 lb)	138,500 for crude, 137,750 for diesel fuel
Coal	1 lb	11,000
Natural Gas	1 cubic ft	1,000

Natural gas is generally purified to the extent necessary to achieve 1,000 Btu/cubic ft. Coal varies from 7,500 Btu/lb for lignite to 15,000 Btu/lb for the best bituminous coal. The energy content of petroleum products can range from 73,380 Btu/gal for ethane to 149,000 Btu/gal for the residual fuel burned in power plant boilers.

COAL AND HOW TO ENRICH IT

Coal is a plentiful fuel, composed of carbon deposited by ancient plants, and gathered by mining with surface or underground techniques. Coal at one time was easily bought in cities for heating of buildings and houses. Clean-air standards in most cities limit the burning of coal to boilers and furnaces that achieve smoke-free combustion. Even more costly are requirements that limit allowable sulfur emissions. Coal now is a common fuel for power generating steam plants. A 1,000 megawatt (MW) power plant running at full load will burn, during one day, about 10,000 tons of coal, the cargo of a unit train having one hundred 100-ton hopper cars.

Energy features of coal are:

- Ordinary bituminous coal releases around 11,000 Btu of heat/lb when burned in air.
- Ash is an undesirable ingredient in coal because it contributes no heat during combustion.
- Sulfur degrades the value of coal because in most states expensive sulfur-removal apparatus is required for cleaning stack gases when high-sulfur coal is used.
- The spot market price of coal was around $40 a ton in 1985. This corresponds to 2 cents/lb. If this coal had 12,000 Btu/lb, then 23 cents would buy 137,750 Btu of heat, the energy content of a gallon of diesel fuel costing around $1.00 at that time.
- Coal is best handled in large quantities. For example, the 1,500 MW base load plant being built by the Intermountain Power Project in Utah will consume four million tons of coal a year.

Coal is important because it is the world's most abundant form of fossil energy. Coal was first used by the Chinese in 1200 B.C. In the U.S. coal is the predominant fuel for electricity generation (Figure 3-1). Of the 2,309 billion kWh generated in the U.S. in 1983, 1,290 billion was generated with coal energy. Coal is a source of liquid and gaseous fuels as petroleum and natural gas resources approach depletion.

Limits in the use of coal are:

- The best semi-bituminous coal has around 5 percent moisture, 2.5 percent ash, 0.55 percent sulfur, and delivers 14,000 Btu/lb. Poorer grades have more ash, moisture, and sulfur. One Montana coal has only 0.5 percent sulfur, but a 25 percent moisture content results in 9,300 Btu/lb.
- United States coal reserves are so great, around 440 billion tons, that their exhaustion need not concern the energy systems engineer. The U.S. consumption is around one billion tons a year.

Coal is sold by the ton, which weighs 2,000 lb (907 kg). Mine production is reported in tons/year, and coal processing and handling equipment is described in tons/hour and tons/year. Coal-carrying freight cars are rated in lb or tons, common ones having capacities of 100 tons. The energy content of coal is commonly specified in Btu/lb. Steaming coal is a fuel burned in steam power plants.

Mine productivity is measured in tons per man per day, a value found by dividing the day's production of a mine by the number of its employees.

Coal and How to Enrich It

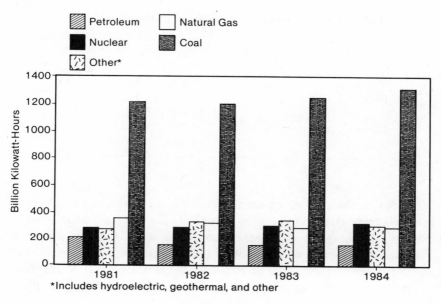

Figure 3-1

Coal availability problems. You would have very little difficulty in having a few hopper cars of any kind of coal you want delivered to your railroad siding. The problems come when you want to invest a billion dollars in a coal-burning steam plant that generates 1,000 MW of electric power. A power plant boiler is usually optimized to burn coal with a specific quality that is available from a particular mine. Changing coal suppliers can involve costly boiler modifications or off-optimum operation. The options and the opportunities for mistakes are many.

Once, Seattle City Light—a municipal utility—considered building a coal-burning steam plant near the center of its loads. An apparent source of coal was the mines in Roslyn and Cle Elum, a hundred miles east of the city. These mines had supplied the Great Northern railroad in its coal-burning days. However, the coal there was in deep underground seams, some only two feet thick, and there was no hope of supplying 5,000 tons every day for a 500 MW steam plant. A group of mines in British Columbia could have delivered the coal by barge, but they could not form a contracting entity that could have assured a 30-year supply. Seattle City Light subsequently abandoned the plans for a downtown coal-burning power plant.

TABLE 3-1 From Wood to Graphite, Heating Values

Fuel	Heating value, dry and clean, Btu/lb
Wood	5,000– 9,000
Peat	7,000–11,000
Lignite	11,000–12,000
Bituminous coal	12,000–15,000
Anthracite coal	15,000–16,000

Coal formation and resources. Coal was formed when giant seed-ferns and non-flowering trees, growing in a tropical rain forest, died in a swamp in which water without oxygen killed the bacteria that would have aided decomposition. The youngest peat is approximately 10,000 years old. Pennsylvania coal is around 250 million years old. The coal-forming process required pressure, but high temperatures would have converted the carbohydrates into gas rather than coal.

Graphite is pure carbon, that yields carbon dioxide—a gas—when it burns, and hence has a high heating value. Bituminous coal, lignite, and peat have ash, moisture, and hydrogen-oxygen compounds that affect heating value (Table 3–1). Coal as burned in power plants is rarely pure, clean, and dry.

World coal resources are shown in Figure 3–2. India is the only developing nation with significant coal deposits. The U.S. accounts for more than one-half of the coal mined in non-communist countries. Most of the U.S. reserves of low-sulfur coal are located in the West. This coal is needed in the East for power plants that are restricted in their release of sulfur (Figure 3–3).

Mines and mining. Many U.S. underground mines, which are based on blasting, gathering, and hauling out the coal, are obsolete. Enforcement of new safety standards and recurring labor troubles have cut production to as low as 10 tons per day per employee, whereas an open-pit mine can deliver 70 tons per employee per day. In a conventional underground mine only about one-half of the coal is recovered, the rest being left for pillars that keep the roof from caving in (Figure 3–4).

Longwall mining is the most productive technique for extracting coal in an underground mine. A 450–600 ft row of jacks supports the roof while a cutting machine moves back and forth across the coal face, shearing off coal and carrying it away on a conveyor. Hydraulic jacks support the roof during the mining (Figure 3–5). The cutter and conveyor are withdrawn after a block of coal has been excavated and the jacks are advanced, one by one. The ceiling behind the jacks is

Coal and How to Enrich It

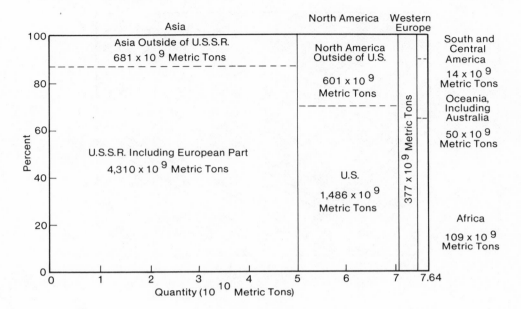

Figure 3-2

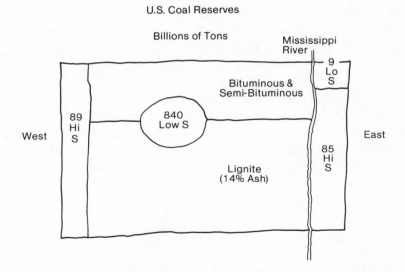

Figure 3-3

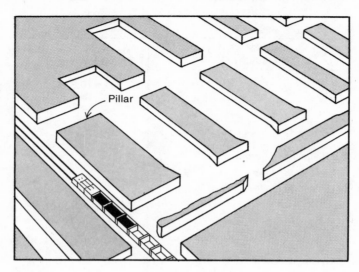

Figure 3-4

left to collapse. This underground shifting does not seem to disturb the surface, according to a Utah Power report.

Open-pit mining is used where removing some 10–100 ft of overburden with huge drag lines exposes the coal seams. The coal is stripped out with power shovels or loaders and trucks, and the land is restored to its initial condition. Most western coal mines are open pit.

An insight into the marginal cost of coal production comes from the Southern Utah Fuel Co. mine at Convulsion Canyon, Utah. An effective employee motivation that includes bonuses has maintained a 30 tons per day per worker for a yearly total of 2.2 million tons.[1] This was claimed to be 2.5 times the average daily rate for miners of coal in underground mines. The starting pay plus bonuses was more than $30,000 a year. The mine had 300 workers.

We might assume that the average pay is $40,000/worker/year, and that the fringe benefits add another 40 percent to this. If the management chooses to continue mining coal without profit or interest on its investment, the cost of labor (Cl) would be:

$$Cl = \frac{\$40{,}000 \times 300 \text{ employees} \times 1.4 \times \text{year}}{\text{employee} \times 2.2 \times 10^6 \text{ tons}} = \$7.64/\text{ton}$$

[1]Matt Moffett, "Great Productivity at a Utah Coal Mine Is Called 'God's Plan,' " *Wall Street Journal,* April 12, 1984, p. 1.

Coal and How to Enrich It

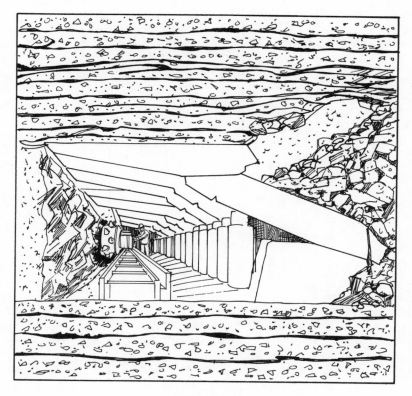

Figure 3-5

Note that 30 tons/day is exceptional productivity in an underground mine. The Pyro mine in Kentucky achieved a production of 25 tons/employee/day by paying miners $29,000/year. An open-pit Western mine can deliver around 70 tons/day/employee. Montana coal is sold for around $10/ton at the mine. Out-of-state buyers pay around $2 severance tax plus freight.

El Paso Electric's 1984 annual report notes that during that year its average coal cost was $14.56/ton. The company paid $0.83/million Btu for coal, compared to $3.62 for natural gas and $7.16 for oil. Minnesota Power's 1984 average cost of fuel for electric generation was $1.15/million Btu. It burned Montana coal and North Dakota lignite.

EPRI has predicted the effect of depletion on the price of coal for the period 1990–2000 (Table 3–3). Note that no depletion is expected for North Dakota lignite. Price rises caused by depletion are not significant compared with the possible inflation in labor rates and the effect of world market conditions.

TABLE 3-3 Impact of Depletion on Minemouth Coal Price Growth Rates (1990–2000)

			Price increase (%/yr)		
Region	Coal type	Sulfur content	Initial base estimate	With hypothetical constraints	Revised base estimate
Western Pennsylvania	Bituminous	High-medium	0.8	1.6	0.8
Southern West Virginia	Bituminous	Low	0.8	1.8	1.3
Southern West Virginia	Bituminous	Medium	0.9	1.9	1.4
Illinois	Bituminous	Very high	1.1	1.8	0.2
North Dakota	Lignite	Medium	0.7	1.0	0
Wyoming (Powder River Basin)	Subbituminous	Low	0.8	1.8	0.9
Texas	Lignite	Medium	1.8	2.3	2.7
Central Utah	Bituminous	Low	0.5	2.5	0.5
New Mexico (San Juan Basin)	Subbituminous	Low-medium	1.7	2.1	0.6

PETROLEUM, A DWINDLING RESOURCE

Petroleum is a mixture of hydrocarbons, inert liquids, gases, and contaminants, lifted from exhaustable underground reservoirs by natural pressure or by various pumping techniques. In a refinery, petroleum is converted into useful products by combinations of distillation, thermal and catalytic cracking, and synthesis.

Petroleum is important because nearly all of our transportation is propelled with fuels derived from petroleum. Many of our buildings are heated with oil. Conversions to alternative fuels are possible, but take years to achieve. Sudden increases in the price of petroleum and interruptions of supply have caused worldwide eonomic crises.

Energy features of petroleum are:

- A petroleum field is exhaustable. For example, the Prudhoe Bay field, which at its best production accounted for 20 percent of the U.S. petroleum production, will pass its peak production around 1987 and then degrade at a rate of around 1 percent/month.
- Elements in the price of petroleum are the costs of exploration, drilling dry holes, drilling production wells, royalties to the owners of the affected mineral rights, lifting the oil to the surface, transporting to markets, taxes, and profits.
- The energy content of liquid petroleum products varies from 149,690 Btu/gallon for residual fuel oil down to 125,070 Btu/gallon for motor gasoline.

Petroleum consists of hydrocarbons of various molecular structures. The simplest is methane, CH_4, normally a gas with an energy content of 1,027 Btu/cubic ft. Ethane, which can be a liquid, has an energy content of 73,380 Btu/gallon. The heavier molecules have more carbon and hence more energy per unit of volume.

The limits for petroleum will be determined by its cost and availability:

- The largest petroleum reserves in the world are in the Mideast oil fields where the actual cost of a barrel of oil lifted to the surface in the 1960s was about 25 cents.
- Taxes and OPEC decisions fixed the price at around $27 a barrel in 1985. A higher price motivates conservation and pumping from marginal wells. A lower price produces less profit for the oil producer.
- Most U.S. oil fields have passed their peak production and are relying on secondary recovery processes. Production will increase only when oil prices rise.
- Little oil is found in structures deeper than 10,000 ft (3,050 m).

A petroleum barrel contains 42 U.S. gallons (159 liters). United States oil consumption is measured in millions of barrels per day (Mbd). In 1984, the U.S. consumed oil at the rate of 16 Mbd, of which 5.5 Mbd was imported. The biggest oil field in the U.S. was at Prudhoe Bay, Alaska, which delivered oil at the rate of 1.6 Mbd through the Trans Alaska pipeline.

The price of petroleum is affected by the sulfur content. For example, when high-sulfur oil was selling for $27 a barrel, the "sweet" Saudi Arabian oil went for $30 a barrel. At one time heavy crude oils sold for $8 less per barrel than light crudes. Light (low density) crudes yield more gasoline with simple distillation. However, availability of refining processes for producing useful products from the heavy crudes and their higher energy content has narrowed this differential.

The density of petroleum is measured in API degrees, which relates to its physical density as follows:

$$\text{Specific gravity} = \frac{141.5}{131.5 + \text{degrees API}}$$

where specific gravity is the density of the oil in grams/cubic cm. API density is measured with a special hydrometer that shows water as 10. The value in API degrees corresponds roughly to the percent of the kerosene-gasoline content of the crude oil. For example, about 30 percent of API 30 degrees crude oil would be kerosene and gasoline that could be recovered by simple distillation.

Future cost: the key energy systems engineering problem. With petroleum supplying 43 percent of the U.S. energy needs, the cost of petroleum enters into over half of the energy systems engineering comparisons. Solutions to energy systems engineering problems require life-cycle cost analyses, in which the future fuel price can be important. For example, a diesel engine driving a generator continuously at full power will consume around 730 gallons of fuel/year/kW of output. At one dollar a gallon that is $730/kW year. To build the diesel power plant costs only $200 to $300 a kW.

The best estimates of future oil prices have not been accurate. For example, on the basis of persumably valid estimates the U.S. Government funded the Synfuels Corporation to contract development of shale-oil recovery and coal-to-oil conversion plants. The design and construction of the plants had to be contracted to those who could develop them, companies such as Exxon, Union Oil, Texaco, Chevron, and Phillips Petroleum. The objective was to deliver 2 million barrels of synfuel a day in 1992. The Synfuel Corporation's contracting technique guaranteed a price for the delivered synthetic oil. Then the world price of oil dropped from $36 a barrel to $29 a barrel, instead of rising to $45 a barrel. The synthetic fuel program began to fall apart. The history of petroleum prices during this period, extracted from Ref. 2, is shown in Figure 3–6.

[2]"Short Term Energy Outlook," U.S. Department of Energy Information Administration, Washington, D.C. (November, 1983), p. 8.

Petroleum, a Dwindling Resource

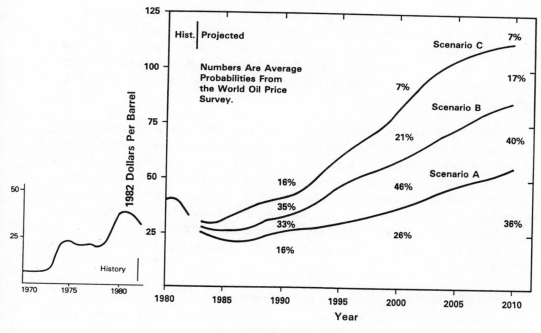

Figure 3-6

An increase in oil prices starts a slow-moving cycle that thwarts the price increase. Examples of the consequences of a price increase are:

- Petroleum consumers conserve petroleum. For example, the U.S. Department of Energy estimates that a $1 per barrel change in the price of petroleum will change the U.S. oil consumption by 3.4 percent.
- Marginal oil fields start secondary production.
- Companies capable of searching for petroleum start geological exploration and drilling of exploratory wells.
- Power and heating plants being designed are changed to burn alternative fuels.

These price-thwarting activities proceed at varying speeds, with unpredictable results.

The U.S. Department of Energy publishes oil price predictions with different scenarios and assigns confidence levels to each scenario.[3] Figure 3–6 is an

---

[3] "Energy Projections to the Year 2000," U.S. Department of Energy, DOE/PE-0029/2, October, 1983, p. 9.

example. Scenario A assumes a high energy-demand reduction and a high energy-supply potential. Scenario B assumes growing energy demand and reducing energy production. The confidence numbers were obtained by surveying world oil experts. The ceiling price of oil will probably be the cost of making oil from coal, a process described in a later section. Oil from coal would cost from $40 to $50 a barrel in 1985 dollars.

Petroleum bearing formations. The most likely zone for petroleum formation was a warm, calm bay where dying microscopic plants and animals were covered by silt that protected them from oxidation. Succeeding deposits and heat from the interior of the earth converted the deposited fats into kerogin, and the kerogin into hydrocarbons, which tend to float on water. These processes occurred some 200 million years ago. Most of the hydrocarbons so created have long since migrated to the earth's surface and have disappeared.

Geological structures in which petroleum might be found must have these features:

- A source rock, usually shale, in which the petroleum had been formed.
- A reservoir rock, like sandstone, that has pores into which the petroleum can migrate when driven by water. The pores of a good reservoir rock might represent 20 percent of its volume.
- A trap that has the reservoir rock covered with an impervious rock so that the petroleum cannot escape upward.

Common traps are anticline, structural, and stratigraphic types (Figure 3–7). An anticline trap is an underground dome-like structure into which petroleum migrates, being lighter than the water in the source rock. Escape of the petroleum is prevented by an impenetrable rock, such as shale. In a fault, the upward migration of oil is likewise prevented by a cap rock that is sealed against the face of a fault. A stratigraphic trap is found when the nature of the sedimentary deposit effectively terminates the layer of reservoir rock. Most of the oil fields around the Gulf of Mexico are in anticlines. The Prudhoe Bay field is in a stratigraphic trap. A well drilled into a promising geologic formation that contains no oil is called a "dry hole."

Finding petroleum. An oil field is discovered in three steps: geology studies, geophysics exploration, and test well drilling. The geologist tries to eliminate unlikely sites from further evaluation. For example, if surface outcroppings show no signs of reservoir rocks at depths where oil could accumulate, then the area is rejected. The expert also knows that oil is not found in igneous or precambrian rocks. The profile of underground structures is mapped, searching for

Petroleum, a Dwindling Resource

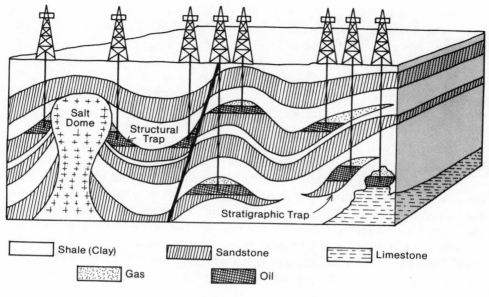

Figure 3-7

possible oil traps in anticlines, faults, and stratigraphic structures. The necessary tools include sensitive instruments that measure variations in the earth's gravity and magnetic fields.

The geophysics team maps underground structure by observing sound reflections from underground rock layers. The sound source can be an explosion in a shallow hole; for example, five pounds of dynamite in the bottom of a 120-foot deep hole. A newer technique uses a hydraulically driven hammer that sweeps the frequencies between 5 and 60 Hz. Sound reflections from underground layers are sensed with hydrophones and delivered to a tape recorder for subsequent computer-generated plotting. From a stack of plots, each representing a cross-section of the structure being investigated, the geophysicist reconstructs the size and shape of underground structures.

Exploratory drilling. An exploratory well is drilled, not to produce oil, but rather to confirm the geologic structure predicted by geological and geophysical work. For example, a well testing an anticline trap would not penetrate the dome of the trap and release the pressurizing gas, but should, rather, penetrate the edge of the oil pool to confirm its extent. An exploratory well is usually not suitable for production, but it can be used later for water injection to sustain the oil flow. The preliminary drilling plan would locate exploratory wells appropriately for such future use.

Drilling oil wells. Most wells are drilled with machines called "rotary rigs." The rig consists of a derrick, draw works, mud handling equipment, rotary table, and power plant (Figure 3–8).

The hole is cut with a turning bit, faced with teeth of hardened steel or tungsten carbide for ordinary rock. The rotary bit has gear-teeth like projections that roll over the rock face. The pressure of each tooth on the rock breaks chips off the rock. This pressure is generated with heavy collars, sometimes extending for fifty feet over the drill stem above the drill bit. Diamond-faced bits are used to grind shale that is too hard for rotary carbide-insert bits. The rest of the oil-well drilling rig supports the operation of the drill bit.

A drilling mud, typically consisting of water, bentonite clay from Wyoming, and additives, must be supplied through the drill stem to the drill bit to cool the bit and carry away the rock chips. The mud is returned to the surface through the space between the drill stem and the outside of the hole. A high-pressure gas pocket, when penetrated, could push the entire drill stem out of the hole. The hydrostatic pressure of the drilling mud, which can weigh 15 to 20 lb/gallon compared to water at 8.34 lb/gallon, prevents this.

Well casing. A well is cased to prevent collapse and ingress and egress of fluids. Casing designations are:

Casing	Diameter		Typical Depth	
	Inches	*Centimeters*	*Feet*	*Meters*
Conductor	20 –30	51 –76.2	100	30.5
Surface	8.6–16	22 –40.6	3000	914
Intermediate	7 –13.37	17.8–34	8000	2438
Production			15,000	4572

The conductor casing supports blow-out preventers, which are hydraulically or pneumatically actuated seals that are automatically actuated if a potential blowout is sensed, from mud pressure, for example. The surface casing prevents contamination of groundwater. The intermediate casing is sometimes run all the way to the production zone at the reservoir. Each section of casing is cemented into place by forcing cement grout past the bottom of the hole into the space between the casing and the rock. Centering springs installed at intervals on the outside of the casing as it is lowered preserve the space for the cement.

Many wells are normally drilled from one arctic or off-shore platform. For example, on Alaskan Cook Inlet platforms, each of the four legs supporting the platform has eight slots for drilling wells. There the wells are drilled straight

Petroleum, a Dwindling Resource

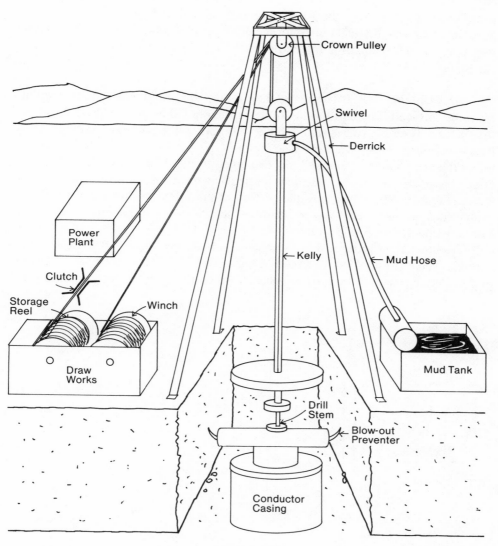

Figure 3-8

down for the first 7,000 feet, and then deviated up to 30 degrees to penetrate the oil reservoir thousands of feet horizontally from the upper axis of the hole. The deflections are obtained with a whipstock or with a special bit that is driven by mud pressure rather than by the rotating drill stem.

After the casing is cemented, the well is completed by firing shaped charges, which not only penetrate the casing and cement, but also bore short holes into the reservoir rock. A string of tubing, 2 to 2.5 inches (5 to 6.3 cm) in diameter, is lowered into the well, with seals above and below the production zone, and then water is substituted for the mud. A valve assembly, called "Christmas tree," is installed at the top end of the string, and production of oil and gas can start.

Artificial lift. Oil is less dense than water, so a column of water extending from the surface of the earth to the oil reservoir should be able to push the oil to the surface. The reservoir is said to have water drive if it is continuously pumped by a water head. The drive could also come from a pocket of gas on top of the oil under the cap rock. Compressed gas dissolved in the oil can also provide a gas drive. These drives can be exhausted, particularly if the oil is extracted at too high a rate (Figure 3–9). Measuring the bottom-of-hole pressure with various flows enables the petroleum engineer to optimize the rate at which the most possible oil can be recovered from the well.

Eventually the natural drive weakens and can no longer lift the oil to the surface. Then artificial lift can recover more of the oil. For example, a piston pump operated by a walking-beam mechanism on the surface, is a common way

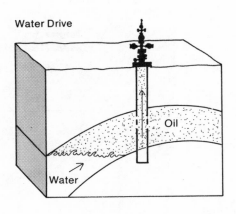

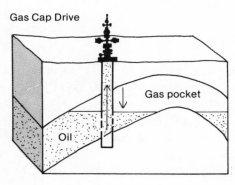

Figure 3-9

of pumping oil. Other mechanisms include injecting gas in the gas cap to restore pressure and to reduce the density of oil. Data and techniques for artificial lift are described in Ref. 4.

Secondary oil recovery. The key to future oil production in the U.S. is recovery of oil from wells that no longer deliver oil even when pumped. An oil reservoir might consist of 85 percent rock and 15 percent space, which when filled with oil and gas, is called "oil in place." Of this oil in place, self-pressurized primary production might extract 15 percent. With secondary recovery techniques extraction of 90 percent of the oil in place has been claimed under favorable conditions, although 33 to 60 percent is doing well. The simplest technique of secondary recovery is to re-pressurize the reservoir with gas or water injected at the correct points. The injected fluid can also absorb oil and carry it to the producing well.

The mechanism of water flooding is to push the oil toward the production well (Figure 3–10). The water flooding of the Prudhoe Bay oil field involves pumping 3 million barrels of water a day through some 200 injection wells distributed over 75 square miles (194 sq km). Seawater, filtered and treated with anti-foam agents, oxygen scavengers, and scale inhibitors, is pumped through 135 miles (217 km) of distribution lines and injected at 2,700 lb/in^2 (18.6 MPa). This primary recovery technique is expected to produce an additional billion barrels of oil before the field is considered exhausted.

Wells having gas drive can sometimes be pressurized with gas to produce petroleum after the natural gas drive is exhausted. A disadvantage of gas pressurization is that the market value of the gas may exceed the value of the oil recovered. Gas not subsequently recovered then represents a loss, whereas loss of water into the strata surrounding the reservoir rock is not serious.

Enhanced recovery. Oil in place that cannot be extracted by secondary recovery methods can be salvaged by enhanced recovery techniques described by H. K. van Poolen and Associates.[5]

Steam stimulation, a thermal process, is the technique that is used in 70 percent of the world's enhanced oil recovery projects. One version, called "huff and puff," involves forcing steam into a well for two or three weeks, shutting off the steam for a few days, and then pumping out the oil. Another technique is to inject steam into alternate wells, spaced like one well per five acres. The quantity

---

[4]Kermit E. Brown and H. Dale Beggs, *The Technology of Artificial Lift*. (Penwell Publishing Co., Tulsa, Oklahoma, 1970).

[5]H. K. van Poolen and Associates, *Fundamentals of Enhanced Oil Recovery*. (Tulsa, Oklahoma: Penwell Publishing Co., 1980).

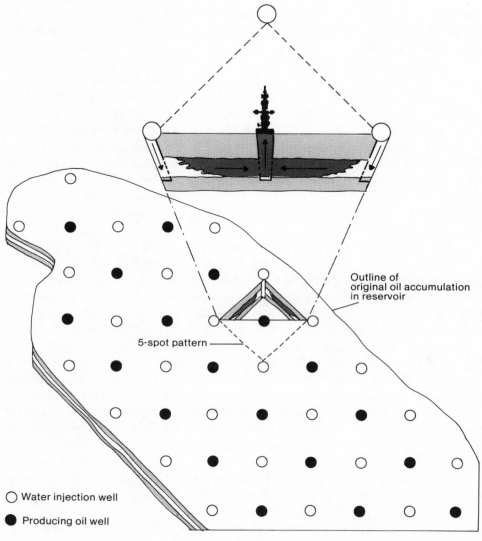

Figure 3-10

of steam for one well is that generated by boiling some 1,000 barrels of water a day. The pressure and temperature must be high enough to deliver steam rather than hot water into the oil reservoir. This techniques doesn't work well for wells deeper than 5,000 feet. Heat loss calculations organized by van Poolen show that oil reservoirs should be thicker than 10 ft (3 m) for this technique to be practical.

The most practical fuel for steam generation is crude oil itself. Some areas have reached their atmospheric pollution limits, so pollution rights would have to be purchased if high-sulfur crude oil is to be burned.

Other enhanced recovery techniques use miscible hydrocarbons and carbon dioxide.

Refining petroleum. A modern refinery extracts useful products from crude oil and converts low-value products into high-value ones. Four processes are used: separation, conversion, purification, and blending. Each refinery is designed for a specific crude oil and a specific mix of products, with flexibility for handling limited variations of crude and product. For example, at Pump Station 6 on the Trans Alaska pipeline is a refinery that extracts 3,300 barrels/day of turbine fuel from the crude oil being pumped. The turbine fuel is burned in gas turbines that drive the 18,200 hp pipeline pumps at that station. All of the refinery by-products are injected back into the pumped oil stream.

The first step in a typical refinery is to distill the incoming crude oil, at a temperature of around 650°F (343°C). The products condense at various temperatures in a distillation column, a conspicuous structure in a refinery (Figure 3–11). The column contains trays that collect products at appropriate temperatures. In a modern refinery, the distillation column is insulated and heat exchangers recover the sensible heat from the products for preheating the crude oil going into the boiler.

Conversion processes are used to make valuable products from less valuable ones. For example, the large molecules in heavy fuel can be broken into smaller ones by heating to a temperature of around 1000°F (538°C), or to a lower temperature in the presence of catalysts. In hydro cracking, hydrogen, available from thermal cracking, is introduced into the process to generate useful products. The output of the conversion process is again distilled to separate liquids called "reformates." The conversion process can be tailored to favor the production of specific reformates; for example, hydrocarbons that improve the octane rating of light gasoline.

Purification usually involves extracting sulfur from the products of distillation. For example, diesel engine fuel having 0.5 percent sulfur is generally acceptable. Crude oils containing 3 percent sulfur require sulfur extraction. One way is a catalyzed reaction with hydrogen, which combines with sulfur to produce hydrogen sulfide, an easily separated gas. From the hydrogen-sulfide elemental sulfur or sulfuric acid can be readily produced.

The final refinery process is one of blending the output to produce the most possible high-value products from the components at hand. For example, gasoline intended for winter use might contain butane for quick starting. Summer gasoline wouldn't contain butane because it evaporates too fast.

Economics of oil: that sets the price of fuels. Prudhoe Bay in Alaska is an example of a petroleum reserve that was marginal at $5 a barrel, but profitable at $30. The development of the field cost around $10 billion, which was more than anticipated. It produces 1.6 million barrels of oil a day. Assuming a 20 percent interest and financing charge, plus 3,000 employees that cost $100,000 each/year, plus 100 new wells per year at $2 million each, the cost of a barrel of oil can be estimated as follows:

Cost element	Annual cost of Prudhoe Bay operation, billion of dollars
Interest on initial investment (I) $I = 0.2 \times \$10\ B$	2.0
Employee cost (E) $E = \$0.1\ M \times 3000 \times 0.001$	0.3
New wells (W) $W = 100 \times \$2\ M \times 0.001$	0.2
Total	2.5

Cost of oil, per barrel (Co)

$$Co = \frac{\$2.5 \times 10^9 \times day \times year}{year \times 1.6 \times 10^6\ bbl \times 365\ days} = \$4.28$$

Of the oil produced, one-eighth goes to the State of Alaska for royalty, so the cost of the owners' remaining share (Co) is:

$$Co = \frac{\$4.28}{0.875} = \$4.89 \text{ per barrel}$$

The cost of delivering the oil through the 800-mile, $8 billion Trans Alaska pipeline to the tanker terminal at Valdez can be estimated from the interest on the capital (Ip) of the pipeline and its $208 million annual operating cost (Op)

$$Ip = \frac{\$8 \times 10^9 \times 0.2 \times day \times year}{year \times 1.6 \times 10^6\ bbl \times 365\ days} = \$2.74/bbl$$

$$Op = \frac{\$208 \times 10^6 \times day \times year}{year \times 1.6 \times 10^6\ bbl \times 365\ days} = \$0.36/bbl$$

$$\text{Total} = \$3.10/bbl$$

Thus the marginal cost of oil delivered to the tanker terminal at Valdez is $3.10 plus $4.89 or $7.99 a barrel. The price of oil must be higher than this to compensate for risk and to provide for reasonable profit. No investment funds would be available for Prudhoe Bay if the price of oil were under $10 a barrel. On the other hand, a $30 barrel oil price makes the Prudhoe Bay development a successful project.

Petroleum, a Dwindling Resource

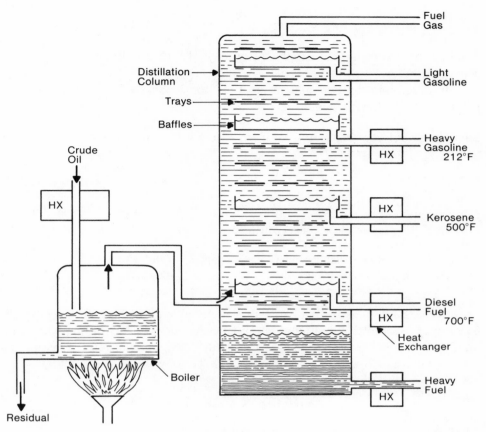

Figure 3-11

An indication of the market price of Prudhoe Bay crude oil came from the price offered to the State of Alaska in an auction of its share of the oil. Texaco bought 90,000 barrels/day and Chevron bought 30,000 barrels/day in December, 1984. Bids ranged from 17 cents to $1.04 a barrel above a base price of $17 to $18 per barrel for various grades of crude.

New petroleum fields and enhanced oil recovery can deliver petroleum, but the available quantity needs to be related to consumption. For example, Prudhoe Bay in Alaska, the largest field in the U.S., in 1983 delivered 1.6 million barrels per day, which corresponds to about 10 percent of the 15.2 billion barrels per day that the U.S. consumed in 1983. Prudhoe Bay had some 9.6 billion barrels that are recoverable by normal means and water flooding. Two billion barrels had already been produced in 1984, leaving 7 billion barrels, which corresponds to supplying the 1983 U.S. consumption for the period (D):

$$D = \frac{7 \times 10^9 \; bbl}{15.2 \text{ million barrels/day}} = 460 \text{ days}$$

A billion barrels of oil can supply for 66 days the 15.2 million barrels per day consumed by the U.S. in 1983.

In describing the potential in the U.S. for enhanced oil recovery, Exxon publishes the following table:

Enhanced Oil Recovery Potential from Already Discovered Fields[6]

	Billions of barrels	
	World	United States
Original oil in place	3000–4000	500–600
Recovery by primary and secondary methods	1100–1400	180–225
Already produced	500	150
Remaining by primary and secondary methods	600–900	30–75
Enhanced oil recovery potential	100–200	20–30

We could optimistically assume that all possible oil in the U.S. is recovered by primary, secondary, and advanced methods to deliver 105 billion barrels of oil. The best possible self-sufficiency (Ss) of the U.S. with this reserve, assuming the 1983 consumption of 15.2 million barrels per day, would be:

$$Sc = \frac{105 \times 10^9 \; bbl \times \text{day} \times \text{year}}{15.2 \times 10^6 \text{ barrels} \times 365 \text{ days}} = 19 \text{ years}$$

Obviously oil imports are essential to the U.S.

NATURAL GAS: FROM WELLS AND ARTIFICIAL SOURCES

Natural gas is mostly methane, which originates in the underground decomposition of plants and animals, and is subsequently trapped in geological formations from which it can be recovered. "Associated" natural gas is found in petroleum and coal deposits. There are also wells that yield only natural gas. Methane is also produced in the treatment of sewage. It can be recovered from garbage dumps.

Natural gas is important because it is the principal source of energy for heating homes in the U.S. Pipelines built after World War II made natural gas available in all large U.S. cities and in most of the small ones. Natural gas fuels

[6]"Improved Oil Recovery," Exxon Background Series, (New York: Exxon Corp., 1982), p. 8.

gas turbines that supply peaking power. It is also used in combined-cycle plants that generate base-load.

Energy features of natural gas are:

- Natural gas is an exhaustable resource, and its cost rises as existing wells run out and marginal new sources are brought on line.
- Natural gas is a good heating fuel because its heat energy can be extracted with greater than 95 percent efficiency in heating apparatus.
- Natural gas is an excellent fuel for gas turbines.

The limitations of natural gas are:

- Natural gas is being used to power vehicles. However, a compressor plant is needed to charge the vehicle fuel tank.
- A coefficient of performance of 3 to 6 has been predicted for natural-gas fueled Stirling-cycle heat pumps. Such a heat source would be hard to beat from an efficiency standpoint by alternatives.

Natural gas is usually tailored to contain 1,000 Btu/cubic ft at standard conditions of 59°F (15°C) and 750 mm of mercury or 1 atmosphere (0.1 MPa). The highest energy content of dry gas is 1,027 Btu/cubic ft. Transactions in natural gas production, transmission, and distribution are conducted in units of one thousand cubic feet, which corresponds to one million Btu. The content of a gas field is quoted in trillions of cubic feet.

Distributors sell natural gas by the therm, which contains 100,000 Btu (105,506 kJ) and occupies 100 cubic ft of volume at standard conditions.

Washington Water Power Company's average residential revenue per therm for natural gas in 1983 was 62.14 cents and in 1984 it was 58.44 cents. A therm has 100,000 Btu, so the 1984 retail price corresponds to $5.84 per million Btu. A million Btu or 1000 cubic feet of gas has energy equivalent to 7.25 gallons of 137,750 Btu diesel oil. Thus the gas price corresponded to 80 cents a gallon for diesel oil, which at the time was selling for around $1 a gallon. In 1984 the utility paid $4 per million Btu for Canadian natural gas and $3 per million Btu for domestic gas.

Formation of natural gas. Natural gas is found with and without association with other fossil fuels. It is 75 to 85 percent methane (CH_4), which is also called "marsh gas." Other constituents of natural gas include ethane (2 to 8 percent), propane (0.5 to 5 percent), and nitrogen (5 to 14 percent). Methane is manufactured in India by depositing cow manure into pits, and collecting the gas as decomposition takes place. Sewage processing plants sometimes supply their own heating energy from collected methane. In San Diego methane is collected from a site where garbage had been buried previously. Methane in coal mines is a nuisance and is blown away to avoid explosive mixtures in the mine.

Petroleum, if subject to high temperature, will eventually decompose into methane. The Earth has a temperature gradient that varies with the geologic structure, but averages about 17 F per 1000 ft. Wells deeper than 15,000 feet are more likely to produce natural gas rather than crude oil. Wells deeper than 20,000 feet are likely to contain only carbon.

Essential to a natural gas field is a trap. Gas that wasn't trapped had diffused into the atmosphere and disappeared millions of years ago. A trap consists of a porous reservoir covered with a impervious rock (see page 79). Often natural gas and petroleum co-exist in the same field with the gas dissolved in the petroleum. Bringing the petroleum to the surface relieves the pressure, and the natural gas bubbles out. The petroleum is valuable, and can be trucked, piped, or tankered to its market. The natural gas doesn't have value unless there is a pipeline to the market. Too often it is disposed of by burning in a flare at the well.

Alaska natural gas. At Prudhoe Bay on the North Slope of Alaska the oil wells produce natural gas as well as oil. The estimated gas content of the field is 2.6 trillion cubic ft. Some of the gas is used for local heating and some is piped to nearby pumping stations of the Trans Alaska pipeline for burning in the gas turbines that drive the pumps. The rest of the gas is pumped back into the ground, to be saved for later export to the 48 contiguous United States.

Uses of natural gas. Natural gas is a common fuel for heating buildings in cities. Previous fuels included artificial gas, and coal which produced disagreeable smoke when burned without sophisticated equipment. Gas manufactured from coal provided only around 400 Btu/cubic ft. Switching to natural gas permitted the gas distributors to carry 2.5 times the former energy in the existing pipes.

In the U.S., 17 trillion cubic ft of natural gas were used in 1983. The important uses were:

Use	Trillion cubic feet/year
Space heating	4.47
Water heating	2.04
Process heat and process steam	5.38
Conversion to electricity	2.94
Petrochemical and feedstock	0.72
Cooking	0.51
Transportation, on and off highway	0.66

All other users consumed less than 0.3 trillion cubic feet total.

Natural Gas: From Wells and Artifical Sources

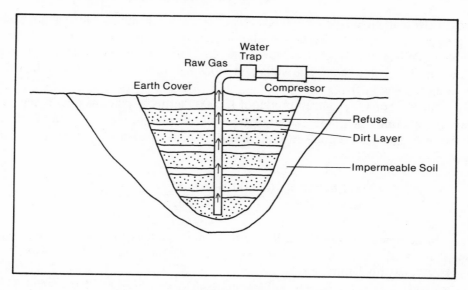

Figure 3-12

Methane gas from garbage. Decomposing garbage in landfills generates methane, which is often collected and burned in flares to avoid odors. The gas as generated doesn't have 1,000 Btu/cubic ft, so it cannot be comingled with natural gas in pipelines unless it is purified.

Landfill gas is easily gathered from wells from 20 to 100 feet deep (Figure 3–12). The American Gas Association reports that in 1983 around 5.6 trillion Btu of gas, roughly equivalent to one million barrels of oil, was recovered from some twenty landfill projects. The largest of these landfills have 2 to 3 million tons of waste in place. Most landfills are smaller, containing less than a million tons of waste. Landfill gas production can be increased by regulating moisture content and acidity of the refuse, and by injecting or recycling nutrients, bacterial matter, and fluids that percolate through the refuse.

The product of bacterial action on biomass is a methane-containing gas that also has water vapor and carbon dioxide, resulting in a low energy content. Water is easily condensed out, and carbon dioxide in the gas can be removed in natural gas processing by dissolving it into a potassium hydroxide solution or in mono-ethanolamine. In a separate chamber the liquid is heated to drive out the carbon dioxide, and is then reused. This process works best at pressures such as 100 lb/in^2, so the raw gas must be compressed by a gas turbine or electric motor-powered compressor. A heated pressure vessel looks like a boiler, so codes can require the purification plant to be manned.

A new development by Monsanto uses a membrane to separate the carbon dioxide from methane. Metal wrapped pipes contain thousands of hollow permeable fibers that separate the carbon dioxide from methane. The equipment was tested at Pelham Bay landfill in New York. In 1984 it was installed in the Oregon City landfill, where gathering wells and pipes had previously been installed for burning the gas in flares.

The Oregon City landfill produces 10,000 therms of pipeline quality gas per day. At that time, in 1985, Canadian gas could be bought for $3.60/million Btu. A therm of gas has 100,000 Btu, so the value of the gas produced per day (Gd) was:

$$Gd = \frac{10,000 \text{ therm} \times 100,000 \text{ Btu} \times \$3.60}{\text{day} \times \text{therm} \times 1,000,000 \text{ Btu}} = \$3600 \text{ per day}$$

In comparison, this landfill's energy production is the same as that of a small oil well producing 173 barrels/day at $30/barrel of oil. Such a well is considered worth pumping.

The Los Angeles Sanitation District operates two 1490 kW Kongsberg KG2 gas turbines on landfill gas having 320 to 390 Btu/cubic ft. A liquid scrubber removes water and a superheater prevents bacteria formation in the fuel supply of the turbines. A compressor pressurizes the gas for injection into the combuster of the turbine.

Solar turbines, Inc. makes gas turbines for low-Btu landfill gas. The firm suggests that the economics of power generation will work out if the site has at least 30 acres containing one million tons of refuse in a thirty-foot layer. Smaller landfills can supply useful heating gas, but they may not justify a gas turbine plant, which should have a life of at least five years.

Methane from cattle manure. Capture of volatile solids is important in recovering methane from cattle manure, according to Frank Liebrock, Project Engineer in the State of Idaho Department of Natural Resources.[7] He calculated the possible methane production from 1,700 head of cattle at the Snake River Cattle Company. He estimates the volatile solids to be 25–50 percent of the waste, and for every pound of volatile solids destroyed, 12 to 16 cubic feet of methane could be generated. From the 9,050 pounds of waste generated each day by the herd, he could produce from 0.8 to 1.0 million Btu/h.

No serious interest, however, has developed in the U.S. for converting cattle manure into methane.

A team from M & E Pacific, Inc. designed a 233 cubic ft (6.59 cubic m)

[7]"Methane Projects Underway in Idaho," *Pacific Northwest/Alaska Bioenergy Bulletin,* Vol. IV, No. 4, June 15, 1983, p. 8.

digester, based on Chinese technology, for use in Mali.[8] Manure, crop residue, and sanitary waste are mixed with water to form a slurry and poured into the biodigester inlet. An equal amount of digested residue is removed from the outlet with a bucket and carried to gardens. Gas produced by anaerobic bacteria is captured above the digester slurry and pressurized to 5 to 10 inches of water by the displaced water. The biogas is used for cooking, lighting, and pumping water.

Water hyacinth, napier grass, and sorghum. P. H. Benson and his associates predict that Florida's dwindling gas wells, which supply 10 percent of state's 2.5 quadrillion Btu per year primary energy consumption, can be replaced with locally generated biogas.[9] Research on biogas at twenty-two research centers in Florida was cofunded by the Gas Research Institute and the State of Florida through its Institute of Food and Agricultural Sciences. The investigators improved the efficiency of bacteria that digest biomass, built pilot plants, and speeded plant growth.

Adding nutrients to a pond produced water hyacinth yields of fifty tons of dry mass/acre/year. Napier grass yields of twenty tons/acre/year have been achieved on land. At Texas A & M, the top yield for hybrid sorghum has been sixteen tons/acre/year. The most efficient producer of biomass is green algae grown in Thailand, yielding over seventy tons/acre/year.

In comparison, the U.S. Forest Service estimates that a fast-growing alder forest will produce 8,000 board ft/year. Assuming that one-third of the growth is wasted, then the biomass growth would be 12,000 board ft or 1,000 cubic ft. At fifty lb/ft^3, the annual production would be 50,000 lb or 25 tons.

The energy content of dry water hyacinth is 16.02 megajoules per kg, according to Donald L. Klass.[10] This corresponds to an energy content (Ewh) of:

$$Ewh = \frac{16.02 \times 10^6 \text{ J} \times \text{Btu} \times 0.4536 \text{ kg}}{\text{kg} \times 1055 \text{ J} \times \text{lb}} = 6888 \text{ Btu/lb}$$

Benson and his associates suggest a theoretical yield of 7.5 standard cubic ft of methane/lb of low-lignin sorghum. This corresponds to 7,500 Btu/lb of input. Water hyacinth doesn't have the low lignin for such a high conversion efficiency. Benson cites achievements of 5 to 6.5 cubic ft of methane/lb of sorghum.

[8]Michael H. Weitzenhoff and James S. Kumagi, "Low Cost Fuel Generated Imported to Mali," *Civil Engineering/ASCE,* October, 1984, p. 14.

[9]P. H. Benson and Associates, "A Regional Approach to Producing Methane from Biomass and Wastes," *Proceedings of the 19th Intersociety Energy Conversion Engineering Conference,* ANS (1984), pp. 683–90.

[10]Donald L. Klass, "Methane from Anaerobic Fermentation," *Science,* 223, No. 4640, 9 March 1984, p. 1024.

If the process could produce 5 standard cubic ft (SFC)/lb from water hyacinth, and the yield were 40 tons/acre, then the energy production/acre (Ea) would be:

$$Ea = \frac{5\ SFC \times 1000\ Btu \times 40\ tons \times 2000\ lb}{lb \times SFC \times acre\text{-}year \times ton}$$
$$= 400 \times 10^6\ Btu/acre/year$$

A gallon of diesel fuel has 137,750 Btu, so the owner of the one-acre pond would be producing the oil-equivalent (Ea) of:

$$Ea = \frac{400 \times 10^6\ Btu \times gallon}{acre\text{-}year \times 137{,}750\ Btu} = 2904\ \text{gallons equivalent per year}$$

This corresponds to sixty-nine barrels of forty-two gallons each. The Oregon City landfill, previously described, delivers 10,000 therms per day, or the equivalent of 173 barrels of oil. The site is about ten acres in size, so its production would be the equivalent of 6,300 barrels/year/acre. Clearly a landfill is a more valuable producer of gas than a hyacinth pond.

Methane is used as fuel for motor vehicles. A compressor is needed for charging the fuel tank, and a special carburetor mixes gaseous fuel with the air. A properly fertilized one-third acre Florida pond and a digester could supply the owner fuel for driving 50,000 miles/year in a 50-miles-per-gallon car.

NATURAL GAS ECONOMICS

The wellhead price of natural gas in the U.S. had been regulated at around 20 cents/1,000 cubic ft (million Btu) to preserve low heating costs in the Northeast. As energy costs climbed with inflation, the producers not committed with contracts to pipeline companies, sold their gas into intrastate channels as power plants and industries converted from fuel oil to low-cost gas. The result was gas shortages in the Northeast during cold winters. The Federal Government then permitted owners of new deep wells to charge up to $9/1,000 cubic ft for gas produced. Then, in 1979, the Government decontrolled natural gas to a limit of $3.73/1,000 cubic ft. Note that 1,000 cubic ft of natural gas is equivalent to 7 gallons of fuel oil.

The price of gas rose from $1.18 to $2.70/1,000 cubic ft, where the supply met the demand of the conservation-motivated consumers. This sent many drillers of deep wells producing $9 gas into bankruptcy.

Texas Oil and Gas Co., in 1984, drilled 925 gas wells at a cost of around $0.6 million each. This firm is an example of an organization that responds quickly and survives in the natural gas business. Brian O'Reilly observed that the firm's engineers analyze drilling and production records filed in state offices, searching for land adjoining proven fields.[11] The firm buys mineral rights for as little as $5 an acre, often adjacent to existing or exhausted gas producing property. Each district manager has authority to drill 100 wells/year, and 65 percent of the drilled wells produce commercial quantities of gas. The average well is 6,500 feet deep.

Natural gas, unlike oil, cannot be conveniently hauled in a tank, so a pipeline must be brought to each new well. Again, quick response by district managers brings the piepline to the well within a few weeks after completion. Other drillers can wait for a year, paying interest on the well investment.

SYNTHETIC FUELS FROM COAL

The chemistry of the route from coal to gas and liquid fuels is simple. To obtain carbon monoxide and hydrogen from the reaction of coal and water requires the addition of energy, which is obtained by oxidizing part of the coal:

$$C + H_2O \longrightarrow CO + H_2$$

With catalysts, the products can be converted to methyl alcohol.

$$CO + 2H_2 \longrightarrow CH_3OH$$

This reaction produced town gas from coal prior to the piping of natural gas to virtually every city in the U.S. Town gas was produced by burning coal in a retort with air to get it hot, and then quenching the hot coal with steam. The product was a gas with a low energy content, around 300 Btu/cubic ft, being diluted with nitrogen and carbon dioxide. By-products were coke, a solid fuel, and coal tar, a raw material for the manufacture of chemicals. Natural gas made town gas obsolete.

EPRI, in cooperation with industrial suppliers, process developers, and domestic and foreign utilities, developed in a $300 million program the Cool Water generating station on Southern California Edison's network. The plant gasifies coal to run a 100 MW combined-cycle power plant. The capacity of the synthetic-fuel plant is the equivalent of 4,300 barrels of oil/day. It uses a second-generation, Texaco-developed gasifier with an entrained-bed coal feeding device that achieves better efficiency than the Lurgi gasifiers developed in the 1930s. A Lurgi gasifier was used in South Africa's Sasol plant that produces 50,000 barrels of synthetic fuel per day.

---

[11]Brian O'Reilly, "A Gasser of a Texas Company," *Fortune*, October 1, 1984, pp. 59–68.

Two problems, cost and efficiency, haunt the synthetic fuels program. The efficiency of conversion can be considered in terms of energy of coal going into the plant and the energy of the fuel leaving. The Sasol plant that occupies hundreds of acres and has sophisticated heat-recovery stages, still requires three tons of coal to produce in liquid output the energy equivalent of two tons of coal. Assuming a 50 percent efficient gas turbine with a steam-turbine bottoming cycle, the overall efficiency (η_{oa}) would be:

$$\eta_{oa} = 0.5 \times 0.67 = 0.33$$

This same efficiency can be obtained easily with less cost in a conventional coal burning steam power plant with flue-gas desulfurization.

Liquid fuel production from coal faces economic hurdles. For example, the Texaco process used in the Cool Water project had also been used by Eastman Kodak Company's chemicals plant at Kingsport, Tennessee, to make chemicals from coal.[12] The first step produced methanol, which is also a liquid fuel that burns easily and cleanly in internal-combustion engines. Eastman Kodak intended to make acetic anhydride, and process chemicals from the hydride, but synfuel tax breaks made it more profitable to sell the methanol in Ohio as a gasoline octane booster.

Methanol can be made more cheaply from natural gas, so presumably Eastman Kodak makes its chemicals from methanol derived from that source.

[12]Eliot Marshall, "The Synfuels Shopping List," *Science*, Vol. 223, 6 January 1984, pp. 31–32.

4

Renewable Power from Water

Hydro power plants extract energy from water as it drops from one elevation to a lower one. The difference in elevation is generally achieved by building a dam or by diverting water from a river to a channel or tunnel that bypasses rapids or falls. The power is extracted from the water with Pelton wheels, or various kinds of hydraulic turbines such as Francis, Kaplan, and propeller.

Energy features of hydro power plants are:

- With no fuel to buy and trivial maintenance expense the main cost of a hydro plant is interest on the original investment, which is a fixed sum each year.
- The choice hydro sites have uniform stream flow, with lots of water or high head obtainable at low cost. In the U.S. these sites already have power plants. Marginal sites become feasible as the cost of fuel for steam plants rises.
- Water flow in a river will vary during the year. The hydro power plant can be rated for maximum flow, minimum flow, or some intermediate flow. The optimum rating, and hence the cost of the plant, is a key energy systems engineering analysis.
- Tunnel boring machines that excavate 10- to 20-foot diameter bores in solid rock for $700 to $2,000 a foot, and apparatus for remotely controlling hydro generating stations, are developments that reduce the cost of hydro power.

Hydro plants are important because they convert a renewable resource into electric power, without releasing sulfur and nitrogen-containing pollutants, and without increasing the carbon dioxide content of the atmosphere. Hydro plants with reservoirs can supply a utility's variable load, allowing the coal and nuclear heated steam plants to run at constant output. Changing steam plant output is a slow process, whereas hydro plant output can be changed in seconds.

The limit of hydro power output (Po) in kW is simple:

$$Po = k \times V \times h \times \eta_p$$

where V = volume rate of water flow, in cubic ft/second
h = head, the difference between the intake and tailrace, in feet
η_p = overall efficiency of the plant
k = 0.0847

For example, if we assume a flow of 100 cubic feet per second, a head of 100 feet, and an overall efficiency of 75 percent, the power output would be:

$$Po = 0.0847 \times 100 \text{ cfs} \times 100 \text{ ft} \times 0.75 = 635 \text{ kW}$$

In the U.S. water flow is measured in cubic ft/second, head is measured in feet of water, and the volume of water stored is measured in acre feet. Turbine power output is quoted in horsepower and generator real power output is in kilowatts. Generators also produce reactive power, which does not represent energy but does add current to the generator windings. The generator is rated in kilovolt amperes (kVA) or megavolt amperes (MVA). The real power output, in kilowatts (kW) or megawatts (MW), is the power factor times the kVA or MVA. A common power factor is 0.8.

Pertinent metric conversions are:

1 foot head of water = 0.4335 pounds per square inch (lb/in^2)
= 2989 Pa
1 cu ft/sec = 0.0283 cubic meters per second
1 acre foot = 43,560 cubic feet = 1233.5 cubic meters

In this chapter the principles of hydro power are described and illustrated with data from completed hydro projects. The last sections describe systems engineering considerations for pumped-hydro storage, the capture of tidal and ocean thermal energy, and geothermal energy.

Hydro plants: big and little. Small hydro plants can be found near abandoned mines in the mountains. A typical unit, often designed by a mining engineer, has a diversion dam, a few hundred feet of penstock, and a Pelton or Francis turbine that drives either a generator or mill. On the other end of the scale of complexity are the multi-gigawatt projects like Grand Coulee where a half century elapsed between concept and final full-power generation. Dozens of years can go into the site selection and design optimization of a big hydro plant. This engineering work, done by specialists, is outside the scope of this chapter.

Hydro plants are not standardized because every site has its unique head, available water flow, geography, and geology. Thus, every hydro project requires a new engineering analysis, which may or may not show opportunities for using adaptations of previous designs of dams and turbines.

Profits have been earned by engineers who have found, acquired, and rebuilt small hydro generating plants that had previously been abandoned because their power cost too much. Sometimes these plants can again be viable because (a) with automatic control an on-site operator is not needed, and (b) the Public Utilities Regulating Policies Act requires utilities to buy power offered to them.

Engineering data for designing and building small hydro plants is now available. For example, the Electric Power Research Institute has identified low-cost equipment for units in the 1 to 15 MW range.* They evaluated even centrifugal pumps for turbines and induction motors for generators.

Water flow and head are essential elements of a hydro power plant. If either is missing, there is no hydro power. With suitable head and flow a hydro plant can be economically viable if a suitable combination of dams, canals, penstocks, turbines, and generators can be developed. For small plants the cost of permits can be significant. The Sultan River Project, described later, required 22 of them. The project also spent $12.5 million for mitigation costs, which included preservation of the environment for fish.

River flow. River flow is the rate at which available water can flow through the hydraulic turbines. For example, rivers flowing to the Pacific Ocean from the Cascade Mountains carry the most water in June and July each year when the snow melts. Fall floods occur when early snow in the mountains is melted by warm rain. When electricity is need for heating homes in the area during winter, the snow-fed rivers are at their lowest.

Some rivers have such wide variations in flow that the only practical hydro plant is one with water storage. The Columbia river at Bonneville is an example (Figure 4-1). Note the huge flood in 1943. When completed in 1936, Bonneville Dam had only 400 megawatts of generating capacity, with water flowing over the spillways most of the time. Many of the subsequently built dams on the Columbia river have storage, and some, like Mica in British Columbia, only store water and have no generators. The result of the managed river flow is that the Corps of Engineers could justify adding 540 megawatts of generating capacity to the Bonneville Dam, and now water rarely runs over the spillway.

Water flow in large rivers, measured at river gaging stations, is compiled by the United States Geological Survey (USGS). Reports are available in large

---

*Fritz Kalhammer, "Small Hydro Activities," *EPRI Journal,* July/August 1982, pp. 52–53.

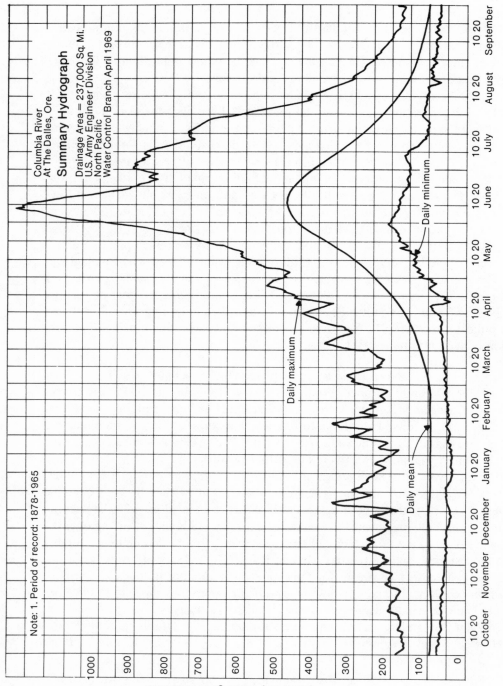

Figure 4-1

libraries.[1] Flows in small streams, which are not monitored by the USGS, can be derived or measured. In Wisconsin, for example, Mahmoud Mahmoud developed a computer program that predicts flows in small streams, based on flows in nearby larger rivers that have gaging stations.[2] He confirmed his program and was able to predict with reasonable accuracy the consequences of major storms of the past.

Flow in streams and rivers can be measured if historic data is not available. The United States Department of Interior publishes instructions on measuring stream flow with weirs, Parshall flumes, submerged orifices, moving floats, and current meters.[3] This 327-page handbook contains extensive tables that simplify interpretation of flow measurements. The book even has an appendix, "Hydraulics for the Novice," in which the methods of measuring stream and river flow are developed.

Estimating hydro power. The power that can be derived from descending water is simple to calculate. First, in English units, consider a stream that releases one cubic foot of water weighing 62.4 lb/second to fall 100 ft down to a lake. Each second this cubic foot delivers in power (Pw):

$$Pw = \frac{62.4 \text{ lb} \times 100 \text{ ft} \times \text{second} \times \text{hp}}{\text{second} \times 550 \text{ ft lb}} = 11.35 \text{ horsepower}$$

Since one horespower is 0.746 kW, the metric equivalent is:

$$Pw = 11.35 \text{ hp} \times 0.746 \text{ kW/horsepower} = 8.46 \text{ kW}$$

Energy losses must be considered. Friction losses occur in the canal, entrance to the penstock, penstock, scroll case, turbine, and draft tube (Figure 4–2). Assuming 0.75 efficiency of the turbine and generator, and friction losses corresponding to 0.7 efficiency of water delivery, the turbine output becomes:

$$Pw = 8.46 \text{ kW} \times 0.75 \times 0.7 = 4.44 \text{kW}$$

An example shows the ease with which metric units can be manipulated. The key to metric calculations is the newton (N), the force when applied to one kilogram of mass will cause it to accelerate one meter/second. We round off our newton equivalent of a kilogram force to 10.

[1] For example, C. C. McDonald and W. B. Lanhbein, "Trends in Runoff in the Pacific Northwest," *Transactions of the American Geophysical Union,* 29 (June, 1948), pp. 387–397.

[2] Mahmoud Mahmoud, "Flood Flow Frequency by SCS TR-20," *Paper Presented to the CEE 919 Hydraulics and Fluid Mechanics Seminar,* University of Wisconsin, Madison, September, 1982.

[3] "Water Measurement Manual," United States Department of Interior, Bureau of Reclamation, United States Government Printing Office (Denver, Colorado), 1981.

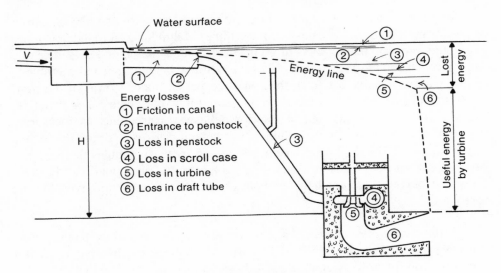

Figure 4-2

Consider the power available (Pm) from a cubic meter per second with a head of 100 meters:

$$Pm = \frac{1 \text{ cu m} \times 1{,}000 \text{ kg} \times 10 \text{ N} \times 100 \text{ m} \times \text{J} \times \text{second-watt} \times \text{kW} \times \text{MW}}{\text{sec} \times \text{cu m} \times \text{kg} \times 1 \text{ Nm} \times 1 \text{ J} \times 1{,}000 \text{ W} \times 1{,}000 \text{ kW}}$$

= 1 megawatt of hydraulic power

Dams make head. The example shows that huge flows are required for producing even moderate amounts of power. Raising our power output from 4.44 kW to 100 megawatts, a useful sized power plant, requires that with our 100 foot head, 703 tons of water must flow through the turbine every second. With a 200 foot head only half the water flow is required. Thus a hydro plant designer looks for the highest head available at reasonable cost.

The simplest dam is the gravity dam, built with enough concrete so that it can't be tipped over by hydrostatic pressure (Figure 4–3). Grand Coulee is an example of this construction. The quantity of concrete required for a dam can be reduced if the base of the dam is made to extend further downstream, and the plate resisting the water pressure is supported by concrete buttresses, hence "buttress dam." Lower Monumental Dam on the Sualeg River is an example.

The plate resisting water pressure can be curved to that it transmits the hydrostatic pressure to the cliffs on the sides of the dam. Such a structure, called an arch dam, is used in Seattle City Light's Diablo Dam on the Skagit river.

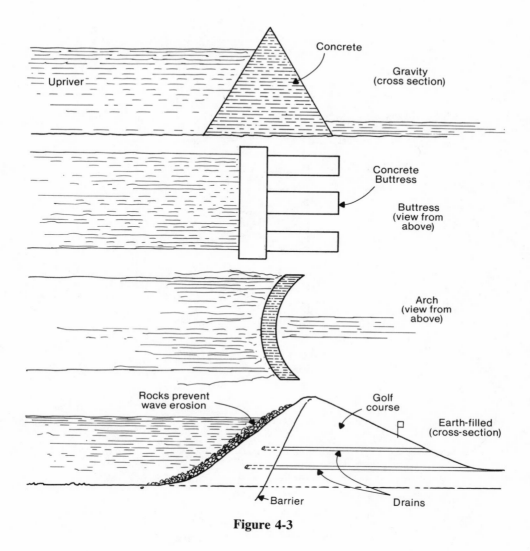

Figure 4-3

A dam can also be made from rocks or earth, with suitable provisions to prevent water erosion of its upstream face. "Piping," a phenomenon in which water develops a channel through the dam and washes it away, must likewise be prevented. Piping destroyed Teton dam in Idaho in June, 1975.

Turbines convert water flow into shaft power. The flow of water under pressure is converted to mechanical power with turbines. The most common turbine, called "Francis," has water flowing inward and downward to make the

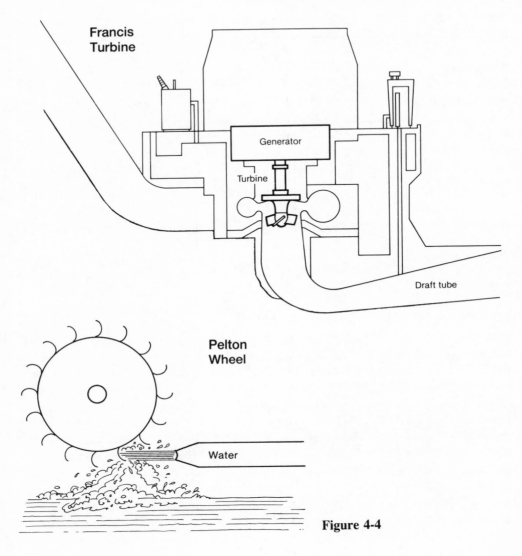

Figure 4-4

turbine rotor turn and drive a generator (Figure 4–4). Francis turbines are generally used on low heads. However, Hitachi developed a Francis turbine that also serves as a pump for Pacific Gas and Electric's Helms plant, where the head is as high as 1,745 ft (532 m).

At Bonneville Dam, where the head and flow vary, Kaplan turbines are installed. This turbine is essentially a huge four-bladed propeller with adjustable pitch to get maximum efficiency under the variable head and flow conditions at the site. In the winter a full seventy-six-foot head can be available, but not much

Renewable Power from Water

flow. In June when the snow melts in the mountains the water is plentiful, but the tailrace backs up to reduce the head.

For heads of over 200 feet the Pelton impulse turbine is often used (Figure 4-4). Water is directed from nozzles into buckets on the periphery of a wheel. Multiple nozzles permit the turbine to run efficiently at part load, with some of the nozzles turned off. Otherwise, throttling the water would waste energy.

For very low heads, such as obtained in tidal power plants, a "bulb" turbine is sometimes used. It has a horizontal turbine shaft that eliminates the sharp change in direction of water flowing through it.

Performance of hydro turbines is described in the *Energy Technology Handbook*.[4]

HYDRO POWER COST IS INTEREST

No fuel has to be purchased when generating power from a hydro plant. The predominant cost is interest on the investment. The following examples illustrate the point.

In 1975, the Copper Valley Electric Association in Alaska found itself paying increasing prices for 3 million gallons of diesel oil/year. It applied for a permit from the Federal Energy Regulatory Commission (FERC) to build a 12 megawatt (MW) hydro power plant at Solomon Gulch. The site had been licensed for a 480 horsepower plant in 1932. In 1976, construction started on the 400-foot rock-fill, asphalt-faced dam that raised the level of Solomon Lake from 610 to 685 feet above sea level, providing a good head for the power house that is a few feet above Prince William Sound. The 615-acre Solomon Lake storage enables the plant to generate 36,000 megawatt hours (MWh) of power per year, which corresponds to running the plant at its full 12 MW for 3,000 of the 8,766 hours of the year. However, the communities of Valdez and Glennallen, which are served by the utility, do not need full power 24 hours a day.

To finance the expected $67 million cost of the plant, including interest during construction, the association borrowed $44.9 million from the Rural Electrification Administration (REA) at 5 percent interest, and $23.3 million from a bank at 13.5 percent interest. The annual interest cost then would have been:

REA loan of $44.9 million × 0.05	= $2.24 million
Bank loan of $23.3 million × 0.135	= 3.15 million
Total	5.39 million
The 1983 operating cost of the plant was	0.86 million
Total 1983 cost, with original financing	$6.25 million

[4]Considine, Douglas M., *Energy Technology Handbook,* Hydropower chapter (New York: McGraw-Hill Book Co., 1977), pp. 8-3 to 8-24.

In 1983, the plant generated 30,800 MWH, so the cost of power (Cp) was:

$$Cp = \frac{\$6{,}250{,}000 \times \text{MWh}}{30{,}800 \text{ MWh} \times 1{,}000 \text{ kWh}} = \$0.203/\text{kWh}$$

This cost exceeded the cost of power generated by the diesel-engine driven power plants previously used. Diesel power was projected to cost $0.119/kWh in 1985.

The Alaska legislature, recognizing the need for a carefully planned and integrated power development, established and capitalized the Alaska Power Authority to coordinate, manage, and finance power generation projects. By injecting capital into the project and by equalizing the cost of power from all of its projects, the Authority expected to bring the cost of hydro power to the users to the same level as diesel power.

Features of hydro plants being built by the Alaska Power Authority are:

Plant	Nearest Alaska city	Head Feet	Head m	Rating, megawatts	Output MWh/year	Cost/kW dollars
Solomon Gulch	Valdez	685	209	12	36,000	4417
Tyee Lake	Wrangel	1400	427	20*	130,000	6280
Swan Lake	Ketchikan	336	103	22	88,000	4226
Terror Lake	Kodiak			20#	136,800	9450

*Vertical axis impulse turbine.
#Two vertical axis impulse turbines, plus provisions for a third.

MEDIUM SIZE HYDRO PLANT: LIFE CYCLE COST

Changing power costs and new developments made feasible a 112 megawatt (MW) hydro plant on the Sultan River in Washington. Part of the Sultan river had been previously diverted into Lake Chaplain to provide drinking water for the city of Everett (Figure 4–5). The Sultan river water comes from rain and melting snow in the river's mountainous watershed. To equalize spring floods and autumn droughts, the city had erected the earth-filled Culmback Dam that created Spada Lake, now a reservoir.

The power generating opportunity came from the possibility of raising the level of Spada Lake by 90 feet to provide 153,270 acre feet of storage, and delivering from 700 to 1,300 cubic ft of water/second (cfs) to a power house.

Medium Size Hydro Plant: Life Cycle Cost

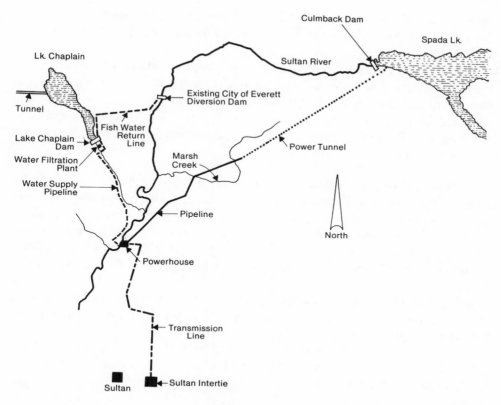

Figure 4-5

Routing the water through four miles of tunnel and four miles of pipe produces a respectable head of 1,170 feet at the power house. A flow of 1300 cfs and 1170 feet head produces in hydraulic power (Ph):

$$Ph = \frac{1300 \text{ cu ft} \times 62.4 \text{ lb} \times 1170 \text{ ft} \times \text{sec-hp} \times 0.746 \text{ kW}}{\text{second} \times \text{cubic ft} \times 550 \text{ ft-lb} \times \text{hp}}$$
$$= 128{,}700 \text{ kW} = 128.7 \text{ MW}$$

Losses reduce this available power to the actual rating of 112 MW. At 6 cents/kWh, 112 MW is worth $6,720/h.

Adopted Sultan River hydroelectric project configuration. Power is generated by two 47.5 MW Pelton-turbine driven generators, and two 8.4 MW Francis-turbine driven generators. Water leaving the Francis turbines is under pressure and is carried in a 6.5-foot pipe up to Lake Chaplain dam, where part of the water is delivered to the city of Everett's drinking water reservoir. The rest is piped to the existing diversion dam and released for the benefit of the downstream fish.

The power house is controlled remotely from the utility's headquarters in Everett. Generated power is transmitted in a 155 kilovolt transmission line for a distance of 3.5 miles to the Sultan substation.

Serious design work on the hydroelectric project began in 1968, and construction started in 1982. Filling of Spada Lake reservoir started in November, 1983, and in 1984 power was produced.

Tunnel diameter option. A trade study resulted in the fourteen-foot diameter tunnel that carries the water through the base of Blue Mountain. A twelve-foot diameter was originally specified. The tunnel contractor proposed to use a tunnel boring machine to excavate the four-mile tunnel in one pass without any intermediate entries. The tunnel-boring machine grips the tunnel walls and pushes a cutting head with many knobbed or sharp wheels against the rock face. The muck consists of rock chips, as large as 4 inches in diameter, which are conveyed or hauled out through the excavated tunnel. The machine consumes 1,000 kW of electric power, and its boring rate varies from 10 to 100 ft/day, depending on the type and competence of the rock.

In his proposal, the contractor offered to bore the four miles of tunnel for $15.5 million, and for an extra half million dollars he would make the tunnel 14 feet in diameter. A study showed that the reduced water-flow friction in the larger tunnel was worth between $6 and $7 million during the life of the power generating project!

Cost of power from Sultan River Project. The decision to proceed with the Sultan River Project was based on the cost analysis summarized in Table 4–1. The engineering services entry represents fees paid to Bechtel Civil and Minerals of San Francisco, and R. A. Beck and Associates of Seattle, who conducted analyses, prepared detailed drawings, and supervised construction.

Interest during construction was paid to banks on short-term notes. The predicted cost of the Sultan project was $180 million. The actual cost of $215 million was reasonably close to the prediction.

The annual cost of operating the Sultan power plant consists mainly of interest on the investment, maintenance labor and supplies, operators' salaries, and administration. The district, being a publicly owned utility, does not pay property taxes as would an investor-owned utility. The interest is related to the strategy of long-term financing. The district, in consultation with its underwriters, decided to sell tax-exempt bonds maturing between 1986 and 2020. This

Medium Size Hydro Plant: Life Cycle Cost

TABLE 4-1 Cost of Construction (From Ref. 5)

	Total estimated cost ($1,000)
Land and Land Rights	$ 1,465[1]
Powerhouse	10,441
Reservoir, Dams, and Waterways	
Spada Lake Facilities	18,090[2]
Blue Mt. Tunnel	17,548
Power Pipeline	22,385
Manifolds	1,827
Lake Chaplain Pipeline	9,437
Turbines and Generators	15,871
Accessory Electrical Equipment	1,917
Miscellaneous Powerplant Equipment	3,247
Emergency Water Treatment Facility	182
Permanent Roads	
Spada Lake Facilities	1,840[2]
Water and Power Facilities	571
Switchyard Structures and Substation	596
Switchyard Equipment	1,299
Transmission Lines	2,251[1]
Communication Equipment	
Remote Terminals	604
System Control and Data Acquisition Equipment	4,797[1]
Total Specific Construction Cost	$114,368
Washington Sales Tax	7,282
Reserve for Contingency	6,909[5]
Engineering and Construction Management	21,879[3]
Indirect Construction Costs	
District's Administrative Costs	11,471[1]
Mitigation Costs	12,523[1,4]
Preliminary Investigations and Engineering	5,713[1]
Total	$180,145

[1] Cost estimate furnished by the District.
[2] Cost estimate furnished by R. W. Beck and Associates.
[3] Cost estimate furnished by Bechtel and R. W. Beck and Associates.
[4] Includes the District's share of cost of Construction of City of Everett's permanent water treatment facility and filtration plant.
[5] Determined by the District, based upon cost data developed by the District, Bechtel and R. W. Beck and Associates.

TABLE 4-2. Estimated Annual Power Costs: 1985[5]

Item	Annual cost, dollars
Operation and maintenance	$1,230,000
Administrative and general	248,000
Insurance	545,000
Taxes (Incremental taxes required to be paid, 0.314 mills/kWh)	96,000
Subtotal	2,119,000
Debt service	28,497,000
Less interest earned on invested funds	3,901,000
Net annual cost	$26,715,000
Estimated average annual energy output	= 450,000,000 kWh*
Estimated cost of energy	= $0.0594/kWh (59.4 mills/kWh)

*During periods of flooding the District may be requested to empty the Spada Lake reservoir to make room for flood waters. Such operations can reduce the annual output by around 4,000,000 kWh.

committed the District to pay during 35 years $28.5 million/year for interest and repayment of principal.

The forty-year average flow of the Sultan River supports generating 50 MW average a year. This corresponds to an annual energy generation (E_g) of:

$$E_g = 8766 \text{ h/year} \times 50 \text{ MW} = 438,300 \text{ MWh} = 438 \text{ million kWh}$$

The cost of power (C_p) delivered to the generating-station bus bars is then the cost of operating the plant divided by the power generated as developed in Table 4-2. The cost of generated power is 5.94 cents per kWh.

After thirty-five years, the plant will still produce power, but all principal and interest on the operating cost will have been paid. Then the cost of power generated for the next thirty years (C_t) will be:

$$C_t = \frac{\$2.119 \times 10^6}{450,000,000 \text{ kWh}} = \$0.0047/\text{kWh}$$

Power cost is generally stated in tenths of a cent, called "mills" ($0.001). Bonneville Power Administration at one time sold hydro power from the Columbia River plants to preference wholesale customers for $18.20 per kilowatt year. This corresponds to 2 mills kWh. On the other hand, some of the new nuclear

[5] "Official Statement, Public Utility District No. 1 of Snohomish County, Washington, Sultan Hydroelectric System Revenue Bonds," Series 1983, November 17, 1983.

plants cost around $4,000/kW at the time the Sultan river project was built. Assuming an 11 percent interest rate and an 80 percent availability on an annual basis, the cost of interest and thirty-year loan payments (Cn) of a $4,000 per kW plant would be:

$$Cn = \frac{(4{,}000 \times 1/30) + (\$4{,}000 \times \text{year} \times 0.11)}{1 \text{ kW} \times 0.8 \times 8766 \text{ h} \times \text{year}} = \$0.082/\text{kWh}$$

To this would be added other costs of operation. Thus, while fifty-nine mills/kWh might not appeal to rate payers used to getting electricity for twenty mills/kWh, the Sultan River project looks good when compared with available nuclear alternatives.

Permits and approvals. A hydro project affects the river that it's built on, the local environment, owners of water rights, and the public in general. The Snohomish County Public Utility District, being an agency of the county, had convenient access to regulating organizations. However it still had to get an impressive array of permits and licenses (Table 4–3).

A smaller hydro project would not necessarily get by with fewer permits and licenses. Arranging these consumes legal, engineering, and administrative time.

PUMPED HYDRO: A BIG RECHARGEABLE BATTERY

Energy can be stored by pumping water to a higher elevation, and later recovering the energy by releasing the water through a turbine into the reservoir from which it was pumped (Figure 4–6). The pumping power as well as the recovered power can be alternating current, and the same synchronous machine can be used to motor the pump and generate the recovered power. Thus the recovered energy can be injected directly into the utility grid without the direct-current-to-alternating-current losses that come with storage of energy in chemicals or superconducting coils. Pumped hydro is useful because it can supply power during peak loads, enabling its owner to avoid building extra fossil or nuclear power plants.

Energy features of pumped hydro storage include:

- Turbines and generators are not 100 percent efficient, so 20 to 30 percent of the invested energy is lost in penstock and pipe friction, and turbine and generator losses.
- A pumped hydro plant must be charged with off-peak power. Nothing is gained by using oil or gas fuels to charge the hydro plant because oil and gas can be used directly in low-cost peaking power plants.
- Pumped hydro plants, like ordinary hydro plants, have long lives but take a long time to construct.

TABLE 4-3. Sultan Hydroelectric System, Major Permits, Licenses, Approvals

---

Federal Energy Regulatory Commission: Amended License to Construct
Federal Aviation Administration: Determination of No Hazard
U.S. Army Corps of Engineers: Section 404 Permit
Federal Communications Commission: Operating License, Microwave Radio
U.S. Department of Agricultre, Forest Service: Special Use Permit
Washington (state) Department of Ecology
   Section 401, Certificate of Reasonable Assurance
   National Pollutant Discharge (NPDES)
   State Waste Discharge Permit
   Short-Term Exception to Water Quality Standards (Multiple)
   Flood Control Zone Permit
   Permit to Appropriate Public Waters
   Reservoir Permit
Washington Department of Social and Health Services: Public Water Supply Approval
Washington Department of Natural Resources
   Application to Purchase Valuable Materials (Multiple)
   Forest Practices Permit
   Right-of-Way Permit (Multiple)
   Surface Mining Permit (Multiple)
Snohomish County
   Shoreline Substantial Development/Conditional Use Permit
   Building Permit/Drainage Plan Approval
   County Road Construction Agreement
   Sewage Holding Tank and Potable Water System Approval
Town of Sultan
   Right-of-Way Permit and Pipeline Easement

Limits of pumped hydro power include:

- To generate one megawatt requires at least one cubic meter of water a second flowing through a turbine with a 100-meter head. Additional water must flow to generate turbine and generator losses.
- In areas without mountains pumped hydro can be achieved only with underground caverns.
- The competition for pumped hydro is the gas turbine. As the cost of gas turbine fuels rises, pumped hydro plants can cost more and still pay off.

Helms pumped storage project. Pacific Gas and Electric Company's 1125 MW Helms project illustrates pumped storage. Water is transferred between

Pumped Hydro: A Big Rechargeable Battery

Pumped Storage

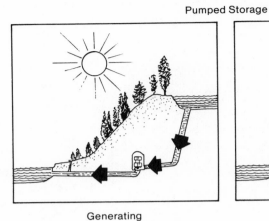

Generating   Pumping

Figure 4-6

two reservoirs, Courtright Lake where the maximum elevation is 8,184 ft (2494 m) and Lake Wishon at 6,550 ft (1996 m).

The significant underground facilities are almost 20,000 ft (6096 m) of 17-foot (5.2 m) diameter concrete-lined pressure tunnel; 3,700 ft (1,128 m) of access tunnel; 1,600 ft (488 m) of steel-lined high pressure conduit, two major underground chambers more than 1,000 ft (305 m) below ground; three deep vertical shafts, and one inclined shaft. Maximum static pressures within the conduits range from 200 ft (61 m) to 1,744 ft (532 m), with hydraulic transients that go higher. These underground works provide 475,000 horsepower of water power to each of three reversible pump-turbines. Each is coupled to a 375-MW generator motor. The units, rated at 1,744 ft (532 m) maximum static head, are unique, considering the size-head combination.

The arrangement of tunnels and plant is shown in Figure 4–7. Pertinent details are summarized in Table 4–4. Arthur G. Strassburger, the manager of the project, describes it in detail in Ref. 6. He enriches his description with interesting anecdotes about some of the troubles, such as ten-foot icicles hanging over the crown of the tunnel and fifteen feet of snow on the ground during the 1978–79 winter.

Pertinent engineering features of the project were:

- New technology tunneling techniques, including the use of tunnel boring machines, plus the rising value of energy storage, made this project economically feasible.

[6] Arthur G. Strassburger, "Helms Pumped Storage Project," *USCOLD Newsletter*, 68 (New York, The U.S. Committee on Large Dams).

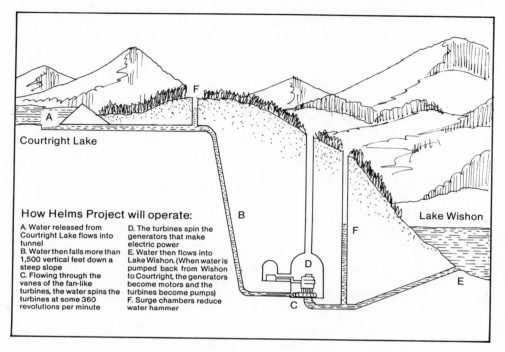

Figure 4-7

- A single stage Francis turbine that can also pump water into a 1,800 ft head is at the edge of technology. A Pelton turbine is normally used for this head in a straight hydro plant. The turbine was built by Hitachi.
- The project was conceived in 1970 and the company applied for a Federal Energy Regulatory Commission license in late 1973. It was completed eleven years later.

The cost of the project rose with inflation. An early 1973 estimate was $212 million, which was revised to $383 million in 1977, and finally completed for $738 million, including all overhead, in 1983. The adopted rating is 1,125 MW, which corresponds to a cost of $656/kW. The lowest cost nuclear and coal power plants being completed at the same time cost around $1,000/kW.

The Sultan River hydro project in Washington is smaller, 115 MW, but similar in that it has tunnels, intake works, and back-pressure turbines. It was built for $180 million or $1,560/kW. Small hydro projects, in the range of one to fifteen MW, are reported by the Electric Power Research Institute to cost around

Pumped Hydro: A Big Rechargeable Battery

TABLE 4-4. Features of Helms Pumped Hydro Project

Reservoirs	
Working volume-Courtright	146.8 million cu m
Wishon	109.8 million cu m
Flow rate, maximum	280 cu m/sec
Range of reservoir level	
Upper reservoir	50 m
Lower reservoir	34 m
Static head at turbines	448 to 532 m

Tunnels	
Upstream conduit-diameter	8.23 m
length	5,430 m
Tailrace conduit length	1440 m

Power generator	
Pump-turbines (3), reversible Francis	Hitachi
Generators (3), 1,195 MW Total	Westinghouse
Synchronous speed	360 rpm
Pump starting	Pony motor

Governor time	
Load on	15 to 60 sec
Load off	30 sec

$2,000/kW.[7] Alaska hydro plants built during the same period cost around $4,000 to $6,000/kW.

Energy storage capacity. The maximum static head at the turbine at Helms power house varies between 532 and 448 m (1745 to 1470 ft). The maximum flow rate is 280 cubic meters/second (9888 cubic ft/second). One megawatt of power can be generated with 1 cubic meter/second flowing with a

---

[7]Fritz Kalhammer, "Small-Hydro Activities," *EPRI Journal,* Electric Power Research Institute, Palo Alto (July/August 1982), p. 52–53.

100 meter head. Assuming an average head of 500 meters, the hydraulic power produced by the turbines (Ph) would be:

$$Ph = \frac{500 \text{ m} \times 280 \text{ cubic meters} \times \text{megawatt} \times \text{second}}{\text{second} \times 1 \text{ cubic meter} \times 100 \text{ meters}} = 1,400 \text{ MW}$$

The electrical output of the plant is given as 1,125 MW, suggesting that the efficiency (η_h) is:

$$\eta_h = \frac{1125 \text{ MW}}{1400 \text{ MW}} = 0.8$$

This is reasonable considering generator and turbine losses, and friction in 7 km (4.3 miles) of tunnel where the water flows at 5 meters/second (16.4 ft/second).

The capacity of the Courtright reservoir, the upper one, is 146,800,000 cubic meters. The duration of continuous operation before the reservoir is completely drained (Dr) would be:

$$Dr = \frac{146,800,000 \text{ cubic meters} \times \text{second} \times \text{minute} \times \text{h} \times \text{day}}{280 \text{ cubic meters} \times 60 \text{ seconds} \times 60 \text{ min.} \times 24 \text{ h}}$$
$$= 6 \text{ days}$$

This is not the intended mode of operation because preserving head in the Courtright reservoir is important for other plants in the Kings River hydro complex. The value of the water in Courtright reservoir for its downstream plant is highest when the water level is at the top of the dam. The other plants are much smaller than the Helms plant, for it would not be economical to build a hydro plant that could drain its reservoir in six days, and then stand idle for months until the reservoir is filled again.

The 6 days of storage capacity could also be obtained with GNB "solar" (lead-acid) batteries costing around $300/kWh. The cost of these batteries (Cb) for 6 days of storage would be:

$$Cb = 6 \text{ days} \times 24 \text{ h/day} \times \$300/\text{kWh} = \$43,200/\text{kWh}$$

The Helms project stores energy for $659/kW for six days. For storing energy for one hour, the batteries cost less. They cannot compete with stored hydro for storage periods of more than a few hours.

If the efficiency of the Helms project is 80 percent, then the round trip efficiency would be around 0.8 for pumping and 0.8 for recovery, or 0.64. This indicates that the energy to be stored would be best generated in other hydro plants during off-peak hours when water otherwise would be spilled.

Gas turbines: the alternative to pumped hydro. A gas turbine power plant could be built, ready to run, for about $300/kW. Assuming that Pacific Gas

Power from Tides and Ocean Thermal Gradients

and Electric pays 20 percent per year for interest and debt service, and that the difference in interest (Di) is available for buying fuel, then the manager of the equivalent gas turbine plant could spend this much for fuel per year:

$$Di = \frac{0.20\ (\$659-\$300)}{kw} = \$71.80/kW$$

If PG & E could get gas for $5/million Btu, and if a new Mitsubishi AGTJ 100A turbine with 40 percent efficiency were bought, then the money saved in interest would buy this much energy (Egt):

$$Egt = \frac{\$71.80 \times 1 \times 10^6\ \text{Btu} \times \text{kWh} \times 0.4}{\$5 \times 3412\ \text{Btu}} = 1683\ \text{kWh}$$

This represents an average of 4.6 hours a day of power generation during an 8,766 hour year, so if the cost of natural gas remained constant, the choice between the Helms pumped hydro station and a 40 percent efficient gas turbine might be a toss-up. On the other hand, the cost of pumped hydro peak power from the Helms project is frozen, and no fuel price escalation can affect it.

Pumped hydro storage without mountains. Pumped hydro energy storage is feasible in underground caverns if the rock is strong enough. The deeper the lower cavern, the smaller the volume of water that needs to be stored and handled for a given energy storage.

Potomac Electric, in research sponsored by the Electric Power Research Institute, has quantified the value of deep reservoirs (Figure 4–8). Their estimated construction cost is in Table 4–5. The work is described in Ref. 8.

POWER FROM TIDES AND OCEAN THERMAL GRADIENTS

Energy deposited on ocean shores by waves and tides has inspired many to speculate on the power that might be captured from these sources. No significant power is being produced from wave energy, and only two large tidal plants are in operation. Useful comparisons with alternative sources of power are hard to make when the required equipment hasn't been designed.

The energy source for a tidal power plant is water, with a low and variable head. The best site in North America is the Bay of Fundy between Maine and Canada, where the average elevation difference between low and high tide is 39 feet and the maximum is 53 feet. A flow of 8 million cubic ft/second is available. However, the estimated cost is around $23 billion for a 5,000 MW plant. The

---

[8]"Underground Pumped Hydro," *EPRI Journal*, October, 1981, pp. 51–53.

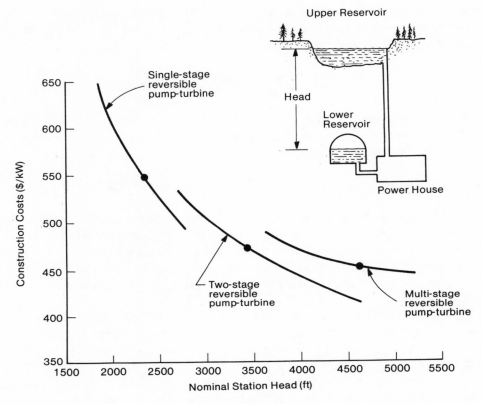

Figure 4-8

tidal plant produces full power only part of the time. James A. Fay and Mark A. Smachlo suggest a capacity factor of 0.25 to 0.35 for a tidal plant because the power generated depends on the difference in elevation of the pond and the sea.[9] In a companion paper, the authors analyze the capital cost of tidal power plants.[10]

Coal and even nuclear power plants can be built with much less risk than can tidal plants. The Bay of Fundy project was first proposed in the 1930s, but getting the U.S. and Canada to adopt it seems unlikely, considering the environmental impact.

Ocean thermal power is generated, for example, with an ammonia Rankine

---

[9]James A. Fay and Mark A. Smachlo, "Performance of Small Scale Tidal Power Plants," *Journal of Energy*, 7, no. 6, November–December 1983, pp. 529–35.

[10]James A. Fay and Mark A. Smachlo, "Capital Cost of Small Scale Tidal Power Plants," *Journal of Energy*, 7, no. 6, November–December 1983, pp. 536–41.

TABLE 4-5. Estimated Construction Costs for 2000-MW (10-Hour) Underground Pumped Hydro Plant

	$ Millions (mid-1979)
Land, site access	6.2
Dams, reservoirs	302.3
Tunnels, shafts	198.1
Powerhouses, civil structures	53.5
Pump-turbines, valves, mechanical equipment	99.1
Motor-generators	80.9
Transformers, electrical and transmission equipment	91.4
Total direct costs	831.5
Administration, overhead, engineering and construction management	249.4
Contingencies	124.7
Total	$1205.6

engine, which uses the surface water at 25°C as a heat source and 5° to 10°C water pumped from a depth of 700 to 900 meters as a heat sink. The Carnot cycle efficiency is low, so huge quantities of water must be pumped through heat exchangers. Experimental plants have produced only enough power run the research ship.

GEOTHERMAL: FREE HEAT FROM THE EARTH

The core of the earth is molten because the thickness of insulation around it limits the escape of heat left by the nova or other event that created the planets. This primordial heat is supplemented by energy released by later radioisotope decay. In escaping through the earth's crust this heat creates a temperature gradient, typically 30°C/kilometer (17°F/1,000 ft). In volcanic regions the gradient is higher, with molten lava sometimes appearing at the earth's surface. No practical way of generating power with heat from molten lava has been found.

A goethermal plant generates power with heat brought from the interior of the earth by a goethermal fluid that can be water or steam. Any geothermal fluid can be used for heating buildings. For generating power, the hotter the fluid, the more useful it is.

Energy features of geothermal power include:

- Geothermal heat is free, so the main cost of generating power is in the capital equipment and in drilling new wells to replace exhausted and plugged ones.
- Very few places in the world have geothermal steam at the temperatures and quantities appropriate for building power plants.
- Unlike solar and wind power, geothermal resources supply firm power which is available continuously.
- The temperature of geothermal fluids is generally less than 350°F (177°C), so the efficiency of geothermal plants cannot match those of fuel-burning and nuclear power plants. This low efficiency means that pipes, pumps and heat exchangers must be large, and a lot of condenser-cooling water is needed.
- Hot water under pressure dissolves solids, which precipitate in pipes as the pressure is relieved. At Wairakei, New Zealand, the wells are redrilled as the pipes clog.

The limiting features of geothermal energy include:

- Use of geothermal energy as heat is limited only by the cost of piping and pumping the fluid from the well to the using facility. For example, for district heating of the city of Reykjavik, Iceland, 86°C hot water is pumped 13 km through pipes insulated with turf and rockwool.
- Geothermal fluids at temperatures below 250°F (121°C) could be used for generating power, but alternative power sources are usually more economical.
- Geothermal water rejected by a power plant is hot and usually contains sulfur. Its disposition into groundwater or a stream may be objectionable.

Geothermal heat output of a well is often measured in kilowatts thermal (kWt), as well as in Btu/h. Petrothermal resources are rocks that are hot. A hydrothermal resource produces steam or hot water.

GEOTHERMAL RESOURCES

Approximately 1 percent of the U.S. geothermal resources are hydrothermal, some 2,000 MWt being identified as steam, 12,000 MWt being water at temperatures above 210°C, and 10,000 MWt being water between 150°C and 210°C. Pertinent features of some of the world geothermal plants are shown in Table 4–6. Resources at Geysers, California; Wairakei, New Zealand; and Milos, Italy are the best. Discovery and proving of a geothermal resource equivalent to any one of these three would justify the design and construction of a new geothermal power producing complex and the transmission lines for carrying out the power. Such new resources have not been discovered.

TABLE 4-6. Some World Geothermal Plants

MW	Location	Reservoir depth, m	Form	Temp., °C	Notes
908	Geysers, Ca.	600 to 2100	Steam, 100 lb/in^2	177	No. 2 heat rate: 21,190 Btu/kWh
198	Wairakei, N.Z.	600 to 1400	Water, 800 lb/in^2	250	Flash to steam
405	Lardello, Italy		Steam		
8	Milos, Italy	500 to 1100	Water	300	100 tonne/hour; 10% salinity
5	Raft River, Id.	1300 to 1700	Water 0 lb/ft^2	135	Isobutane Rankine-cycle engine

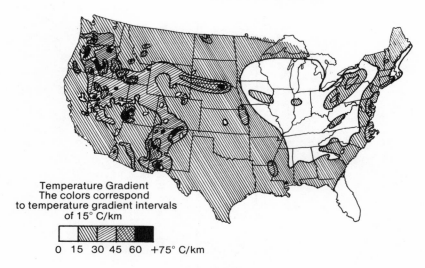

Figure 4-9

Most U.S. geothermal resources are in the western states. Figure 4–9 shows some of the locations.

From hot water to power. A steam turbine generator can convert geothermal heat to electric power if you have a resource that delivers hot steam. For example, wells at Geysers, California, deliver 350°F (177°C) steam, and the ones at Laidello, Italy, deliver 430°F (221°C) steam. However, steam-producing geothermal wells are rare, with the usual resource delivering hot water.

Part of a hot water flow can be flash-boiled into steam if the geothermal water is hotter than 212°F (100°C). Flash boiling is not an efficient way of extracting heat from water. For example, consider the source at Raft River, Idaho, which produces water at 280°F (138°C). Assume that the desired steam is at 240°F, which corresponds to a gage pressure of 25 lb/in². Then the spent water can be no colder than 240°F, and every pound of water contributes only 40 Btu to the generation of steam. More heat could be extracted from the water if steam is generated at 220°F, but then the pressure would be 17 psig, which is rather weak for running steam turbines. At 212°F the steam would be at atmospheric pressure, and would have to be sucked into the turbine by cold condensing water.

Flash boiling is simple, but discarding huge quantities of hot spent water is costly, particularly in the U.S., where it must be injected back into deep wells if it contains minerals or salts. The geothermal resource may be free, but the cost of drilling and maintaining wells is proportional to the quantity of geothermal fluid delivered.

More efficient use of geothermal fluid results if the working fluid in the Rankine cycle has a lower heat of vaporization than that of water. For example, isobutane working fluid is boiled in a heat exchanger with hot geothermal water. The isobutane vapor goes to the turbine and condenser. A boiler feed pump sucks the liquid isobutane from the condenser well and sends it through liquid-isobutane heaters ("feedwater heaters") and on to the boiler. These heaters are heat exchangers that extract more heat from the geothermal water as it leaves the boiler. In one Raft River, Idaho, plant the geothermal water, originally at 290°F (143°C) is reinjected at 144°F (62°C) after delivering around 146 Btu/lb during its flow through the plant. The price paid for efficiency by this means is complexity and heat exchangers.

Heat exchangers for low-temperature geothermal power plants are large. For example, the condenser for a 500 kWe 300°F (149°C) isobutane plant is 26 feet high, 7 feet in diameter, and it weighs 20 tons. It condenses a million lb of isobutane an hour.

5 MWe binary geothermal plant. At Raft River, Idaho, three geothermal wells 1,300 to 1,700 ft deep deliver water at 275° to 300°F (135° to 149°C). The adopted power generating cycle has a high-pressure and low-pressure turbine with isobutane as a working fluid (Figure 4–10). The plant was built by the Idaho National Engineering Laboratory as an integral part of the Department of Energy's plan for commercial development of geothermal energy.

The spent water is 144°F, having contributed some 136 Btu/lb to the heat engine. This heat extraction could not have been achieved in a simple Rankine cycle because of the amount of geothermal water needed for boiling the isobutane at 240°F. This quantity of heat for boiling is 50 percent greater than the amount that could be used in heating the liquid isobutane from the condensing temperature to 240°F. Instead, around one-third of the isobutane is boiled at a lower pressure and temperature, and the heat flows are balanced. In this way all practical heat is extracted from the geothermal water before it is pumped back into injection wells. The benefit of using a binary cycle rather than flashing hot water to steam is shown in Figure 4–12 from Ref. 11.

Carl J. Bliem and his colleagues described the performance of the 5 MWe Raft River plant by tracing the destinations of the geothermal heat (Figure 4–11). They achieved a 4.010 MWe electric power output, and identified the following losses, which when corrected would result in a power output of 5.223 MWe.[12]

[11] "Heber Geothermal Demonstration Plant," *EPRI Journal,* December 1983, p. 34.

[12] Carl J. Bliem and others, "Performance of a 5 MW(e) Binary Geothermal-Electric Power Plant," *Proceedings of the 18th Intersociety Energy Conversion Engineering Conference* (AIChE, 1983), pp. 274–279.

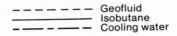

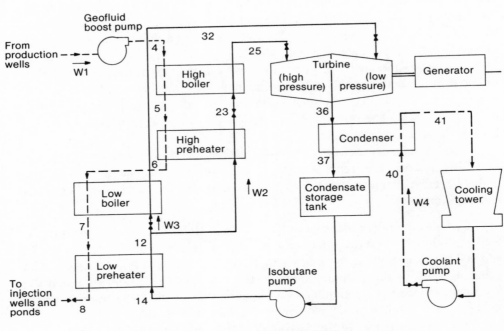

Figure 4-10

Losses that could be corrected	Power, kilowatts
Generator output	4010
Failure to utilize design geofluid flow	110
Moisture in turbine	144
Cooling water pumps not able to produce specified flow	380
Cooling tower unable to produce specified cold water temperature	454
Other components including heat exchangers and turbine generator	125

The important loss of 454 kW was attributed to the cooling water pumps delivering only 78 percent of rated flow. The pump pit was too small for proper pump inlet flow.

Note that if the power required for supply and injection pumps were subtracted from the 3.4 MWe power generated, the net power output would have been 1.6 MW. However, the supply and injection pumps had not been designed for supplying the 5 MWe plant only, so they were not optimum.

Geothermal Resources

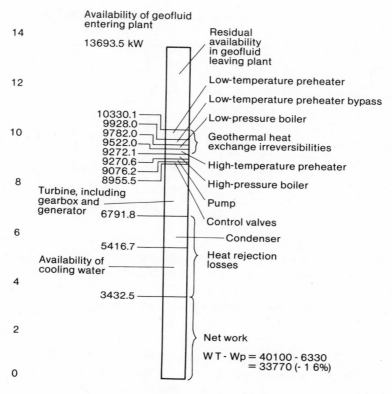

Figure 4-11

45 megawatt binary cycle power plant. A 45 MWe binary cycle power plant was built by a team of utilities, the State of California, the U.S. Department of Energy, and the Electric Power Research Institute, at the Heber geothermal field in the Imperial Valley of California. The purpose of the $122 million project was to scale up the binary-cycle technology and demonstrate economic viability.[13] Geothermal fluid is supplied by the Union Oil Company of California and Chevron USA, who drill the wells and sell the heat to the project. The spent fluid is returned to the heat suppliers for reinjection into a suitable part of the reservoir.

The wells are expected to produce 1.0 cubic meter/second (7.9 million lb/hour) of brine at 182°C (360°F) when the plant starts. The temperature is expected to decline by 17°C by the end of the plant's thirty-year life.

[13] J. E. Bigger, "Heber Geothermal Binary Project," *EPRI Journal,* October 1984, pp. 44–45.

TABLE 4-7. Estimated Annual Average Performance of Heber Geothermal Plant*

Performance parameter	Gross power	Net power to grid
Capacity	65 MW (e)	45 MW (e)
Heat rate	23.7 MJ/kWh (22,500 Btu/kWh)	32.5 MJ/kWh (30,900 Btu/kWh)
Thermal efficiency	15.6%	11%
Brine rate	49 kg/kWh (107 lb/kWh)	69 kg/kWh (147 lb/KWh)
Availability	82% (potential for improvement)	

The turbine is a dual axial-flow machine rated 70 MWe at full load. The working fluid is 90 percent isobutane and 10 percent isopentane. Characteristics of the plant are summarized in Table 4–7. The benefit of the binary-cycle installation for Heber is related to a flashed-steam plant in (Figure 4–12).

Rotary separator turbine. Wells at the Roosevelt Hot Springs near Milford, Utah, produce 421 psia (2.9 MPa) water at around 290°F (143°C). Utah Power Co. operates a 20 MWe power plant there and has a second 14 MWe unit under construction. Utah Power estimates that the southwestern part of Utah has a potential of 200 to 400 MWe of geothermal power, the equivalent of one of the company's coal-burning power units.

Electric Power Research Institute and Utah Power, with other sponsors, tested at this site a fifty-four-inch diameter rotary separator turbine, which is an alternative to flashing hot water into steam.

Hot water pumped out of the ground has only heat energy corresponding to

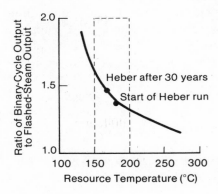

Figure 4-12

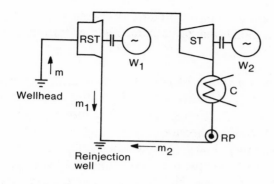

RST	= Rotary separator turbine
ST	= Steam turbine
C	= Condenser
RP	= Reinjection pump
W	= Generator output power

Figure 4-13

its temperature above an available heat sink. Hot water that comes out of the ground under pressure also has potential mechanical energy, just like the water in a penstock arriving at a hydro plant. A rotary separator turbine directly captures the mechanical energy in pressurized geothermal fluid. The pressure is reduced, not by throttling, but in carefully designed nozzles that convert the hydraulic head into a moving water jet. At the same time, part of the water flashes into steam within the nozzles. The steam is normally routed to a steam turbine (Fig. 4–13). Delaval Transamerica has designed an efficient nozzle that handles this changing two-phase flow.

The rotary separator turbine tested at Roosevelt Hot Springs ran for 4,004 hours during an endurance test. Its efficiency was 33 percent and it generated 1.3 MWe.[14] From the 527,000 lb/hour of geothermal fluid it produced 121,000 lb/hour of steam at 43 psia. The steam, which was not used during the test, could have generated 5.75 MWe. Nozzle inlet pressure was 326 psia (2.24 MPa). The spent water was discharged at 102 lb/in^2 (0.7 Pa). The power availability was 94 percent, which was good for an experimental unit.

The rotary separator turbine weighed 85,000 lb, and an accompanying lube-oil skid weighed 8,000 lb. By changing nozzles and diffusers the unit could be modified to handle flow rates up to a million lb/hour.

Utah Power estimates that the new 14 MWe plant will cost $14.7 million.

[14]Evan E. Hughes, "Geothermal Rotary Separator Turbine: Wellhead Power System Tests at Milford, Utah," *Proceedings of the 18th Intersociety Energy Conversion Engineering Conference* (AIChE, 1983), pp. 280–285.

The utility expects the rotary separator plant to produce 30 percent more electricity than could be produced by a geothermal plant extracting power from steam alone.

Geysers, California, geothermal steam plant. The most successful geothermal field in the U.S. is at Geysers, California, where the wells produce steam at 100 lb/in² (0.7 MPa) and 350°F (177°C). Conventional steam turbines are used to drive generators. The heat rate of No. 2 generator is 21,190 Btu/kWh, which is only twice the heat rate of an efficient coal burning power plant. Pacific Gas and Electric Co. is building the twentieth geothermal generating unit, costing an estimated $216 million, to generate 120 MWe. This will bring the power production from the field to 1,139 MWe.

Unfortunately, Geysers is the only steam-producing geothermal field in the U.S. with the quality and quantity of steam needed for efficient power generation.

Hot dry rocks. At any given place the most probable geothermal heat source is underground rocks, from which heat can be extracted by fracturing the zone between two wells and pumping hot water through the fracture. Scientists at the Los Alamos National Laboratory have investigated this resource, measured the temperature of 1,700 wells, analyzed the geology of potential sites, drilled exploratory wells to depths of 5 km (16,000 ft), and generated 60 kWe of electric power. This work is described in part by Roland A. Pettitt.[15]

The team at Los Alamos has shown that the hot-dry-rock concept is feasible. Realistic cost estimates can be made after demonstration power plants are in operation.

Alternatives to geothermal: nuclear and coal plants. For utilities having proven geothermal resources the alternatives are generally importing power or building nuclear or coal steam plants.

Pacific Gas and Electric has in its territory the Geysers field, which is probably the best geothermal resource in the world. In Chapter 1 you will find the analysis of the alternatives that led to the building of the twentieth Geysers unit, 120 MWe, at a cost of $1,800/kW. Importing power into its territory was an important alternative. Constructing nuclear and coal plants had not been seriously considered because they are essentially prohibited by California laws.

Utah Power found that geothermal power from Roosevelt Hot Springs would be competitive with power generated in coal burning power plants.

---

[15]R. A. Pettitt, "Development of Man-Made Geothermal Reservoirs," *Proceedings of the Conference on Energy in the Man-Built Environment*, Colorado, August 3–5, (ASCE 1981).

Where not to build geothermal plants. Not all geothermal plants are successful. The Seattle Times on October 13, 1983, carried a story about how the General Accounting Office discovered that the Department of Energy had sunk nearly $45 million into a geothermal plant in New Mexico before finding that there was not enough underground heat to run it. The $133 million project was started in 1978 and abandoned in 1982.

5

Nuclear Fission: Produces No Acid Rain, CO_2, or Nox

In a nuclear power plant, heat produced in a reactor generates steam that drives a Rankine-cycle turbine, which turns a generator. In most power reactors uranium atoms break up or "fission" into smaller atoms. The heat comes from the kinetic energy of the fission products, radiation they release as they decay to stable elements, and from absorption of released neutrons. The accompanying electromagnetic radiation is absorbed within the reactor and in surrounding shields.

Energy features of nuclear power include:

- The biggest cost element in U.S. nuclear power plants is the interest during construction, particularly when essentially complete plants are prevented from operating.
- Fuel cost is an almost trivial part of the price of delivered power.
- The low cost plants have been built by teams that had previously built successful plants.
- Nuclear power plants produce no carbon dioxide, nitrogen oxides, or acid rain. Reprocessing of spent reactor fuel elements, prohibited in the U.S., but being adopted elsewhere, can provide a practical solution to the waste problem.

Some limits of nuclear power plants include:

- Reactors using water for a moderator are limited to 706°F (374°C) steam temperature. The best obtainable efficiency is around 30 percent.
- Higher efficiency is available with gas-cooled, graphite-moderated reactors. The St. Vrain unit near Denver, Colorado, attained 39.2 percent efficiency.

Nuclear Fission: Produces No Acid Rain, CO_2, or Nox

- Most nuclear power plants are rated around 1,000 to 1,100 megawatts output. Economics of construction and operation are attained if several are built in one complex. New USSR reactor power plants are rated 1,500 megawatts.

Even the preliminary design of nuclear power plants is best done by experienced engineering organizations with management and staff who understand how to work with the complex regulations. Nuclear, solar, and hydro power will become the preferred energy sources if reducing the 5 billion tons of carbon dioxide being added each year to the atmosphere becomes necessary.

Nuclear measurements. Nuclear power plants generate electric power measured in megawatts (MW). The electric power is sometimes designated in megawatts electrical (MWe) to distinguish it from the heat generated in a reactor, which is designated as megawatts thermal (MWt). In this chapter, megawatts will always mean electrical output unless otherwise stated. Nuclear phenomena are generally measured in metric (SI) units, but dimensions of power plants are usually in feet and the energy in steam and cooling water is usually stated in British thermal units (Btu). One Btu is 1055 joules.

Uranium ore is processed into an oxide called "yellow cake," which is sold by the pound. Uranium fuel, which has been enriched in its U 235 isotope content, is measured in kilograms when it is loaded into reactor fuel bundles.

Nuclear power reactors and how they work. A phenomenon known as "binding energy" is the glue that holds the nucleus together. Consider the fusion reaction, where we force a hydrogen nucleus to combine with a lithium nucleus.

$$Li + H \longrightarrow 2\ He$$

Nucleus mass is expressed in atomic mass units (AMU). One AMU is one-twelfth the mass of a carbon-12 nucleus. Assigning to a hydrogen atom a mass of 1.0081 AMU, we have on the left side of the equation:

$$7.0182\ \text{AMU} + 1.0081\ \text{AMU} = 8.0263\ \text{AMU}$$

On the right side we have:

$$2 \times 4.0039\ \text{AMU} = 8.0078\ \text{AMU}$$

The difference in mass, 0.0185 AMU, can be converted to energy using Einstein's equation where:

$$E = M \times C^2$$

where E is energy, M is mass, and C is the speed of light. The equivalence in energy units is 931 million electron volts (MeV) = 1 AMU. Thus the fusion reaction releases:

$$E = 0.0185 \text{ AMU} \times 931 \text{ Mev/AMU} = 17.2 \text{ MeV}$$

A simple calculation shows the significance of this binding energy. One gram atom contains 0.6023×10^{24} AMUs. Therefore the fusion of seven grams of lithium with one gram of hydrogen would release:

$$E = 17.2 \text{ MeV / reaction} \times 0.6023 \times 10^{24} \text{ reactions} = 1.036 \times 10^{25} \text{ MeV}$$

The thermal energy available for power generation is:

$$E = \frac{\text{MWh} \times 1.036 \times 10^{25} \times 10^6 \text{ eV/MeV} \times J}{3600 \times 10^6 \text{ J} \times 6.242 \times 10^{18} \text{ eV}} = 461 \text{ mW h}$$

Binding energy per nucleon is plotted as a function of atomic mass in Figure 5–1. Combining nuclei such as hydrogen and lithium produces the element helium, which has less binding energy per nucleon, as shown in the example.

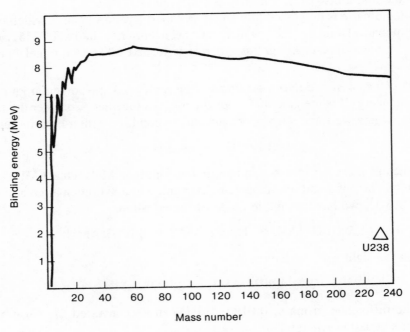

Figure 5-1

TABLE 5-1. Energy Released in U 235 Fission

Process	Energy released, MeV
Instantaneous energy	
Fission fragments, kinetic	166
Prompt neutrons, kinetic	5
Prompt gamma rays	8
Delayed energy from decaying fission products	
Beta particles	8
Antineutrinos	12
Delayed gamma rays	7
Total	206

Similarly, breaking a uranium nucleus into two pieces releases binding energy. However, nothing can be done to an element such as iron to release binding energy. Hence there are only two useful energy-producing nuclear reactions, fusion and fission, the combining of two light nuclei and splitting a heavy nucleus.

The power-producing reactors contain fissionable fuel in the form of an oxide of slightly enriched uranium. The fuel, often uranium oxide, is formed into pellets that are encapsulated in tubes made of metal such as zirconium to form the reactor core. Reactor physics is the science that tells engineers how to design the reactor core so that the chain reaction continues in a controlled manner as neutrons are released and absorbed in fission and non-fission processes.

The energy released when a thermal neutron induces fission in a U 235 atom is shown in Table 5-1. Most of this energy is collected in the water or gas coolant flowing through the reactor. The neutrino energy escapes, with trivial attenuation, even when passing through the earth and out the other side.

Enrichment of natural uranium. Natural uranium contains 99.3 percent of U 238 and 0.7 percent of fissionable U 235. Enrichment raises the U 235 content to 2 to 3 percent for power plant fuels. Both isotopes have identical chemical characteristics so the isotopes must be separated by a process based on the mass difference of (238 AMU − 235 AMU) = 3 AMU per atom. Processes used are summarized in Table 5-2.

Electromagnetic separation, which produced U 235 for the first nuclear weapons during World War II, was soon replaced with gaseous diffusion, which has produced most of the enriched uranium for power plants. A disadvantage of gaseous diffusion is its power consumption. For example, in 1984–85 when the need for enriched uranium for power plants diminished in the U.S., the U.S.

TABLE 5-2. Processes that Enrich the U 235 Content of Reactor Fuel

Separation process	Principle
Electromagnetic	A mass spectrograph magnetically deviates accelerated and ionized atoms into separate cups.
Gaseous diffusion	Lighter atoms diffuse through membranes faster than do heavier ones. Thousands of stages are required.
Gas centrifuge	Centrifugal force on the heavier U 238 atoms is greater than on the lighter U 235 atoms.
Laser	A carefully tuned laser beam vibrates U 235 atoms, stripping off electrons, so that they can be electrostatically separated from neutral U 238 atoms.

Department of Energy had to pay over $300 million in penalties to the Tennessee Valley Authority for power not purchased for its thermal diffusion plants.[1]

Development of centrifugal separation processes in Europe enabled consortiums to undercut the Department of Energy prices for enrichment. This resulted in a secondary market that supplied two-thirds of the uranium needs of U.S. power plants in 1984. The U.S. Department of Energy, in 1985, stopped development of centrifugal separation because existing diffusion plants could supply all needed enrichment.

A measure of the enrichment is the separation work unit (SWU). In 1985, when the Department of Energy was charging $144 for a SWU, the spot market was $80.

NUCLEAR POWER PLANT CONFIGURATIONS

Three configurations of nuclear power-producing reactors are used in U.S. power plants. Most have either boiling water reactors, built by General Electric, or pressurized water reactors, built by Westinghouse or Combustion Engineering. Only two plants have high-temperature gas-cooled reactors. The U.S. government has encouraged multiple designs to promote competition and innovation.

Common to all power-producing nuclear plants are steam turbines, generators, condensers, and means for dissipating Rankine-cycle losses. Each plant also has a containment structure with thick walls of reinforced concrete, designed to prevent release of radioactive materials during a worst-case failure. The highest internal pressure comes from a break in the high-pressure steam or water pipes.

[1]Colin Norman, "Hard Times in Uranium Enrichment," *Science,* 9 March 1984, pp. 1041–1043.

Nuclear Power Plant Configurations

Such a break might pressurize the structure to only 50 lb/in^2 as the steam expands into the volume of the structure, but this pressure means a hoop stress of 533,000 lb/foot of height in a cylindrical building 148 feet in diameter.

In some containment structures the hoop stress is absorbed by tensioned tendons around the outside of the structure. In others the earthquake scenario requires a wall so thick that the internal reinforcing can carry the hoop stress.

Pressurized water reactors. In a pressurized water reactor (PWR) circulating water moderates the neutrons and cools the core. The fuel is typically in the form of uranium oxide pellets, inserted into zircalloy tubes. In the Trojan plant in Oregon these fuel rods are 0.374 inches in diameter and 12 feet long. The rate of heat release at the tube surface can be 55 kWt/square ft. In comparison, sunlight, the source of heat in a solar power plant, arrives on earth with an intensity of less than 0.1 kW/square ft. The 50,952 fuel rods in the reactor are organized into 193 fuel element assemblies with 264 rods per assembly. These assemblies form the power-producing reactor core that contains ninty-five tons of enriched uranium.

The reactor core is housed within a heavy-walled pressure vessel. The vessel in the Trojan plant is 14.5 feet in diameter and 43 feet 10 inches long. Wall thickness varies from 8.5 to 10.5 inches. A 2,250 lb/in^2 pressure is required to keep the 617°F (325°C) water in the reactor from boiling. The steel vessel must be thick enough to carry the hoop stress from this pressure with a margin of safety when hot.

The requirement to sustain high pressure in a big hot tank is unique to nuclear power plants. In fossil fuel heated boilers the boiling water and superheated steam are in pipes, and the steam drum is only a few feet in diameter.

The pressure vessel of the pressurized water reactor has a removable cover, which can be opened for installing and replacing fuel bundles. Fuel bundle manipulation is generally done under water. On the cover are the control rods that establish the reactivity and hence power output of the reactor.

A heat exchanger, called steam generator, is where the heat generated in the reactor is converted to steam. The hot water is moved through the reactor and steam generator by motor driven circulating pumps (Figure 5–2). Multiple pumps and steam generators provide redundancy. The steam from the heat exchanger goes to turbines that drive the electric power generator. Condensed steam is delivered to the steam generator by feedwater pumps. Power is supplied to circulating pumps by standby diesel generators if normal power fails.

A requirement of pressurized water reactor operation is that the circulating water that moderates fast neutrons and transfers heat be in its liquid phase at all times. This requirement is achieved with a pressurizer where electric heat maintains a steam bubble at an appropriate pressure.

136 Nuclear Fission: Produces No Acid Rain, CO_2 or Nox   Chap. 5

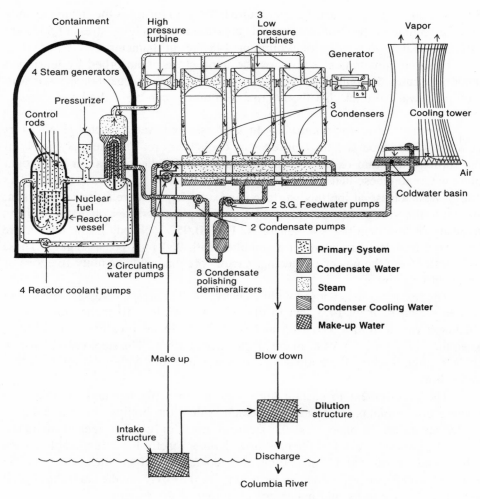

Figure 5-2

Pressurized water reactors have been successful. However, they do have limitations:

- At a temperature of 706°F, where the pressure is 3,226 psia, the latent heat of evaporation of water is zero. At higher temperatures the fluid is not clearly defined water, so it cannot be used to moderate neutrons. A steam bubble, for example, would not slow neutrons as much as water would, so the reactivity and energy release in the reactor would drop.

Nuclear Power Plant Configurations

- At 706°F and with 100°F condensing-water temperature, the Carnot cycle efficiency is 52 percent, and the best plants achieve around 30 percent overall efficiency. Note, however, that nuclear fuel costs only one-third as much as coal, so the lower efficiency is tolerable.
- Generation of superheated steam is not practical. As a result, the low pressure stages of the turbine have wet steam.
- At least three large-diameter pressure vessels are required. They are fabricated from low alloy steel, clad with stainless steel in the reactor vessel, and all welds are radiographically inspected.
- To keep pressure-vessel size reasonable, the core most be compact and high rates of heat release are required, for example, 55 kWt/square ft of fuel rod surface.
- Multiple and redundant safety features and controls assure orderly shutdown if continued operation becomes unsafe. However, for the ultimate accident there isn't much heat-absorbing material in the reactor compared with its heat producing capability.

The pressurized water reactor technology was adapted from propulsion power plants for submarines, which were the first users of nuclear power.

Boiling water reactors. In a boiling water reactor steam is generated in the reactor core. Feedwater is pressurized by feedwater pumps and circulated through the core with other pumps. Part of the water boils into steam, which then rises into separators and driers in the top of the pressure vessel. The dry steam is piped to the turbine (Figure 5–3).

The steam drying apparatus above the core leaves no space for control rods, so reactivity is controlled with rods that move within tubes that penetrate into the pressure vessel through its bottom. In a pressurized water reactor with control rods at the top, some of the rods can be magnetically held and driven by gravity into the core to shut down the reactor if power fails. Equally reliable mechanization of shut-down rods are provided for control of the boiling water reactor.

A maintenance advantage of the boiling water reactor is that the control rods need not be disturbed when the reactor is opened for refueling. Other details of boiling water reactors are in Ref. 2.

A temperature drop across the heat exchangers of the pressurized water reactor has been avoided in the boiling water reactor. However, boiling water reactors operate at temperatures of around 600°F, so the plant efficiency has not been much different from that of pressurized water reactors.

Stress corrosion cracking has been a development problem in internal components of boiling water reactors. The cause of failure has been trace elements in

[2] "General Description of a Boiling Water Reactor," General Electric Co., Nuclear Energy Division, San Jose, California, 1975.

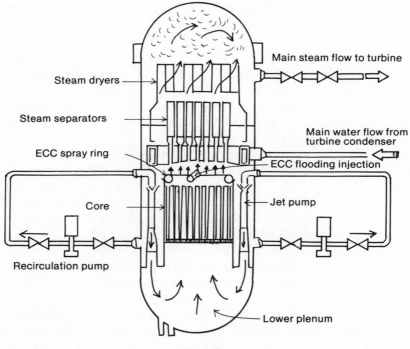

Figure 5-3

the circulating water, which is the same water that becomes steam and flows through the turbine and condenser. The Electric Power Research Institute has developed techniques for estimating the rate of corrosion cracking, and cracking resistant alloys for replacing failed parts have been developed.[3]

HIGH TEMPERATURE GAS COOLED REACTOR FOR HIGH EFFICIENCY

The high temperature gas cooled reactor (HTGR) is an inherently fail-safe reactor that uses graphite for a moderator and an inert gas, helium, for a coolant. Not applicable to the gas cooled reactor is the 706°F temperature limit of water-moderated reactors that keeps their efficiency in the 30 percent range.

Energy engineering features of the HTGR's include:

[3] "Influence of BWR Chemistry on Pipe Cracking," *EPRI Journal,* June, 1984, pp. 58–60.

High Temperature Gas Cooled Reactor for High Efficiency

- Relatively few have been built, but those that have, have been successful.
- In the U.S. HTGR development is at a standstill, but the Germans are building and operating new plants.
- Efficiencies as high as 39.2 percent have been attained.

HTGRs are important because they provide an approach to making reactors absolutely fail safe.

How the HTGR works. The key to a high temperature gas cooled reactor is the fuel particle, a ceramic uranium dioxide core surrounded by layers of silicon carbide and graphite. These particles are formed into fuel blocks. In one reactor design, the block is a graphite ball that has the fuel particles protected by a graphite shell. The ball, around 2 inches in diameter, is so sturdy that it can be bounced off of a concrete floor without breaking! The balls are stacked into the reactor to become the heat-producing core, with the uranium being the fuel and the graphite being the moderator. The fuel particles are designed to retain the fission products.

The gaseous heat transfer fluid, usually helium under pressure like 1050 lb/in^2 is circulated past the core and through heat exchangers. The reactor pressure vessel consists of a welded steel diaphragm, surrounded by a reinforced concrete structure that carries the structural load. A design developed by GA Technologies is shown in Figure 5-4. The steam generated by the hot gas flowing through the heat exchangers in the pressure vessel is piped to steam turbines outside of the pressure vessel.

A HTGR is considered inherently safe because it can withstand any kind of failure without releasing radioactivity.

The Fort St. Vrain plant in Colorado uses a steam temperature of 1000°F (538°C), resulting in 39.2 percent efficiency, compared with around 30 percent for pressurized water and boiling water reactors. Other features of the Fort St. Vrain power plant include:

Thermal power	842 MW thermal
Electrical output	330 MW electrical
Steam pressure	2400 lb/in^2 (16.55 Mpa)
Core exit helium temperature	1445°F (785°C)
Average fuel temperature	1830°F (1000°C)
Maximum fuel temperature	2372°F (1300°C)
Power density	6.3 watts/cubic centimeter
Average burnup	100,000 megawatt days per tonne
Moderator	1,500 tons of graphite

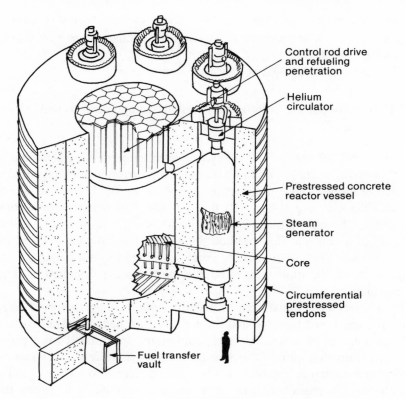

Figure 5-4

In a HTGR, the internal pressure is reacted with tensioned steel tendons pressing against a massive reinforced concrete structure that is erected on the site. The tensioned steel that carries the pressure is at low temperature, and furthermore can be loaded lightly and made redundant. The high pressure water and steam are contained in small pipes and drums, as in conventional coal burning boilers. This contrasts with the boiling water and pressurized water reactors where the pressure is reacting against the pressure vessel containing the core.

Canadian Candu reactor. Enriching the U 235 content of natural uranium is a costly process because the two isotopes to be separated are identical in chemical properties but only slightly different in atomic weight. To avoid enrichment, Canadian engineers developed the Candu reactor, which is fueled with natural uranium. Core size has been made manageable by using heavy water for a

High Temperature Gas Cooled Reactor for High Efficiency

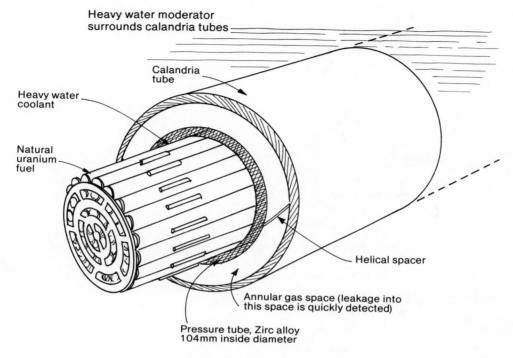

Figure 5-5

moderator. In heavy water the hydrogen atoms are replaced by "deutrons," hydrogen isotopes with one proton and one neutron in each nucleus. Heavy water has a lower capture cross-section to fast neutrons than does ordinary light water or carbon.

A key feature of the Candu reactor is its pressure tube design (Figure 5–5). The 3,100 psig pressure in the 700°F water is resisted by a pipe of only around 4 inches in diameter. This means that the walls have to withstand a hoop stress of only 6,000 lb/inch (Figure 5–6). American boiling-water and pressurized water reactors have the pressure retained by the pressure vessel, some 15 feet in diameter, requiring thick steel walls to withstand the hoop tension.

Cost of nuclear power. The major elements in the cost of nuclear power are the interest on investment for the plant, operating staff, and nuclear fuel. Of these the smallest is fuel, which seems to be dropping further on a cents/kWh basis. Staff and interest cost are affected by plant availability, with low availability distributing these fixed annual costs among fewer kilowatt hours.

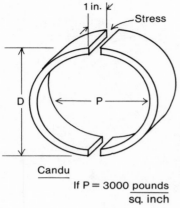

Candu

If $P = 3000 \dfrac{\text{pounds}}{\text{sq. inch}}$

$D = 4$ inches

Stress $= 4$ in. $\times$ 1 in. $\times$ 3000 $\dfrac{\text{pounds}}{\text{sq. inch}} \times$ ½

$= 6000$ lb. per linear inch

United States Pressure Vessel

$P = 3000 \dfrac{\text{pounds}}{\text{sq. inch}}$

$D = 180$ inches

Stress $= 180 \times 1 \times 3000$ pounds $\times$ ½

$= 270{,}000$ pounds per linear inch

Figure 5-6

The interest on investment depends on the cost of the plant. The range of costs of power plants completed during 1984–85 were:

			Cost, in dollars	
Owner	Plant	Rating, MW	Plant	Per kW
Duke Power	McGuire No. 2	1,180	1.1 B	941
Florida Power	St. Lucie	802	1.4 B	1746
Arizona Public Service	Palo Verde No.1	1,270	2.3 B	1811
Middle South Utilities	Grand Gulf No.2	1,250	3.065 B	2452
Washington Public Power Supply System	No. 2	1,100	3.2 B	2909
Long Island Lighting	Shoreham	820	4.1 B	5000

High Temperature Gas Cooled Reactor for High Efficiency

The effect of power plant cost on power price can be illustrated by assuming that the interest and debt service cost 20 percent per year and that the power plant has an availability of 75 percent. If it runs at full power 75 percent of the time, the cost of interest (Ci) for power from Duke Power's Mcguire No. 2 plant would be:

$$Ci = \frac{\$941 \times 0.20 \times \text{year}}{\text{kW} \times \text{year} \times 8766 \text{ h} \times 0.75} = \$0.0286/\text{kWh}$$

It's worth noting that Duke Power had achieved, at its Oconee No. 3 unit, a 94.2 percent capacity factor, the highest in the nation in 1983

On the other end of the scale, the interest in power from the Shoreham plant would be:

$$Ci = \frac{\$5000 \times 0.20 \times \text{year}}{\text{kW} \times \text{year} \times 8766 \text{ h} \times 0.75} = \$0.152/\text{kWh}$$

The cost of the staff might be estimated from Portland General Electric's Trojan plant, which has around 120 employees and generates 1,100 MW. We again assume that the plant runs at full load for 75 percent of the time. Assuming an average of \$40,000/employee/year for salary, fringe benefits, and overhead, the cost of staff (Cs) would be:

$$Cs = \frac{\$40,000 \times 120 \text{ persons} \times \text{MW} \times \text{year}}{\text{year} \times 1,100 \text{ MW} \times 1,000 \text{ kW} \times 0.75 \times 8766 \text{ h}} = \$0.00066/\text{kWh}$$

The staff cost, even if twice the \$40,000 per year per person, would still be trivial compared with the interest for new plants.

The cost of nuclear fuel was quoted as one-third that of coal for the Peach Bottom plant by John H. Austley, President of Philadelphia Electric Co., in his 1983 report to stockholders. During the following year the price of nuclear fuel dropped as imported fuel entered the secondary markets in the U.S.

A. D. Rossin and T. A. Rick, in 1978, published a breakdown of the cost of nuclear fuel in mills per kWh (Table 5–3). Their analysis showed that the refined uranium ore, called yellow cake, and enrichment were the biggest components of the cost of nuclear fuel. Some recent prices have been \$80/SWU for enrichment and \$20/lb for yellow cake. Even at 7.1 mills/kWh shown by Rossin and Rick, the price of uranium fuel is a very small component of the cost of generated power, compared to the interest on investment.

Interest rate, the key cost of nuclear power plants, varies with economic conditions. The prime rate varied between 1.5 percent and 15.8 percent between 1940 and 1980. Interest on funds used during construction contributes to the cost of nuclear power plants. For example, assume that a \$2 billion plant is complete

TABLE 5-3. Fuel Cost Assumptions (1977 dollars)*

Assumptions	Cost (mills/kWh)
Yellow cake, $40 per pound	3.5
Uranium for conversion to UF6, $2.75 per pound	0.1
Enrichment (0.20 percent tails assay), $75 per SWU	1.8
Fabrication, $110 per kilogram of uranium	0.7
Net salvage	1.0
	7.1

*A. D. Rossin and T. A. Rick, Economics of Nuclear Power," *Science,* 201, August 1978, p. 587.

with all construction cost paid, but power generation is delayed for two years. If the finance cost is 20 percent, then the final plant cost (Cf) is:

$$Cf = \$2 \text{ billion } (1.2)^2 = \$2.88 \text{ billion}$$

The cost of interest during construction can be reduced by speedy construction. For example, Florida Power started construction of St. Lucie Unit No. 2 in June, 1977. It went into service in August, 1983. Innovative construction techniques included slipforming, a continuous concrete casting process in which the forms move upward as concrete is poured. At a rate of 11.5 vertical feet per day, the contractor in 16.5 days poured 10,000 cubic yards of concrete to complete the 190 foot high, 74 feet radius containment-building wall. Traditional construction would have taken 98 days. De-bugging of components started just as soon as a component or subsystem was installed, rather than waiting until the plant was completed. Through constant communication with the Nuclear Regulatory Commission, the safety checkoff was completed and an operating license issued in nine months, instead of the usual thirty months.[4]

Nuclear power plant costs in other countries. Edwin L. Zebroski of the Electric Power Research Institute observed that power generated in the higher-cost half of nuclear power plants completed in the U.S. during the 1980s couldn't compete even with oil and gas fired fossil fuel plants in cost of delivered energy.[5] However, he noted that the cost of power generated at a nuclear plant goes down

[4] Annual Report, Florida Power and Light Co., Miami, Florida, 1984.

[5] Edwin L. Zebroski, "U.S. and World Nuclear Energy Growth and Developments," *Address at the 19th Intersociety Energy Conversion Engineering Conference,* San Francisco, California, August 21, 1984.

as the loans used to build the plant are paid off, and interest expense declines. At the end of its 40-year-life the cost of power generated by a nuclear power plant would be less than the cost of power from a coal plant because of the lower fuel cost.

Zebroski also observed that most plants that had been completed before 1980 were producing electricity that cost less than electricity from coal burning plants. In Table 5–4 he compares the cost of nuclear-generated power with coal-generated power in other countries. The costs are in cents per kWh at the station bus. He cautioned that the cost of power generated in different countries cannot be compared directly. Power generated in overseas coal-fired plants costs an average of 49 percent more than the power generated in overseas nuclear plants. Power from pre-1980 U.S. coal plants costs 10 to 20 percent more than power from nuclear plants built at the same time. The cost of generating power from gas and oil fired plants range from 40 to 70 percent more than the cost of power from nuclear plants built before 1980.

Nuclear power plant maintenance . . . a money saving opportunity. A power plant that is generating power 80 percent of the time is better than one that's generating power only 60 percent of the time. The predominant cost of nuclear power is interest on investment, which persists even when the plant generates no power. Thus if power from a 100 percent available plant costs 5 cents/kWh, then power from an 80 percent available plant would cost 6.25 cents/kWh, and from a 60 percent available plant, 8.3 cents/kWh. Furthermore, a network of fifteen 80 percent available plants can supply as much energy as twenty 60 percent available plants.

In the U.S. the better nuclear power plants achieve around 85 percent availability, and the worst ones have 55 percent, according to Dr. Hal Booher, Chief of the Licensing Division of the Nuclear Regulatory Commission (NRC).[6] The highest in the nation in 1983 was Duke Power's Unit 3 in its Oconee station, which achieved a 94.2 percent capacity factor. Unit 2 at this same plant on January 30, 1985 began its 418th full day of continuous power operation, breaking a world record at the time. Another measure of availability is the mean time between failures during the operating period between scheduled shutdowns of the reactor. Such failures are also called forced outages or "scrams." These periods vary from 25 days to 43 days in the U.S. The average forced outage in the U.S. lasts 2.5 days. Again the rate varies. The ten best United States plants have 1.5 scrams a year, compared with an average of six for all plants.

The forced outage rate of Japanese power plants is one-tenth that of U.S.

---

[6]Hal Booher, *Address at the IEEE Power Engineering Society Conference*, Seattle, Washington, July, 1984.

TABLE 5-4. Summary of Levelised Discounted Electricity Generation Costs

Discount rate 5% 10^{-2} ECU per kWh at January 1st 1981

Country	Nuclear				Coal				Ratio
	Investment	O & M	Fuel	Total	Investment	O & M	Fuel	Total	$\frac{Coal}{Nuclear}$
Belgium	1.26	0.57	0.68	2.51	0.59	0.32	2.59	3.50	1.39
France	1.02	0.36	0.69	2.07	0.83	0.29	2.50	3.62	1.75
Germany, F.R.[a]	1.58	0.47	0.82	2.87	0.79	0.60	3.32	4.71	1.64
Italy	0.99	0.22	0.78	1.99	0.56	0.19	2.38	3.13	1.57
Japan	1.34	0.47	0.76	2.57	0.95	0.42	2.51	3.88	1.51
Netherlands	1.61	0.37	1.02	3.00	0.79	0.41	2.68	3.88	1.29
Norway	1.26	0.44	0.78	2.48	0.82	0.43	2.27	3.52	1.42
Sweden	1.75	0.45	0.85	3.05	0.84	0.49	2.74	4.07	1.33
United Kingdom[b]	2.85	0.34	0.93	4.12	1.73	0.35	3.82	5.90	1.43
United States[c]	1.85	0.37	0.67	2.89	1.03	0.37	1.52	2.92	1.01

Note - The figures for different countries are not directly comparable.
[a]Fuel cost of coal based on 50% domestic and 50% imported coal.
[b]Data from CEGB Sizewell Case.
[c]Mid-case projection for Chicago region (mid-west or mid-continent region).

plants. However, a Japanese plant is scheduled to be shut down for ninty-two days/year for refueling and thorough maintenance, making the best possible availability 75 percent. In the U.S. annual refueling and maintenance takes an average of fifty-five days, giving a possible availability of 85 percent. With such variations in availability there are obviously opportunities for improvement with big payoff.

Booher observed that 47 percent of the maintenance shutdowns resulted from personnel errors. He described a troublesome power plant in which a revised maintenance program cost $9 million to implement and $1 million a year to operate, but produced savings of $2.5 million per year from reduced shutdowns.

Nuclear waste contains useful elements. Nuclear waste is classified as high-level and low-level. Low-level waste, such as contaminated glassware, lab coats, and waste paper, requires little if any shielding and is buried in shallow trenches. In 1980, Congress assigned low-level waste responsibility to states or regional coalitions that are responsible for designing and operating disposal sites. This legislation was necessary because states that generated much of the waste in hospitals and laboratories chose not to permit waste disposal in their own territory.

High-level wastes generated in nuclear power plant reactors contain long-life fission fragments, trans-uranium elements, and their daughter elements that accrue by radioactive decay after fission. Responsibility for disposal of high-level waste rests with the United States Government, which plans to have a waste disposal site selected by 1990. The utilities are assessed one mill ($0.001) per kWh of nuclear power generated to pay for the cost of the repository for high-level waste. Until the ultimate repository is in operation the spent fuel elements accumulate at power plants.

As a reactor is operated its inventory of fission products builds up until eventually the fission products absorb so many neutrons that reactivity to support full-power operation cannot be achieved. Thus a reactor is shut down, commonly once a year, for a period ranging from forty to eighty days, for maintenance and refueling. The fuel elements at the center of the core, which has the greatest burnup and hence accumulation of neutron-absorbing fission products, are transferred to a storage pool. Outer fuel elements are moved toward the center, and new fuel elements are installed in the peripheral spaces to restore reactivity sufficient for another year's operation.

The need for a repository for spent reactor fuel results from a U.S. decision to not reprocess spent fuel for fear of developing a technology that would make plutonium available for bomb-building terrorists. The underlying thinking was that making weapons grade U-235 from 2 percent enriched reactor-grade uranium required a huge gas diffusion facility. Extracting plutonium from spent reactor

148   Nuclear Fission: Produces No Acid Rain, CO_2 or Nox   Chap. 5

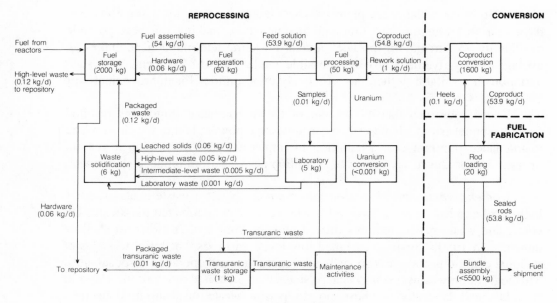

Figure 5-7

fuel required only chemical separation in a laboratory equipped to handle highly radioactive materials. This requirement became obsolete as alternative accesses to nuclear materials became simpler for terrorists. For example, the laser process of enriching uranium is straightforward and requires neither a huge plant nor access to high power.

The volume of high-level radioactive waste placed in a repository could be reduced with reprocessing. Figure 5–7, showing the major steps and the flow, is from an EPRI report.[7] Note that of 54 kg/day of fuel assemblies processed, only 190 grams/day has to go to the repository.

European countries are not restricted by the U.S. prohibition of reprocessing spent fuel, so they can relieve U.S. utilities of their need to store spent fuel elements. Spent nuclear fuel elements contain valuable ingredients that can be recovered by reprocessing. For example, plutonium is a good reactor fuel, and even the uranium is richer in U-235 than is natural uranium. The fission products include strontium-90, a 28.6 year half-life radioactive element, which when shielded can be a good continuous source of heat for buildings where accidental

[7]Ray W. Lambert and Robert F. Williams, "Recycling of Spent Fuel," *EPRI Journal*, November, 1982, p. 53.

access can be prevented. It is used with thermoelectric converters for powering remote ocean buoys.

Spent fuel elements contain useful elements that aren't radioactive. For example, George P. Dix of the U.S. Energy Research and Development Administration predicted that the U.S. need for platinum-type metals for catalysts could be met by extracting them from spent nuclear fuels.[8] He predicted that 60 million grams of stable palladium could be recovered from nuclear waste in 1990.

BREEDER REACTORS AND FUSION POWER

A breeder reactor converts stable elements into fissionable ones at the same time it produces heat for power generation. The usual fissile element is plutonium 239, and the blanket element, uranium 238, is ultimately converted into plutonium 239. The Clinch River liquid-metal fast breeder reactor (LMFBR), which was under development in the U.S. for many years, would have operated at 883°F (473°C), rather than at under 700°F, the common temperature for water-moderated reactors.

A breeder reactor differs from pressurized-water and boiling-water reactors in that fast neutrons are used in the breeding reaction, rather than fission neutrons that have been slowed to thermal velocities by the water moderator. Since water is not required, the heat can be carried from the reactor to the steam-producing heat exchanger with a liquid metal such as sodium. Using a liquid-metal coolant avoids having in the reactor vessel the high pressures that are required in water-moderated reactors. For example, we once evaluated a 1300°F (704°C) reactor that was cooled with a sodium-potassium eutectic alloy. The highest pressure in the reactor was 30 psig.

Breeder reactors could supply the U.S. Energy requirements for 700 years by using only the stockpiled tailings from our uranium-enrichment plants.[9] Thus fuel efficiency is not as important as it would be in other power plants. Reliability and safety become the important considerations in selecting the operating temperature and steam cycle. H. W. Buschman and R. J. McConnell describe the four candidate steam cycles and inconclusive efforts in selecting the best one.[10] The cycles were saturated-steam, Benson, Sulzer, and controlled-recirculation.

---

[8]George P. Dix, "Beneficial Utilization of Nuclear Waste—1977," *Proceedings of the 12th Intersociety Energy Conversion Engineering Conference*, 1977.

[9]Elie Hollander, "DOE Enriches the Nuclear Option," *EPRI Journal*, June 1983, (Electric Power Research Institute), p. 32.

[10]H. W. Buschman and R. J. McConnell, "Issues in the Selection of the LMFBR Steam Cycle," *Proceedings of the 18th Intersociety Energy Conversion Engineering Conference*, AIChE (1983), pp. 48–54.

For a given amount of energy delivered, a breeder reactor requires only 1 percent of the uranium required by an equivalent conventional fission reactor. However, fuel cost in a conventional nuclear power plant is less than 1 cent/kWh. For example, Ontario Hydro's Pickering station produces electricity costing 1.64 cents/kWh, of which the nuclear fuel is only 0.4 cents. Interest on investment during construction, including the period in which the plant is finished but not producing power, drives up the cost of power from new American nuclear plants. Interest on a $2000/kW plant can cost 4 cents/kWh. A breeder reactor will cost more than a conventional one, so no pressing incentive for developing breeder reactors has appeared. No cost experience is available for breeder-reactor power plants in the U.S. and development of a full-size breeder reactor in the U.S. before the year 2000 is unlikely.

About twenty-five breeder reactors are in operation or under construction throughout the world. For example, the Phenix plant in France ran for seven years before developing a leak between the sodium and water.

Fusion for the future. In a fusion reactor fuels such as hydrogen and lithium would be combined to produce heavier elements, and also heat for power generation. Achievement of fusion requires a temperature of about 100 million K, and a plasma density of around 1,000 times normal. The power would be produced in pulses occurring several times a second as pellets of fuel are brought to fusion pressure and temperature. The fuel pellets are small and react only at fusion temperature and pressure, so an accidental reaction is impossible. The secondary radiation produced in a fusion reactor would be low-level and short-lived.

The principles and problems of fusion reactions are described by P. L. Kapitza.[11] Progress in fusion development has been described by W. J. Hogan.[12] His laboratory had not yet achieved controlled fusion.

Most forecasters estimate that fusion power plants will be operating only after the year 2000. Estimates of power cost are not meaningful until a prototype plant is in operation.

---

[11]P. L. Kapitza, "Plasma and the Controlled Thermonuclear Reaction," *Science*, 205, No. 4410 (Sept. 7, 1979), pp. 959–964.

[12]W. J. Hogan, "Inertial Confinement of Fusion: Present Status and Future Potential," *Proceedings of the 19th Intersociety Energy Conversion Engineering Conference*, ANS (1984), pp. 1560–1565.

6

Solar Power: Renewable But Watch the Cost

Solar power, which comes as electromagnetic radiation from the sun, can be converted into heat, energetic chemicals, or directly into electricity. Solar energy is important because most of our energy sources, such as coal, oil, wind, and hydro are residuals from previously received solar radiation from the sun. Nuclear and geothermal resources may be exceptions. Also, plants convert water and carbon dioxide into carbohydrates, using solar energy to power the process.

For many applications, such as mountain-top repeaters, solar power is the best choice. Other applications, such as large fields of arrayed solar-cell panels, become economically sound because of favorable tax treatment. Many solar power installations have been made without the benefit of energy systems engineering evaluations.

Energy features of solar power include:

- The source temperature is so high, around 6,000 K, that Carnot-cycle efficiency can be nearly 100 percent. The visible-light and ultraviolet photons in solar radiation carry enough energy to break most chemical bonds.
- Solar radiation received on earth has low density, around 75 to 100 watts/square ft (70 to 93 mW/square cm) at best. In space in the vicinity of the earth the intensity is 135.3 mW/square cm.
- For most spacecraft operating in the vicinity of the earth, solar cells are the best power source.
- A solar power plant converts no sunlight to electricity at night, and very little on cloudy days. Interest on the money invested in the plant is paid, whether or not the plant generates power.

- Solar power plants that must provide continuous power generally have enough energy storage to carry the load for five days. Mountain-top communications repeaters, where loss of function cannot be tolerated, are sometimes provided with stored energy for several months.
- An array of sun-oriented panels, built with 13 percent efficient solar cells, in bright sunlight will produce one megawatt (MW)/12.5 acres. To generate 1,100 MW of peak power would require 13,700 acres or 21 square miles of solar array.
- Optimized solar heated houses are superbly insulated and equipped with triple glazed windows. The cost of heating such a house with other heat sources would be low.

Some of the limits of solar power include:

- The best predicted efficiency for capturing solar energy is around 80 percent, for gasifying coal with concentrated sunlight.
- Solar heat engines rarely produce higher than 15 percent overall efficiency.
- About 28 percent is the highest predicted efficiency for solar cells. Gallium arsenide solar cells have delivered over 20 percent of the received radiation as electrical output.

"Air-mass zero" intensity is that of sunlight above the earth's atmosphere. The following expressions are useful in solar power calculations involving intensity (I):

I in mW/square cm = 0.929 times the intensity in W/square ft

I in kW/square m = 0.01 times the intensity in mW/square cm

The intensity for calibrating solar cells intended for earth surface use is 100 mW/square cm. The energy in a photon of light in electron volts (Ep) is also the maximum voltage that could be recovered from light of a given wavelength. This energy in electron volts is:

$$Ep = 1.24 / \lambda$$

where λ is the wavelength in micrometers.

QUANTITY AND QUALITY OF SUNLIGHT

The most important energy characteristics of sunlight are its intermittent presence and low energy density. The energy content of solar radiation on earth, around 1 Kw/square m at best, is good for the earth's fauna and flora, which have learned to live with it. A higher energy density would make more desert. However, for energy gathering, a 1 kW/square m is a lower density than almost any other

Quantity and Quality of Sunlight

source except ocean thermal. Even windmills and geothermal wells produce more energy per square meter of producing area. In a pressurized water nuclear reactor the fuel element can produce 55 kW/square ft (592 kW/square m).

Peak power, the basis on which most solar apparatus is sold, is the output at noon on a bright cloudless day when the intensity is 100 mW/square cm. This corresponds to 1 kW/square m. Intervening air mass attenuates the intensity of sunlight, the thickest air mass being at sunrise and sunset.

The spectrum of solar radiation. Nuclear fusion within the sun releases energy, some of which escapes in neutrinos, and the rest eventually reaches the chromosphere, which radiates like a black body at 6,000 K. Gases in various states of ionization absorb discrete wavelengths of radiation, leaving the transmitted sunlight in space with the spectrum shown in Figure 6–1. On the vertical axis is plotted the intensity of light in each narrow wavelength band. Plotted on the horizontal axis are wavelengths of the light, expressed in micrometers. For reference, the visible spectrum is between blue light, at about 0.4 micrometer, and red light at 0.7 micrometer.

The sunlight arriving at the earth's atmosphere first encounters Rayleigh

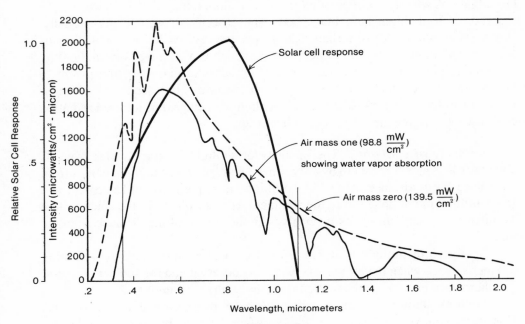

Figure 6-1

scattering, which makes the sky blue. Gas in the atmosphere absorbs in discrete wavelengths. Water vapor absorption in the infrared bands leaves few clear views of the sun there. The resulting spectrum on the earth depends on the thickness of the air mass between the observer and the sun and the amount of water this air mass contains. When watching the short-current of a solar cell with a sensitive milliammeter you can see changes as invisible clouds of water vapor move by. An example of the spectrum at earth is also shown in Figure 6–1.

The spectrum of sunlight is important to developers of solar cells because a given semiconductor will have a cutoff wavelength. Light photons having a wavelength longer than the cutoff won't have enough energy to generate hole-electron pairs. Photons having a shorter wavelength will have plenty of energy, but one photon can contribute only one hole-electron pair. For example, silicon has a cutoff at 1.1 micrometers, and light photons having longer wavelengths pass right through the silicon (Figure 6–1). A photon of blue light, having an energy of 3 electron volts will generate one hole-electron pair that is worth around 0.45 volts at the cell terminals. The remaining 2.55 volts is dissipated as heat as the hole and electron rattle to a stop.

Fortunately for heat absorbing, the solar spectrum comes from a 6,000 K black body. A solar heat collector might be at a temperature of 333°K (140°F). An absorber of heat might be a black body that absorbs all wavelengths equally. However, all black body absorbers also radiate emergy. At 333°K the black body absorber radiation peaks at 8.5 micrometers, which is a long ways from the 0.53 micrometer peak of the solar radiation. Materials like glass are transparent to visible light but opaque to infrared light, so layers of glass separated by air or vacuum can form heat traps into which sunlight can enter, but infrared radiation can't escape. A greenhouse is such a heat trap.

Intensity of sunlight. The intensity of sunlight in space varies from its nominal 135.3 mW/square cm by only some 3 percent. This variation comes from the Sun-to-Earth distance being 147.1 million km at perhelion in December and 152.1 million km at aphelion in the summer. At other places in the solar system the intensity of sunlight varies inversely with the square of the distance from the sun.

On the earth sunlight is at its brightest at local noon on a clear day during summer solstice when the air mass between the observer and the sun is minimum. At other times greater air mass attenuates the sunlight (Figure 6–2).

Solar radiation for weather records has been measured with a pyranometer. This instrument has a black and white surface bridged by thermocouples for measuring temperature differences. The instrument is usually positioned with the plane of the surfaces horizontal, so early-morning and late-evening sunlight pro-

Quantity and Quality of Sunlight

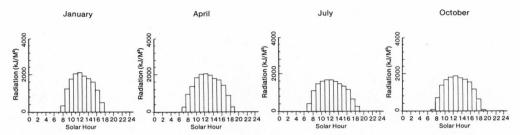

Figure 6-2

duce small readings because the energy absorbed is related to the angle of incidence (Figure 6–3).

The solar collector designer needs to know how the solar measurements provided to him were made. For example, pyranometer readings are appropriately applied to fixed flat heat-collectors. On the other hand, ARCO solar engineers, in designing a 1.0 MWe solar-cell array, found that the generated power was most valuable in the summer afternoons when the customer, San Diego Gas and Electric Co., had its peak air conditioning load. This justified mounting the panels on trunions so that they could be turned by computer control to follow the sun and intercept all available sunlight. Thus data from normal-incidence pyrheliometers is more useful for designers of such sun-following solar arrays. A normal-incidence pyrheliometer is kept pointed at the sun, and it measures only light coming from the sun. It doesn't detect sunlight coming from the blue sky or clouds.

In solar technology the word, "insolation," means the rate of deposition of direct solar radiation. This excludes sunlight that arrives from clouds by reflection or scattering. The insolation can be in terms such as watts/square ft, kW/square m, mW/square cm, watt-hours/square cm/day, or kWh/square m/year. Insolation is shown in a general way in maps such as in Figure 6–4. The Department of Energy publishes more detailed insolation data that can be procured from the Superintendent of Documents, U.S. Government Printing Office, Washington, D.C.[1,2,3] Ref. 1 provides long term geographical, monthly, daily, and hourly distributions of available solar radiation energy resources in the U.S. as well as

[1] "Solar Radiation Energy Resource Atlas of the United States," Stock No. 061-000–00570-6, $18.00.

[2] "Insolation Data Manual," Stock No. 061-000-00489-1, $7.50.

[3] "Direct Normal Solar Radiation Data Manual," Stock No. 061-000–00593-5, $4.75.

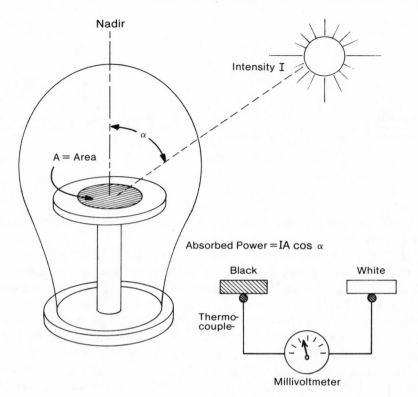

Figure 6-3

relevant meterological data. Ref. 2 has long-term monthly averages of solar radiation, temperature, degree days, and cloudiness indexes for 248 National Weather Services stations. Ref. 3 is an addendum to Ref. 2. Insolation records are also available from the Department of Energy (DOE) in computer-readable form.

Using solar energy with and without storage. Energy storage isn't necessary where solar power is injected into a huge power grid like the one in the western U.S., extending from British Columbia to the Mexican border. The generating capacity available to the grid is around 100,000 MW. Frequency is kept constant by varying the output of the biggest generators, with major utilities contributing or subtracting generation when frequency drifts from 60.00 Hz. The grid can easily respond to a cloud drifting over a 1,100 MWe solar array.

On the other hand, the contractual relationships aren't simple. The solar-generated power must be sold to an entity, usually a public utility. It becomes classified as non-firm power because it is not continuously available. The utility

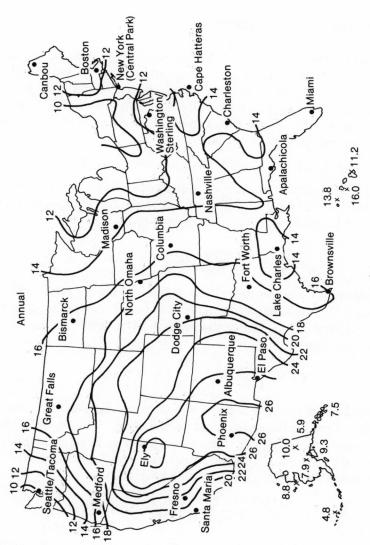

Figure 6-4

COST OF STORING SOLAR ENERGY

An example will illustrate the cost of stored solar energy. Assume that the solar array is at Yuma, Arizona where the sky is rarely clouded, and the plant has oriented solar panels. The plant must deliver 1 kW continuously, 24 hours/day, but it can generate power only between 8 AM and 4 PM, one-third of each day. To produce 1 kW of power continuously, without storing for cloudy days, requires that the stored energy (E_s) each day be:

$$E_s = (24 \text{ hours} - 8 \text{ hours}) \times 1 \text{ kW} = 16 \text{ kWh}.$$

The solar array, when sunlit, would need to charge batteries as well as supply its 1 kW load. If the batteries have a 70 percent charge-discharge efficiency, then the solar panel must be rated with a power output (P_o) of:

$$P_o = 1 \text{ kW} + \left(\frac{16 \text{ kWh}}{8 \text{ hours} \times 0.7} \right) = 3.86 \text{ kW}$$

More important, assume that Globe "Solar Reserve Cells" with gelled electrolyte cost \$235/kWh, but are designed for cycling only to 20 percent depth of discharge if a life of 2,000 charge/discharge is to be obtained. The cost of the battery (C_b) would be:

$$C_b = \frac{\$235}{\text{kWh} \times 0.2} = \$1175/\text{kWh}$$

A GNB starved-electrolyte solar battery costs around \$400/kWh, but it can withstand 80 percent depth of discharge continuously. Its cost would be only \$500/kWh. Using this GNB unit, the cost of batteries alone to store 16 kWh would be \$8,000. Allowing a mere 20 percent/year for battery interest and depreciation makes the power cost (C_p) on a kWh basis:

$$C_p = \frac{\$8,000 \times \text{year} \times 0.2}{1 \text{ kW} \times 8766 \text{ h} \times \text{year}} = \$0.182/\text{kWh}$$

Even if the cost of the solar array were zero, the cost of continuous solar power would be more than public-utility power. For example, in 1983 customers supplied by Utah Power Company paid an average of 5.67 cents/kWh, and those supplied by Minnesota Power paid 5.38 cents. Both are predominantly coal-burning utilities. Utah Power generates 83 percent of its power in coal-burning plants, and Minnesota Power generates 91 percent.

SOLAR CELLS, FROM MILLIWATTS TO MEGAWATTS

A p-n junction in silicon can separate hole-electron pairs, forming a voltage barrier. Feeding the junction with photon-generated charge carriers makes it possible for the junction to drive a current through an external circuit, generating power.

The words *solar cell* generally refer to a slab of high-purity single-crystal silicon, 0.05 to 0.5 mm (2 to 20 mils) thick, and up to 10 cm in maximum dimension. It is doped to produce a n–on–p junction, some 0.25 to 1.0 micrometers under the illuminated surface. Alternative photovoltaic energy converters have been developed and tested, but none have yet been as useful as single crystal silicon solar cells (Table 6–1).

Solar cells are important because they convert sunlight directly into electric power with high reliability and fair efficiency, 10 percent to 18 percent for silicon. Solar cells are the predominant power source for spacecraft, and alternatives are not often seriously considered. For generating terrestrial electric power the alternatives have not been as successful as silicon solar cells.

Energy features of solar cells include:

- Single crystal solar cells are generally made in sizes under 10 cm in maximum dimension, so many must be connected in series and parallel to generate the required voltage and current. For example, a 12-volt battery charger may have 30 cells in series.
- By-products of spacecraft technology are techniques for making, connecting, and mounting solar cells. These techniques, plus the rock-like nature of silicon, make the solar cell array one of the most reliable power generating devices ever developed.
- Solar cells produce virtually no power when sunlight is intercepted by clouds, even though the sky is bright.
- Solar arrays are being used to replace diesel engines and propane-fired thermoelectric generators for powering mountain-top radio relays. Avoiding the cost of hauling fuels is the important criterion.
- Solar cell arrays and power plants are usually described in terms of kilowatts of peak power. This is the power that would be generated on a bright day when the sun's rays are normal to the solar panels and the sunlight has an intensity of 100 mW/square cm.

Limits that pertain to the use of solar cells include:

- A 1 MW array requires about 6 acres of solar cell panels. The 1 MW would be generated only when normal-incidence sunlight with an intensity of 100 mW/square cm is shining on the panels.

TABLE 6-1. Photovoltaic Energy Converters

Semiconductor material	Advantage	Disadvantage	1985 status
Silicon	Silicon is plentiful. Efficiencies over 19% achieved.	Degrades in radiation. Efficiency drops as cell heats. Manufacturing cost is around $5 per watt.	Commonly used in spacecraft. Low cost manufacturing methods are under development.
Cadmium sulfide (copper sulfate)	Low cost, flexible. Large area cell is feasible. Low lb/ft^2	Efficiencies of over 6% are unstable.	Still under development after 15 years of intensive research.
Gallium arsenide	Operates at high temperature. Resists radiation degradation.	Efficiency generally <10%.	Some work being done on thin-film cells.
Aluminum-gallium arsenide	High efficiency (up to 23% with solar concentration)	Costs more than silicon cells.	New development.
Polycrystal silicon	Low cost.	Efficiency <10%.	Being developed for terrestrial use.
Copper indium-selenide	Low cost.	Efficiency <12%.	Being developed.

Solar Cells, from Milliwatts to Megawatts

- Solar panels can be oriented so that they face the sun at all times during the day. However, they must be spaced apart so that they don't shade each other. About 12 acres of land/MW is required for oriented panels.

Measurements of solar cell performance are derived from spacecraft conditions. The output of spacecraft cells is based on illumination having the spectrum of sunlight outside of the earth's atmosphere and an intensity of 135.3 mW/square cm. A 6,000 K black body light source is not a convenient tool for testing solar cells, so a simulator specification with acceptable tolerances was developed at a NASA Lewis Research Center workshop on terrestrial photovoltaic measurements.[4] Performance of solar cells intended for earth surface use is measured under 100 mW/square cm illumination furnished by a filtered xenon lamp. The efficiency of a solar cell is its maximum power output divided by the energy content of the intercepted light, which is also called irradiance.

Making silicon solar cells. Wafers for solar cells are sawed from single-crystal boules of high-purity silicon, commonly doped to produce p-type material with a volume resistivity between one and 10 ohm-cm. Diffusing phosphorous into the sun-facing surface produces an n-layer within 1 micrometer of the surface. Metallic electrical contacts are formed on the top and bottom surfaces by vacuum deposition and sintering. Plated and silk-screened contacts are sometimes used. Peter A. Iles describes the manufacture of solar cells in Ref. 5.

Cell operation. Under equilibrium conditions the majority carriers, electrons in n-type and holes in p-type silicon, will migrate by thermal diffusion. Some carriers migrate across the p-n junction. The electrons that migrate into the p-type silicon and the holes that migrate into the n-type silicon are called minority carriers and soon combine with the majority carriers in the new region. Each migrating electron uncovers a positive charge and each migrating hole uncovers a negative charge. Thus a potential difference or electrostatic field is created across the junction.

When light strikes the cell, photons with energy more than 1.1 electron volts, the energy gap of silicon, will transfer their energy to electrons in the silicon. These electrons then have enough energy to escape from their bound states, creating hole-electron pairs (Figure 6–5). Most of the photon-released minority carriers will drift to the junction where the electrostatic field will sort the

---

[4]Henry Brandhorst, John Hickey, Henry Curtis, and Eugene Ralph, "Interim Solar Cell Testing Procedures for Terrestrial Applications," NASA TM X-71771, 1975.

[5]Peter A. Iles, "Manufacturing Trends in Space Solar Cells," *Proceedings of the 17th Intersociety Energy Conversion Engineering Conference,* IEEE, 1982, pp. 1624–1630.

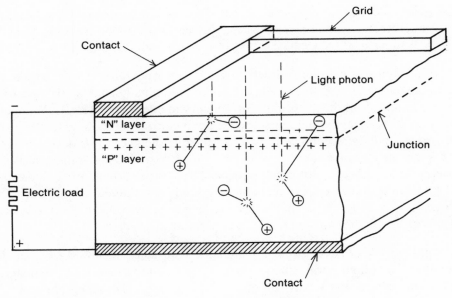

Figure 6-5

electrons to the n side and the holes to the p side, making a voltage appear at the output terminals. Power is generated if the electrons travel through an external load to recombine with holes in the p region of the cell.

Spectral response. Because of the non-linear relationship between absorption coefficient, wavelength, and minimum ionization energy (1.1 electron volts for silicon), the solar cell does not respond equally to all wavelengths of light. The energy of a photon (Ep) in electron volts is:

$$Ep = 1.2398 / \lambda$$

where λ is the wavelength of the photon in micrometers. Thus infrared photons having a wavelength longer than 1.1 micrometers cannot break loose a hole-electron pair, and photons in the visible and ultraviolet parts of the spectrum have excess energy that goes into heat.

The spectrum of sunlight in space and on the surface of the earth, as well as the response of a silicon solar cell, are plotted in Figure 6–1.

Equivalent circuit. The n-region and p-region of the solar cell, to which the terminals are connected, also form a diode, and by Murphy's law, the diode is in the wrong direction. The result is that the solar cell can be represented as a current source with a diode in parallel with it (Figure 6–6). A complete equivalent

Solar Cells, from Milliwatts to Megawatts

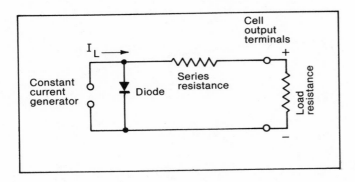

Figure 6-6

circuit would show junction capacitance, which is not pertinent in dc circuits, and a shunt resistance. References 6 and 10 contain explanations of how the physics of photovoltaic energy conversion relates to the current-voltage curve of solar cells.

Power output of solar cells. The power released to the external load is the product of the voltage and current at the cell terminals. With the solar cell short circuited the voltage is zero, and so is output power. With the cell open circuited the power is likewise zero. Between these limits is a condition at which the cell produces maximum power. Solar cell performance is therefore defined with a curve on which voltage is plotted as a function of current for varying load resistance (Figure 6–7). The maximum power point is where the product of voltage and current is the greatest. Engineers working with solar cells use an overlay that has hyperbolas of constant power with which they can easily identify the maximum power point. The hump in the curve is usually close enough.

The ideal volt-ampere curve would be a rectangle, with the top line being open-circuit voltage, and the right hand vertical line being the short-circuit current. Manufacturers, in trying to make their solar cells more valuable, approach this with various techniques. They measure their success with a term, "fill factor," which is the maximum power of the cell divided by the product of open-circuit voltage and short-circuit current.

The performance of a solar cell for a given illumination is then defined by its open-circuit voltage, short-circuit current, and maximum power. Usually current and voltage at maximum power are also specified. The cell efficiency is the power output of the cell divided by the power in the intercepted illumination.

Solar cell performance is usually specified at $28° \pm 2°C$. Designers of solar arrays usually have a computer program that will calculate the cell volt-ampere curve at any temperature, if given the short-circuit current, open circuit voltage, and voltage and current at maximum power. The equations for doing this are

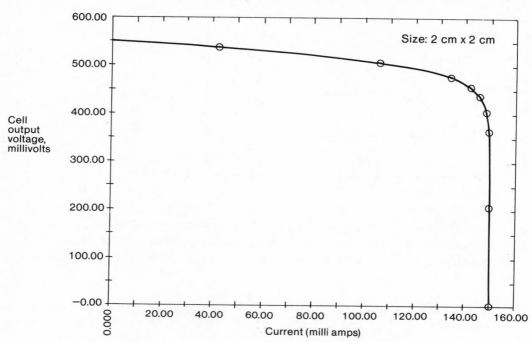

Figure 6-7

published in the Solar Cell Array Design Handbook.[6] A family of current-voltage curves from this source is plotted in Figure 6–8.

As temperature of a cell increases to about 80°C the open circuit voltage decreases linearly about 2.2 mV per degree C. In fact, the temperature of a cell can be derived from its open circuit voltage. In space sunlight the short circuit current increases about 0.05 percent/degree C. In tungsten light the increase is much greater, about 0.1 percent/degree C. The voltage at maximum power will change in proportion to the change in open circuit voltage, thus decreasing power output as temperature rises. To keep the temperature down in spacecraft cells, a cover of fused silica or other glass is bonded to the cell. This substitutes a material with a high emittance in infrared for silicon, which has a low emittance. The cover also attenuates low-energy protons and electrons. Backs of solar cell panels, if provided a view for radiating heat, are usually painted with a high emittance coating.

[6]"Solar Cell Array Design Handbook," *Jet Propulsion Laboratory JPL SP 43-38,* Vols. 1 and 2, October, 1976 for National Aeronautics and Space Administration.

Solar Cells, from Milliwatts to Megawatts

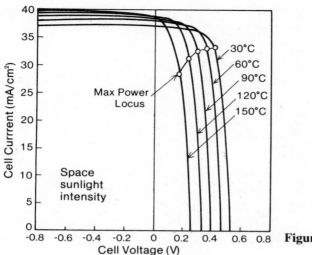

Figure 6-8

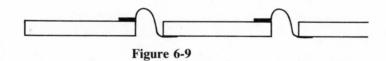

Figure 6-9

Installing and interconnecting solar cells. At one time we shingled solar cells with the bottom contact of a solar cell soldered to the top contact at the edge of the next solar cell. The result was a weak structure that usually broke at the interface between the contact metal and the silicon. The common way of connecting cells in series is with Z-shaped interconnectors which have strain relief (Figure 6-9). Terrestrial cells are usually bonded with silicone adhesive such as Dow Chemical's DC 93-500 to a covering glass sheet that protects the contacts from water and wind-borne sand and dust. Resilience is important because silicon has temperature coefficient of 2×10^{-6} cm/cm/°C, compared to 23×10^{-6} cm/cm/°C for aluminum, the common solar panel substrate.

Routes to higher efficiency in silicon cells. The mass of a spacecraft is limited to that which the selected booster can raise to the desired orbit. The incentive for developing efficient solar cells came from the extra payload that

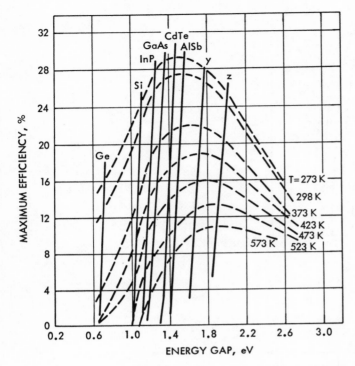

Figure 6-10

could be put into orbit if the size and mass of the solar array is reduced. Clever ways of making solar cells more efficient are described in Ref. 7 and 8.

Solar cells other than silicon. Semiconducting materials other than single crystal silicon can be configured into solar cells. In fact, silicon's energy gap of 1.107 electron volts is too low to collect maximum energy from the solar spectrum. Germanium, with an energy gap of 0.7 electron volts is even worse, although it recovers more energy from infrared than does silicon. A better photovoltaic material is a binary compound of gallium arsenide, which has almost the best energy gap. Figure 6-10 from Jet Propulsion Laboratory's handbook[9] shows

[7] J. Lindmayer and J. Allison, "An Improved Silicon Solar Cell-the Violet Cell," (Conference Record of the 9th Photovoltaic Specialists Conference) *IEEE*, 1972, p. 83.

[8] R. A. Arndt and others, "Optical Properties of the COMSAT Non-Reflective Cell," (Conference Record of the 11th IEEE Photovoltaic Specialists Conference, 1975), p. 40.

[9] "Solar Cell Array Design Handbook," *Jet Propulsion Laboratory JPL SP 43-38*, Vols. 1 and 2, October, 1976, National Aeronautics and Space Administration.

that cadmium telluride is the best, with a potential efficiency approaching 30 percent. Alternative solar cell materials are described in Hans Rauschenbach's book.[10]

Higher efficiency has been attained with gallium arsenide solar cells. The best is about 23 percent with concentrated sunlight. The original gallium arsenide cells weren't spectacular until investigators succeeded in incorporating a 0.5 micrometer thick gallium aluminum arsenide layer on the top cell surface. This layer effectively captures blue photons and improves surface conductivity. Gallium arsenide solar cells are being made for spacecraft where 18 percent efficiency pays off.[11]

Not all solar cells that promise low cost and high efficiency turn out to be practical. We once spent three years testing cadmium sulfide solar cells that had a projected cost of about $1 a watt and efficiency of higher than 10 percent. We never found a way to stabilize the efficiency at its high value.

Many have worked on polycrystal silicon solar cells, the objective being to avoid the costly crystal growing and slicing process. Efficiencies of over 10 percent have been obtained. However, the cost of solar array structure and real estate have generally favored the higher efficiency of single crystal cells even though they cost more.

Two approaches to better solar cells may become practical. One is the multiband-gap cell, which recovers energy from red photons with one junction and from blue photons at a higher voltage with another junction. For this approach G. S. Kamath predicts a possible 30 percent efficiency (see Ref. 11). Such cells would cost even more than gallium arsenide solar cells, but would pay off for spacecraft because a smaller and lighter array would be possible.

The other promising approach is a tertiary copper-indium-selenide compound that has demonstrated over 10 percent efficiency and durability appropriate for terrestrial solar arrays. A cost as low as $1/square ft has been predicted for this plastic-sandwich type of cell.

SOLAR CELLS FOR SPACECRAFT

For generating power for spacecraft, trade studies have in the past compared solar cell arrays with alternatives. These have included nuclear reactors with a variety

---

[10]Hans S. Rauschenbach, "Solar Cell Array Design Handbook," Van Nostrand Reinhold, Florence, Kentucky, 1980.

[11]G. S. Kamath, "Recent Development in GaAs Solar Cells," *Proceedings of the 18th Intersociety Energy Conversion Engineering Conference,* AIChE, 1983, pp. 1224–1228.

TABLE 6-2. Alternatives to Solar Cells for Spacecraft

Power source	Principal reason why rejected
Chemical fuels and oxidizer	Weight, and cost of developing heat engine.
Radioisotope heated heat engine	Cost and handling of isotopes. Weight of shielding and radiators.
Nuclear reactor and heat engine	Cost of developing reactor and engine. Weight of radiators and shielding. Radiator vulnerability to micrometeroids.
Batteries	Weight

of heat-to-electricity converters, chemically heated engines, and radioisotope heated engines and thermoelectric converters. Most of these alternatives are not now being seriously considered for most spacecraft for the reasons summarized in Table 6–2.

The power density of sunlight in space varies inversely with the square of the distance from the sun. For example, at Jupiter the sunlight has only 5 mW/square cm, compared with 135.3 mW/square cm in the vicinity of the earth. Thus a thermoelectric converter with a radioisotope heat source is used for deep-space missions. Mars oribters are solar-cell powered, but the radioisotope-thermoelectric-generator has worked well for Mars landers because the rejected heat is useful for keeping other components warm during the cold Mars nights. Spacecraft carrying men have been fuel-cell powered because power is needed during launch and re-entry when deployed solar arrays are not practical. Once a SNAP 10 reactor with a 500 watt thermoelectric generator was flown in space, but it operated for only forty-three days before a voltage regulator failure ended the mission. A duplicate ground-test reactor ran for a year.

In a trade study in which power sources for a spacecraft are compared, reliability and weight are generally the key criteria. Without power the spacecraft becomes useless, so the cost of achieving reliability is acceptable. Weight is important because any weight not needed by the power source can be used to provide extra capability, for example, extra channels in a communication satellite.

Sun-oriented paddles are usually the source of power when hundreds of watts are required. Many spacecraft with smaller loads have been built with solar-cell panels mounted on the body of the spacecraft. Only a portion of the array is illuminated and producing power at any instant. As the spacecraft rotates or tumbles the non-illuminated cells are cooling by radiation into space.

DESIGN OF SOLAR ARRAYS FOR SPACECRAFT

Charged particle radiation is the important degrader of solar cells on spacecraft. In low-Earth orbits, such as 700 km altitude, protons and electrons trapped in Van Allen radiation belts vary in fluence, depending on the spacecraft altitude and orbit with respect to such anomalies as the magnetic poles and the magnetic distortion over Brazil. In geosynchronous orbits at 35,768 km (19,313 nautical miles) altitude, the predominant degrading agent is solar flare protons that are emitted from the sun during intense solar magnetic activity. These protons, if energetic enough, can penetrate the earth's magnetic shield. Protons from one solar flare can cause a 5 percent reduction in the power output of a solar array. Solar flare protons are also the degrading agent in solar-cell arrays on interplanetary spacecraft.

All factors affecting the performance and degradation in solar-cell arrays in space are well understood and quantified. The design technique is to achieve an optimum combination of cell type, thickness, and glass cover thickness that delivers required power at the end of the mission. Engineers designing spacecraft solar arrays have computer programs that quickly optimize the performance of an array for any given power requirement and orbit. Solar arrays can also be designed with a calculator using a handbook developed by the Jet Propulsion Laboratory.[12]

Solar power satellite. A solar power satellite is the energy converting portion of a power source that delivers into Earth-surface utilities power generated from sunlight in geosynchronous orbit, 35,768 km altitude. Other elements are the microwave beam that transmits the power to Earth, and the receiving station with its array of antenna elements, rectifiers, and equipment for converting the collected power to high-voltage 60 Hz alternating current for delivery to distributing utilities.

To be practical, a solar power satellite must be large, in the order of 100 sq km. This is because the microwave beam should be sharp enough to focus its energy on a reasonably-sized receiving station, say sixty square km in area. An antenna generating such a beam would have to be about a kilometer in diameter. It can then transmit enough power to support on Earth a receiving station that delivers from 5,000 to 10,000 megawatts of power, the equivalent of 5 to 10 nuclear power plants.

---

[12] "Solar Cell Radiation Handbook," Third edition, *JPL Publication 82-69* (Jet Propulsion Laboratory, California Institute of Technology, Pasadena, California, November 1982).

We once compared solar cells with heat engines for the solar power satellite. A comprehensive preliminary design and comparison could not establish either photovoltaic energy conversion or heat-engine energy conversion as being clearly superior.[13] We selected the photovoltaic solar power satellite for in-depth costing studies because of its 5 to 10 percent lower cost. Other intangible considerations also affected the choice.

The resulting solar array had an area of 115 square km, and its mass was 909,713 metric tons.

When will solar power satellites be built? The key to building solar power satellites turned out to be the heavy-lift freighters for hauling components and personnel into low-Earth orbit. A space shuttle type of vehicle could not haul the huge tonnages of material economically. The new freighters would be several times the size of the shuttle. A fleet of such freighters supporting the construction of three satellites per year was required to make the concept economically sound. Developing such a fleet of freighters requires a commitment of U.S. Government funds, and a 7- to 10-year development period.

TERRESTRIAL SOLAR ARRAY DESIGN

A terrestrial solar array must produce the required power at appropriate voltage, and survive in an environment that has rain, snow, hail, high winds, dust, sand, and bird droppings. The array needs enough solar cells in series to produce the required voltage when the array is illuminated on the hottest expected summer day. The cells may not have to operate at their maximum-power point on this hot day, particularly if the energy storage is sized for the short days at winter solstice. Optimization of the number of cells in series requires consideration of (a) the length of time the array will be at the highest temperatures, (b) the value of the power during the hottest periods, (c) the cost of the extra cells required to support array voltage at the maximum-power point during hot periods, (d) voltage drop in the feeders between the array and its load, (e) the adaptability of the load to voltage variations, and (f) the type of voltage regulation adopted.

Once the number of cells in series in a string is adopted, the array can be designed with enough strings in parallel to produce the required power output.

Solar cells for terrestrial use are often made round to avoid wasting the silicon that would otherwise be trimmed off the disks that are sliced off of a round single-crystal rod, often 3 inches in diameter. Each cell has extending across its

---

[13]H. Oman and D. L. Gregory, "Solar Power Satellites—Heat Engines or Solar Cells?" *Journal of Energy*, Vol. 5, No. 3, AIAA, May–June 1981, pp. 132–137.

TABLE 6-3. Solar Array History Features Falling Cost

Year	Sales KW	Module watts	Average module price	Average cost $/watt	New application first introduced
1972	1	5	250	50.00	Aids to navigation
1973	2	7	245	35.00	Radio repeaters
1974	8	7	175	25.00	Cathodic protection
1975	3	19	300	15.79	Recreation vehicles
1976	8	20	300	15.00	Deep well pump
1977	20	20	270	13.50	Microwave repeaters
1978	80	20	250	12.50	Navaho homes
1979	220	30	300	10.00	10 kW microwave
1980	850	33	290	8.79	7 KW cathodic prot.
1981	1300	35	290	8.29	65 kW Standalone
1982	2500	37	290	7.84	1 megawatt grid tie

front center a conducting bus, and current-collecting fingers are prependicular to the bus. Connections to the front surface of the cell are usually made by soldering wires to both ends of the bus, for redundancy. Redundant connections are likewise soldered to the back side of the cell.

Early terrestrial solar panels were built by users who bought the cells, connected them together, and assembled the panels. Later the Jet Propulsion Laboratory, under the U.S. Department of Energy sponsorship, contracted with firms for the development of reliable solar arrays for demonstrations. An important result was the technology for manufacturing reliable weather-resistant solar panels, so few users assemble their own panels now. The progress of ARCO Solar during this period is described by J. W. Yerkes.[14] Table 6-3 extracted from his paper, shows the growing sales of terrestrial solar panels.

1-megawatt peak solar plant. The ARCO one megawatt photovoltaic power plant located on 20 acres adjacent to Southern California Edison's Lugo substation has features pertinent to energy systems engineering. The solar cells are mounted on panels, each 32 by 32 ft (9.75 by 9.75 m) in size, and supported on steel pedestals with trunions for tracking the sun. Sun orientation provided a 55

---

[14]J. W. Yerkes, "Ten Years of Commercial Photovoltaic Experience," *Proceedings of the 18th Intersociety Energy Conversion Engineering Conference,* AIChE, 1983, pp. 1284–1291.

percent gain of energy delivered over a fixed array of similar size, according to Ralph Megerle.[15] The power plant consists of 108 of these panels, spaced 75 ft (23 m) apart in the north-south rows, and 90 ft (27.4 m) apart in the east-west rows. This arrangement was the best compromise of land area versus shadowing of panels during sunrise and sunset. Further separation to catch early morning and late evening power was not worthwhile. The solar cell strings had diodes to protect the cells from the effects of partial shadowing of strings.

The solar cells were connected in series strings to generate 525 volts dc, and the strings in each panel delivered 18.25 amperes at the maximum power point. Thus each panel would deliver 9580 watts peak. The direct current is converted to 60 Hz ac with two 500-kW Helionetics solid-state inverters, with a 1,000 kW Garrett-AiResearch unit as a backup. Sun-orientation of the panels is controlled by redundant Hewlett Packard 9825 computers and an ARCO orienting mechanism. The power plant is unmanned during operation.

Detailed design of the solar power plant started in April, 1982. The first power was delivered to the customer in November, 1982. This short construction time avoided significant interest on investment during construction, in contrast with a nuclear plant that might take ten years to build.

Power output in during January, February, and March of 1983 was below predictions because ash in the stratosphere from a Mexican volcano reduced solar insolation by some 20 percent. By late March and April 1983 the peak dc power exceeded 1 MW, and by June it was running at its design level of 3 million kWh of dc power per year. Power availability from the solar panels was better than 95 percent during the first year, and availability from the inverter it was 98 percent. The losses were as follows:

Inverter loss	4 to 5 %
Tracking load	2 to 3 %
Wiring loss	1.5 %

The average daily ac energy produced by the plant during one year of operation is plotted in Figure 6–11.

Megerle's report includes an analysis of the actual energy generated and the energy that would have been generated had the panels been stationary rather than sun-following. On June 30 the sun-following panels delivered 45.6 percent more energy, and on July 1, 59.9 percent more energy.

1.2 MW solar cell power plant. Mark Anderson and Robert Spencer describe a 1.2 MW peak power solar plant that Sacramento Municipal Utility

---

[15]Ralph Megerle, "ARCO Solar Lugo 1 MW Photovoltaic Power Plant," Paper 84 SM 523-7 (IEEE Power Engineering Summer Meeting, July 15–20, 1984), p. 3.

Terrestrial Solar Array Design

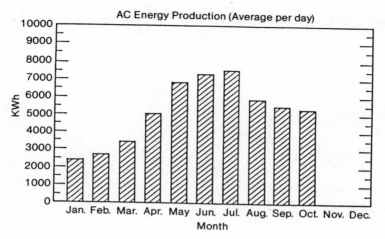

Figure 6-11

TABLE 6-4. Project Cost Forecast for the SMUD Phase I Photovoltaic Power Plant

	($K)
PV Panel Procurement	7,167
PCU Procurement	351
Potential Cost Growth (Scope Addition)	50
General Construction Subcontract	1,460
Construction Contingency	400
A&E Subcontract Expense through 5/1/83	1,191
Estimate to Complete Work thru 9/84	742
SMUD Cost through 9/85 (including operation)	459
Operational Contingency	180
Total	12,000

District (SMUD) built for $12 million on land adjacent to its Ranch Seco nuclear power plant.[16] The utility expects to expand the plant to 100 MW during the next twelve years. The solar cell panels are costing $5.60 a watt, and the power conditioning unit $0.29 per watt.

Table 6–4, from their paper, shows the breakdown of the $12 million cost of the solar project. Table 6–5, also from their paper, shows the balance-of-plant costs for elements other than solar cell panels and inverters. The "PCU" is the

---

[16]Mark Anderson and Robert Spencer, "The SMUD Photovoltaic Power Plant," Paper 84 SM 523-7 (IEEE Power Engineering Summer Meeting, July 15–20, 1984), p. 3.

TABLE 6-5. Project Balance of System Cost Comparison ($/M)

	Battelle	Hughes	SMUD
Site Prep.	1.83		23.18
Foundation & Installation	71.96	93.21	44.02
Structure	31.55	21.66	38.93
Electrical	29.21	22.88	14.72
Fence	11.97	17.69	3.85
Total "BOS"	146.12	155.44	124.70
Gravel	—	—	8.71
Control Bldg. & PCU	—	—	48.75
In Field Wiring & Equipment	—	—	24.12
Inspect Startup & Testing	—	—	17.36
TOTAL			223.64

Notes: The SMUD site preparation cost is high because the site is not level and requires construction of a road. The SMUD project is rated 1.2 MW peak power. The Hughes and Battelle arrays are each rated 30kW.

inverter that makes 60 Hz ac out of the dc output from the solar array. The inverter uses a unique 12-pulse circuit topology that produces high quality 60 Hz 12.5 kV power from the 570 to 700 volt dc supplied by the solar array.

The authors decided to install bypass diodes around every twenty-four cells in series. Otherwise shadowed cells would try to resist the flow of current from the rest of the string, and be damaged by the absorbed heat.

SOLAR VERSUS COAL FOR POWER GENERATION

A useful energy systems engineering task is to compare the cost of power from the Ranch Seco solar power plant with power from a coal burning power plant. We assume that the solar plant is rated to deliver 1.2 MWe when sunlight is available. We consider only fuel and interest. A pound of coal delivers about 10,000 Btu, enough to generate a 1 kWh in most steam plants. Coal costs around $40 a ton, so the cost of coal for generating one kW for a year (C_f) is:

$$Cf = \frac{\$40 \times \text{ton} \times \text{lb} \times 8766h}{\text{ton} \times 2{,}000 \text{ lb} \times 1 \text{ kWh} \times \text{year}} = \$175.32/\text{kW-year}$$

To this is added interest on investment. A modern coal burning steam plant with pollution controls might cost $2,000/kW of capacity, so at 15 percent per year the interest (C_i) would cost:

Solar Versus Coal for Power Generation

$$Ci = \$2{,}000 \times 0.15 = \$300$$

Then the cost of interest and fuel (Cky) would be:

$$Cky = \$175.32 + \$300 = \$475.32/\text{year}$$

The \$12 million 1.2 MWe solar power plant needs no fuel, but interest at 15 percent would represent:

$$Ci = \frac{\$12 \times 10^6 \times 0.15 \times \text{MW}}{1.2 \text{ MW} \times \text{year} \times 1{,}000 \text{ kW}} = \$1{,}500/\text{kw-year}$$

However, the solar plant doesn't produce power continuously. For example, Megerle notes that his 1 MWe (1,000 kW) plant at Lugo substation is rated 3 million kWh/year. The ratio of peak output to average output (Pa) then becomes:

$$Pa = \frac{3 \times 10^6 \text{ kWh}}{1{,}000 \text{ kW} \times 8766 \text{ h}} = 0.342$$

Thus for these assumptions the interest for the solar power plant would cost 9.2 times the cost of interest and fuel for the coal plant. A more rigorous analysis would discount the price of coal bought in the future. It might be argued that no power need be generated while the customers are sleeping, so the coal plant would not be running at full power all year either. However, the cost of interest for each plant persists, whether it produces power or not.

An expense of the coal burning plant is operating labor, for the plant must be manned twenty-four hours a day when it is running. For example, the Centralia, Washington, steam plant has an operating staff of around 120 persons. Assuming that each employee works 1,700 h/year and costs \$25/h for salary and fringe benefits, then the cost of labor (Cl)/kW year would be:

$$Cl = \frac{120 \text{ persons} \times 1{,}700 \text{ h} \times \$25 \times \text{MW}}{1{,}400 \text{ MW} \times \text{person} \times \text{year} \times \text{h} \times 1{,}000 \text{ kW}} = \$3.64/\text{kW-year}$$

The solar plant is not continuously manned, so assuming that a maintenance staff of four is employed full-time, the labor cost at \$25 an hour would be:

$$Ss = \frac{4 \text{ persons} \times 1{,}700 \text{ h} \times \$25 \times \text{MW}}{1 \text{ MW} \times \text{person} \times \text{year} \times \text{h} \times 1{,}000 \text{ kW}} = \$170/\text{kW-year}$$

Anderson and Spencer reported that the Sacramento Municipal Utility District received \$8.8 million in supporting funds. These came from the U.S. Department of Energy (\$6.8 million) and California Energy Commision (\$3.2 million). This left \$3.2 million or \$2,700 per kW as the cost to the District. This District, being municipally owned, has access to funding with tax-exempt bonds, hence enjoys a lower interest rate than would be available to a stockholder owned utility.

However, the owner of a solar power plant receives valuable income-tax credits. The municipally-owned utility, having no taxable income, would not normally get these benefits.

SOLAR HEAT COLLECTION FOR BUILDINGS AND WATER HEATING

Domestic hot water can be efficiently heated with sunlight in non-concentrating collectors (Figure 6–12). Every pound of water heated from say 50°F to 140°F requires 90 Btu of energy. Heating the water from 50°F to just 95°F can be achieved with virtually no losses in a simple solar collector, so half the energy for going to 140°F is free. If further tempering is needed to get the water to 140°F, the fuel for heating would cost one-half that required if solar energy were not used.

A typical solar water heater has two tanks, one heated by sunlight, and the other by electricity or gas as necessary to achieve required water temperature. The need to prevent freezing of the water in the solar collectors during cold nights introduces complication. Collecting solar heat with a non-freezing liquid requires

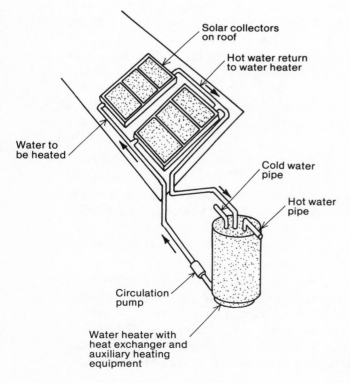

Figure 6-12

heat exchangers. Draining the collector requires valves and valve control. Some building codes require that solar-heated water not be delivered to the faucets in the building. A water-to-water heat exchanger is then required.

Alternatives to solar water heaters are gas and electric water heaters, and heat pump water heaters. Low first cost is a feature of fuel and electricity consuming water heaters. A heat pump that extracts heat from within a building merely transfers the water-heating load to the heating plant of the building during the winter. During the summer it cools the interior of the building, depositing the heat into the hot-water tank. Operating cost relates to the amount of water used.

The economic feasibility of retrofit solar hot water heating in public buildings in Virginia was assessed by Robert R. Sommers II and his associates.[17] This study concluded that the optimum combination of solar and conventional heating of water would pay back the installation cost in less than twenty-five years in seventy-seven of the 177 buildings analyzed. The mean discounted payback period was 17.7 years. These units would provide 78.8 percent of the facilities' hot water loads.

Hospitals were better candidates for solar water heating than were schools. The hospitals needed hot water all year long, whereas schools were closed in the summer. The life-cycle cost was based on a nominal interest rate of 11 percent, which presumably is available to the Commonwealth of Virginia from tax-exempt bonds, and escalation of gas, electricity, and oil was assumed to be 18 percent. A different interest rate and a different fuel escalation cost would have altered the conclusions of this 1979 study. However, the technique used is a valid and useful one.

Heating of buildings with solar energy. Solar heating will be installed in buildings if it makes the building more interesting or easier to sell. Many other costly architectural features are incorporated into buildings for the same reason. Solar heating of a building may also be a sound investment, but the soundness can be determined only by a careful energy systems engineering analysis. Before starting such an analysis, the engineer should determine whether it will be worthwhile. For houses, a rigorous analysis technique has been worked out at the Oak Ridge National Laboratory, and some results have been published.[18] The study concluded that the route to energy efficiency is effective insulation. Once this is

---

[17]Robert R. Sommers II and associates, "Economic Feasibility of Retrofit Solar Hot Water Systems," *AIAA Journal of Energy,* 77, no 4 (July–August 1983), pp. 325–331.

[18]T. A. Vineyard and Associates," Analysis of Conservation and Renewable Options for New Single Family Residences," *Proceedings of the 18th Intersociety Energy Conversion Engineering Conference,* AIChE, 1983, pp. 2050–2055.

attained the required heat is so little that it can be supplied best with a simple gas furnace. This analysis is described in the chapter on heating buildings.

Comparing solar heat with alternatives. The cost of gas, oil, and electric heat can be easily developed because the heating equipment is in production and costs are easy to get. Heating equipment and installation costs can be extrapolated from existing installations. Even heat pumps have been developed to where costs can be accurately estimated.

Solar heating is different. Equipment quotations can be obtained, but installations made by craftsmen inexperienced with solar apparatus can develop costly surprises. Many construction elements are standard. For example, ducts that carry 150°F (65°C) air from a furnace are sized to carry a given amount of heat per hour with a given air velocity. If solar heated air is at 110°F (43°C) then for the same amount of heat transported the duct size must be doubled or the air velocity must be doubled. Other elements peculiar to solar heating installations are:

- Collectors must contain non-freezing fluids or they must be drained when cold nights are likely.
- Power consumed in pumps and blowers is electricity, which adds to the operating cost. In one installation this electricity cost exceeded the fuel savings.
- The cost of constructing the volume occupied by the heat storage must be added to the solar heat cost.
- Allowances for future repairs are hard to estimate when experience is not available.
- Emergency heating for long periods of sunless days may be needed.

At one time the U.S. Department of Energy commissioned architects and builders throughout the U.S. to design and construct local solar-heated demonstration homes. These homes were to be tested for performance. The test report, when published, should provide useful guides on the performance that can be expected.

SOLAR HEATED ENGINES

John Ericsson, in 1872, designed a sun motor for pumping water. In the 1880s thousands of sun-powered Stirling-engine driven pumps were sold by De Lamater. In the early 1900s solar powered irrigation pumps were built for Egypt. The latest solar-heated power plant is Solar One, a 10 MW solar thermal generating station located in Daggett, California. There 1818 sun-tracking heliostats, each 400 square ft (37 square m) in area, reflect sunlight into a central receiver on top of a 300 ft (91 m) high tower. Steam generated at the receiver is piped to a turbine generator on the ground.

Solar heat engines have not been practical in the past for these reasons:

- Sunlight has low energy density, around 70 mW/square cm. Unconcentrated sunlight can't even boil water for a steam engine.
- A concentrator, which must be oriented toward the sun, is needed if higher temperatures are to be used for engines. The more the concentration, the more precise the orientation and concentrator surface must be.
- Heat engines require heat sinks. For irrigation pumps a convenient heat sink is available in the water being pumped.
- Hot fluid output is available only when the sun shines. Storing high-temperature solar heat has not been practical.

Often overlooked in the analysis of solar heat-engine power plants are the energy losses that occur when the sun's rays are collected and converted into heat. In Ref. 19 we provide data for estimating these losses. In general, higher engine efficiency comes with high temperature, such as 1000°F (538°C), which is attainable with precisely oriented, highly accurate concentrators. Also, a 1000°F absorbing cavity radiates energy back to the sun, 260 watts through every square inch of its opening. With a lower temperature the heat collection losses are less, but so is engine efficiency.

Transporting heat or power. In a field of solar heat collectors the engines can be distributed, one at each concentrator, or all at a central location. Distributing the engines at the concentrators complicates maintenance and operation. Installing the engines at a central plant and transporting heat involves difficulties. We found, for example, when using helium, an inert gas, the power required for pumping helium was a good share of the power generated.[20] Liquid metals needed insignificant pumping power, but having miles of insulated pipes, with trace heaters to keep the metal from freezing at night, introduced many unrecognized possibilities for trouble.

At Solar One, heat transport was solved by having a field of heliostats reflect sunlight to an absorber at the top of a 300 ft (91.5 m) tower where steam is generated and sent down to a steam turbine plant on the ground.

[19]H. Oman and G. Street, "Experimental Solar Thermionic Converter for Space Use," *Electrical Engineering,* Vol. 79, No. 12, December, 1960 (IEEE), p. 967–972.

[20]H. Oman and C. J. Bishop, "Feasibility of Solar Power for Seattle, Washington," International Congress—The Sun in the Service of Mankind, Paris, France, July 2–6, 1973.

THE FUTURE OF SOLAR ENERGY

In the late 1970s the U.S. government started a well-funded solar energy development program. Among its objectives were to reduce dependence on imported oil and avoid construction of nuclear power plants by substituting solar energy for nuclear fuel. The program produced varying success. Least successful was the use of solar energy for air conditioning.

With respect to displacing nuclear power, the solar program proved that solar power plants require about ten times the area of nuclear and coal plants. Furthermore, power from solar power plants, if without government subsidy, will cost about ten times the cost of power from nuclear plants built in a normal construction environment.

Among successes of the U.S. solar program were these:

- The cost of silicon solar cell panels dropped to around $5/watt in 1985. Further reductions seem possible.
- Studies established that solar power satellites can deliver power at a cost lower than could be obtained from nuclear power plants. However, no funding was given to the heavy-lift launch vehicles needed to build solar power satellites.
- Effective insulation, a key to solar heated houses, also works in buildings heated with fuels and electricity.

Remarks by J. W. Yerkes of Arco Solar, when describing the future of solar power[21] are summarized as follows:

- Production of solar-cell panels grows each year. In 1983 the production was 20 MWe; of this 60 percent went to the U.S., and the rest was exported mostly to Europe and Japan.
- As long as energy storage costs 6 to 8 cents/kWh, solar power is best used where storage is not needed. He cited such applications as water pumping, cathodic protection, refrigeration, and desalination.
- Solar power pays off in remote places as a supplement to diesel power that can cost $0.60 to $2/kWh.
- There are a million villages in the world that have no electricity. They can benefit from even part-time solar power if they can afford to buy it.
- Generating solar power and using a power grid for storing the energy is not in the best interest of the grid.

[21]J. W. Yerkes, "Photovoltaics for the Marketplace," *Panel Discussion at the 18th Intersociety Energy Conversion Engineering Conference*, AIChE, 1983.

Jay Chamberlin of Scandia Laboratories observed that powering mountain-top repeaters with solar-cell arrays had been successful. The batteries are normally sized for five days of no sun. The batteries at a repeater station can be monitored from the owner's headquaters, and a low-battery warning can be handled in various ways.

Solar coal gasification. D. W. Gregg and associates at Lawrence Livermore Laboratory in an experiment concentrated sunlight to gasify coal.[22] Alternative gasification processes require coal to be heated by burning part of it. The combustion consumes air, and the product gas is diluted in energy content by the carbon dioxide and nitrogen from the combustion. Using pure oxygen avoids the nitrogen dilution, but requires more coal to be burned to power the oxygen-making plant. Furthermore, the burning coal pollutes the atmosphere with sulfur and carbon dioxide.

Gregg notes that there are three possible reactions for producing gas from coal. The first, pyrolysis, requires only a little more energy than is needed to bring the coal to a temperature where it decomposes into coke, carbon dioxide, carbon monoxide, hydrogen, methane, and tar. This was the reaction once used in ovens that produced coke. The other products were discarded. Two other reactions that occur at 1100 ± 100 K, being endothermic, require energy:

Reaction	*Energy consumed*
$C + H_2O \rightarrow H_2 + CO$	32.4 kcal/mole of carbon
$C + 2H_2O \rightarrow 2H_2 + CO_2$	24.6 kcal/mole of carbon

In his experiment Gregg delivered endothermic energy to coal from concentrated sunlight (Figure 6–13). He arranged the steam jet so that it would continuously clean the window. He found the reaction to be well behaved. When the sun set in the evening the reaction stopped. In the morning it started. The product gas is easily stored. He found in the product gas as much as 48 percent of the intercepted solar energy (Figure 6–14). He predicted that efficiencies of 70 percent or greater can be achieved. His product gas had an energy content of 300 Btu/cubic ft (11.2 MJ/cu m). A typical moisture-free gas had 54 percent hydrogen, 25 percent carbon monoxide, 16 percent carbon dioxide, 4 percent methane, and 1 percent higher hydrocarbons.

[22]D. W. Gregg and associates, "Solar Coal Gasification," *15th Intersociety Energy Conversion Engineering Conference,* AIAA, 1980, pp. 633–636.

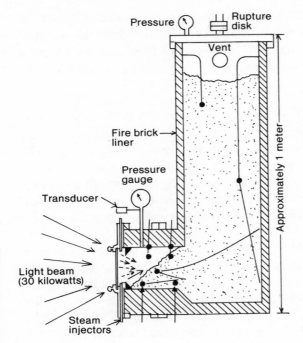

Figure 6-13

The alternative to solar coal gasification is the Lurgi coal gasifier. In making the comparison Gregg used these values:

Item	Cost		
Coal	$1.14 per GJ	or	$1.20 per million Btu
Solar energy	$1.07 per GJ	or	$1.13 per million Btu
Oxygen	$23.00 per Mg	or	$25.00 per ton
or Oxygen	$2.09 per GJ	or	$2.20 per million Btu

His conclusions are summarized in Table 6–6.

Heating the coal through a window with concentrated sunlight, and injecting steam in a way that keeps the window clean gasifies the coal without the need for air or combustion. The sulfur in the produced hydrogen and carbon monoxide can be extracted by processes used to clean natural gas before it is transmitted in pipelines. The product gas can be used for heating, converted to methane for vehicle fuel, or burned as fuel in the gas turbines in a combined cycle power plant.

In their experiments, Gregg and his associates used solar radiation for what it does best. For a billion years sunlight has been breaking down water and carbon dioxide in plants, generating more energetic products. Perhaps there are many

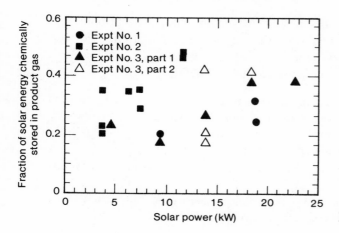

Figure 6-14

Table 6-6 Solar gasifier uses less coal than does a Lurgi gasifier.

Gasifier	Coal cost	Product-gas cost: Oxygen cost	Solar cost	Plant cost	Total
Lurgi	1.4($1.14) +	0.6($2.09) +	none +	$2.28	=5.12/GJ (=$5.40/$10^6$ Btu)
Solar	0.8($1.14) +	none +	0.6($2.14) +	$2.28	=$4.47/GJ (=4.72/$10^6$ Btu)

Coal Consumption:
Solar coal gasifier/Lurgi gasifier = 0.57
Capital Cost:
Solar coal gasifier/Lurgi gasifier = 1.4-1.6

ways of using the energetic photons in solar radiation, other than converting them to low-temperature heat or 0.5 volt electrons.

WIND POWER: DERIVED FROM SUNLIGHT

The intensity of sunshine at a surface is reduced by the cosine of the angle of incidence. For example, on a clear noon at equinox when full sun is overhead at the equator the sunlight on a horizontal surface at the north pole is essentially (0), the cosine of 90 degrees. Winds are produced by these and other temperature differences, plus Coriolis forces on north–south air mass movements. In the past energy has been captured from winds to propel sailboats and drive mills and pumps. Availability of continuous electric power has made obsolete most wind-driven pumps.

Energy can be extracted from a moving air stream with a wind turbine with aerodynamically shaped blades that slow the flowing air. Windmills with flat wood or metal blades are less efficient.

Energy features of power generation from wind include:

- The cost of wind power depends on the cost of the turbine and the average wind power available at the site. Power from a site with 240 watts/square m average will cost twice as much as power from a 480 watts/square m site.
- Available wind power varies with the cube of the wind speed.
- Most wind turbines discard energy from low-speed and high-speed air. Recovering this energy would be too costly.
- Clustering many wind turbines on a farm reduces construction and maintenance costs, but the turbines must be spaced far enough apart to avoid occupying each others tailstream.
- Tax benefits influence cost optimization.
- Wind installations have been failures because the sites were adopted before needed wind data had been gathered.

Limits in extracting power from wind include:

- The available power varies with the square of the rotor diameter. The weight of the rotor varies as the cube of the diameter. Somewhere under 10 megawatts is the economic limit for a single turbine.
- The power in an air stream flowing through a given hoop is related to the mass of the air and its velocity. The Betz limit states that no more than 59.2 percent of this air stream power can be extracted by a wind turbine.
- Wind power can displace oil or coal used to generate power in a public utility. It cannot displace peaking capacity because there may be no wind when the peak occurs.

The key measurement for wind turbine evaluation is the air velocity, in meters per second or miles per hour, recorded from anemometers. The annual wind energy in terms of MWh/square m, is plotted as isopleths on a map. Areas having less than 1 MWh/square m/year are not particularly useful for wind power generators.

The air density also affects available energy. Sea-level air at a given velocity contains more energy than air in the mountains at the same velocity.

WINDMILLS ON THE FARM

Windmills pumped water for American farms before rural electrification. Dempster Industries, Inc. in Beatrice, Nebraska, makes them. For example, their ten-foot diameter horizontal-axis windmill will pump up to 418 gallons/h against a fifty-five-foot head with a 15 miles/h wind, if equipped with a 3.25 inch diameter cylinder having a 7.5 inch stroke. The windmill costs around $1,300 without tower and installation. Its galvanized steel rotor weighs 150 lb.

Water pumping was a good use for windmills because the pumped water can be stored for the time when the wind was not blowing. However, electrically pumped water can be stored in the well, so no tower-mounted tank is needed. A one-quarter horsepower motor would give more than enough power for a pump that delivers 418 gallons/h against a fifty-five-foot head.

WIND TURBINE POWER OUTPUT

A key characteristic of a wind turbine is its output at different wind speeds. The performance of a Boeing 2.5 MW wind turbine, from Ref. 23, is plotted in Figure 6–15. Note that no power is generated when the wind speed is under 14 miles/h at the hub. A 14-miles-per-hour breeze will bring the alternator to its synchronous speed. At higher wind velocities the turbine extracts nearly one-third of the power available in the airstream.

At twenty-eight miles/h, the alternator is at its rated capacity, and at higher velocities the blades of the turbine are feathered to avoid overloading the alternator. A higher rating of the alternator and speed-increasing gear would enable the capturing of energy from higher wind speeds, but the wind blows over twenty-eight miles/h so rarely that the extra cost wasn't worthwhile. Similarly, capturing energy when wind velocity exceeds forty-five miles/h isn't worth the cost of stronger structure, so on such occasions the plant is shut down.

The annual energy output of a wind turbine is the sum of its output at each wind velocity increment during the year. A wind speed spectrum is plotted in Figure 6–16 for Goodnoe Hills near Goldendale, Washington, where three Mod-2 wind turbines are installed. Both design distribution and measured distribution are plotted. Note that an insignificant operating time is available at wind speeds more than 40 miles per hour.

The wind velocity at the surface may not be the same as the velocity at the turbine. The Mod-2 rotor extends from 50 ft to 350 ft above ground (Figure 6–17).

Boeing analysis of wind turbines. A Boeing optimization of wind turbines was based on preliminary designs in which the structure, turbine, and

---

[23]"Mod-2 Wind Turbine System Development Final Report, Volume 1—Executive Summary," U.S. Department of Energy, NASA CR-168006, September, 1982, pp. 5–1 to 5–7.

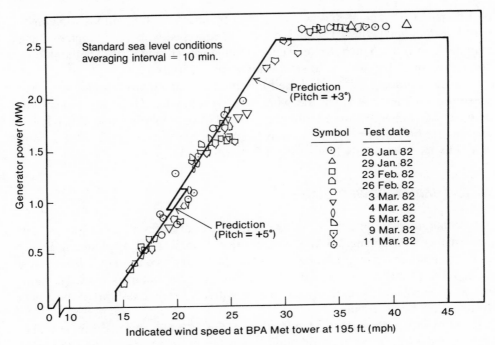

Figure 6-15

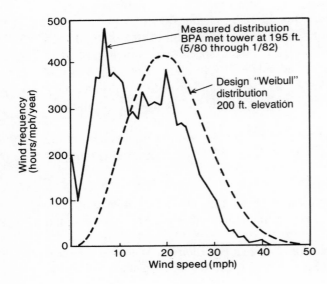

Figure 6-16

Wind Turbine Power Output

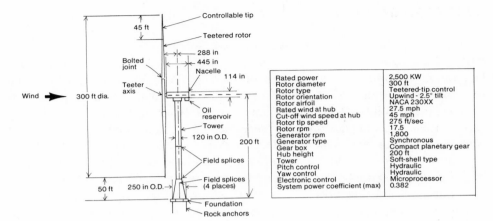

Figure 6-17

electrical and mechanical equipment were optimized to achieve the lowest cost of generated power. For example, the turbine could be supported on a rigid tower or a tubular one that had some flexibility. With computer-based structural analysis, such as done for airplane wings, it was feasible to design a tubular tower with resonances suitably separated from turbine resonances, thus achieving the lowest weight and cost with adequate factors of safety. The adopted design is summarized in Figure 6–17.

Engineers have asked why the rotor has two blades rather than three. Two blades are used in order to get as high an rpm as possible, and thus reduce the torque for a given power output. This lowers the cost of shafts and gears. Power output is proportional only to the rotor diameter, not the number of blades.

Wayne Wiesner and his associates in testing of the Mod-2 wind turbine found that its annual energy output could be increased by 43 percent.[24] Increasing the rotor diameter by 20 feet contributes 11 percent and rotor aerodynamics gives another 12 percent. Other improvements include use of a variable speed induction generator rather than a synchronous unit (8 percent). Contributing to aerodynamic improvements were vortex generators, pitch-change axis seal, rotor tip fairings, and small trailing-edge tabs. The peak power is increased to 3.2 MW, which contributes 9 percent.

[24]Wayne Wiesner and associates, "Mod-5B Advanced Megawatt Scale Wind Project," *Proceedings of the 19th Intersociety Energy Conversion Engineering Conference*, AIChE (1984), pp. 2357.

COST OF WIND POWER

Bechtel Group, Inc. analyzed for the Electric Power Research Institute (EPRI) the cost of wind power plants in Amarillo, Texas; Holyoke, Massachusetts; and San Gorgonio, California. Instrumentation of these sites had produced the actual wind data that was needed for accurate cost estimates. Total capital requirements included plant facilities investment, land, organization and startup costs, working capital, royalties, and allowance for funds used during construction. These capital costs for single-turbine power plants ranged from \$3,000 to \$4,000/kW.[25] The estimates were in mid-1982 constant dollars, with no escalation, but with construction wages and labor productivity experienced at the selected sites.

For estimating the cost of generated power the analysis assumed that the power plant would be built by an investor-owned utility, and that the levelized fixed rate charge is 16 percent, with an 8 percent tax credit. An 8.5 percent annual inflation was used, and a thirty-year plant book life was assumed.

The lowest cost of wind-generated power was 11.8 cents/kWh for a wind farm having 25 Mod-2 wind turbines with a 62.5 MW peak rating. A wind farm costs from 17 to 25 percent less on a per-kWh basis than does a single machine. The highest cost power was 22.57 cents/kWh from a single Mod-2 turbine at Mt. Holyoke, Mass. Table 6-7 from Ref. 25 provides details.

The study concluded that lower capital costs would result from larger machines (Table 6-8).

SMALL WIND TURBINES

At the end of 1984, on Altamont Pass in California some fifty 200 kW machines and ten 500 kW units were contributing to California's 500 MW of wind generation capacity. EPRI Journal noted that in 1983 more than 100 active companies were building wind power stations. These plants sell power to local utilities at a rate related to the cost of avoided generation. The Public Utilities Regulating Policies Act requires utilities to buy this power.

The benefits to the investor in windmills are substantial. As much as 40 percent of a project's cost can be recovered in the first year's operation because of tax rules. Rapid depreciation of capital costs under the Energy Recovery Tax Act adds further benefit.[26]

---

[25]S. M. Kohan, "Cost Estimates for Large Wind Turbines," *EPRI Journal,* May 1984, pp. 41-43.

[26]Tony K. Fung and Kirby C. Holte, "Investment Value Analysis for Renewable Energy Projects," *IEEE Paper 84 SM 520-3,* Summer Power Meeting, 1984, pp. 1-2.

TABLE 6-7. Estimated Levelized Revenue Requirements for All Cases (A-L)

	Amarillo, Texas								Holyoke, Massachusetts		San Gorgonio, California	
	A	B	C	D	E	F	G	H	I	J	K	L
Machine (MOD-2 or WTS-4)	2	4	2	4	2	4	2	4	2	4	2	4
Number of machines	1	1	25	25	1	1	25	25	1	1	1	1
Power plant rating (MW)	2.5	4.0	62.5	100	2.5	4.0	62.5	100	2.5	4.0	2.5	4.0
Wind machine production rate	—[1]	—[1]	—[1]	—[1]	—[2]	—[2]	—[2]	—[2]	—[1]	—[1]	—[1]	—[1]
Capacity factor (%)	26.4	18.9	24.7	17.6	26.4	18.9	24.7	17.6	19.1	14.8	35.0	29.1
Total capital requirement ($/kW)[3]	4775	3703	3496	2700	3978	3004	3233	2497	4717	3610	4865	3775
Levelized revenue requirements for investor-owned utility (mills/kWh; constant mid-1982 dollars)[4]	168.5	182.1	127.1	136.0	142.8	150.1	118.1	126.4	225.7	223.4	132.6	123.4

[1]Nominal.
[2]High volume.
[3]Includes plant facilities investment, land, allowance for funds during construction, organization and startup expenses, and working capital.
[4]Levelized, using before-tax average cost of money (3.76%/yr in constant dollars).

TABLE 6-8. Single Wind Turbine Total Capital Requirement (existing production methods)[1]

	2.5 MW (case a)		4.0 MW (case b)	
	$/kW	%	$/kW	%
Rotor	1238	26	728	20
Nacelle[2]	2166	45	2016	55
Tower	506	11	389	10
Ground support	91	2	51	1
Balance of plant	352	7	219	6
Other[3]	422	9	300	8
Total	4775	100	3703	100

[1] Constant mid-1982 dollars, including contingencies.
[2] Includes the drive train, gear box, and lubrication and hydraulic systems.
[3] Land, organization and startup costs, royalties, working capital, and allowance for funds during construction.

The need for quick benefits has been more pressing than the collection of two-year wind records at operating altitudes. For example, Oregon's largest cluster of twenty-five automated 50-kW windmills at Bandon generated only 100 megawatt hours during its first forty-two days of operation. The combined generating capacity is 1.25 MW. A 20 percent lighter than normal wind speed was cited. A wind powered generator on a 140-foot tower at Agate Beach, Oregon, was torn down because its power cost 16 cents/kWh at a time when the local Lincoln County Public Utility District was buying power for 2 cents/kWh.

GAS TURBINES: THE ALTERNATIVE TO WIND POWER

Wind power can displace oil and gas for the generation of electric power in a big public utility like Pacific Gas & Electric. In fact, PG & E had in 1984 thirty-two contracts with wind projects greater than 100 kW. These contracts call for PG & E to purchase as much as 848 MW of wind-powered generation. The utility's announcement continues to state that, "The projects contribute to PG & E's capacity without risk to ratepayers or stockholders."

Wind power plants cannot be depended on to generate power when demand is at its peak. The wind might not be blowing then. A gas turbine is available when needed, and it costs around $300/kW, compared to around $3,000/kW for a wind plant. Laws restrict continuous operation of gas turbines running on oil and

Gas Turbines: The Alternative to Wind Power

natural gas. Hence all of the annual interest on the investment on gas turbines must be carried during the 2,000 or so hours that the turbine is permitted to operate. However, the interest on a wind turbine must likewise be paid, whether the wind is blowing or not.

Gas turbines are beneficial because they carry peak loads, leaving the base load plants more heavily loaded at other times. A wind turbine coming on line reduces the load at base load plants, increasing the cost per kWh of power generated at these base-load plants.

7

Storing Energy for Nights and Days with No Sun or Wind

Energy has been stored by pioneers in woodpiles, by homeowners in oil tanks, by power-generating utilities in coal piles and reservoirs behind dams, and by the U.S. Navy in oil reserves. These are energy reserves. It is also useful to be able to accept heat, electricity, and mechanical energy when available and store them for later use; for example, electricity from a solar array for use at night. Approaches for evaluating mechanisms for repeatedly storing and extracting heat, electricity, and mechanical power are the topic for this chapter.

Energy is repeatedly stored in raised fluids and objects, stressed springs, spinning flywheels, processed chemicals, and compressed gases. For any requirement, the energy systems' engineering task is to compare alternative techniques that meet the needs of efficiency, life, and cost. The analysis becomes complicated when promising new techniques offer advantages over proven commercial products, and when electrochemical storage is compared with thermo-mechanical storage.

Important engineering considerations in storing energy are:

- A common alternative to energy storage is to generate energy with a prime power source when needed.
- Energy stored as heat tends to escape.
- The heat released during compression of gases is energy that may not be available when the gas is expanded.
- A twenty-year lifetime of new energy storage apparatus is hard to establish in a one-year test.

Storage Batteries

Some limits in energy storage technology are:

- Nickel-iron and pocket-plate nickel-cadmium batteries can be built to last at least thirty years.
- The starting-lighting-ignition battery used in automobiles is about the lowest cost mechanism for storing electricity. Its life when discharged deeply, however, is only around 100 cycles.
- New high-strength materials make practical flywheels that can store the braking energy of vehicles for subsequent starts.
- Superconducting magnets offer the highest efficiency of energy storage. Around 95 percent of the deposited energy can be recovered. Flywheels supported by magnetic levitation and turning in vacuum also offer high efficiency.

The efficiency of energy storage is the energy recovered from the storage mechanism divided by the energy delivered to it. For example, in a public-utility peak-shaving battery, the losses must include the transformer and rectifier that charge the battery, the inverter that makes ac out of the dc battery output, as well as the battery and control losses.

Coulombic efficiency is the ampere hours delivered by a battery divided by the ampere hours required to recharge it.

Heat rate in compressed air storage is the heat energy added to the air to generate 1 kWh. The compressed air is generally heated before it is admitted to the turbine. A perfect engine produces 1 kWh with 3412 Btu of heat. The best of steam power plants need around 9,000 Btu. Having compressed air available for the gas turbine reduces the fuel consumption to around 4,000 Btu/kWh at best.

Sections that follow are in the order of practical importance, starting with storage batteries. Pumped hydro, a practical way of storing thousands of megawatt hours of energy, is described in Chapter 4, "Renewable Power from Water." Compressed air storage is covered in the Brayton-cycle section of Chapter 2.

STORAGE BATTERIES

A storage battery is a combination of positive electrodes, which are sometimes called "cathodes" during discharge, negative electrodes, which are called "anodes" during discharge, and an electrolyte, arranged to accept charging current from a dc power source and deliver power to a load when discharging. A battery cell is a unit that has one or more negative plates in parallel, and one or more positive plates in parallel, all immersed in a common electrolyte. Two or more cells connected in series is called a "battery." A single cell is often also called a "battery." Among the smallest storage batteries are those mounted on circuit boards of computers for keeping alive random access memories. Some of the largest batteries are installed on submarines to supply propulsion power while submerged.

Storage batteries have evolved to match requirements of applications (Table 7–1). Batteries designed for one application are not necessarily suitable for another one. For example, a starting-lighting-ignition battery wouldn't last long in traction service. Important energy features of batteries are:

- A battery can have close to 100 percent coulombic efficiency in that nearly every coulomb of charge is recovered during discharge. However, batteries frequently need to be overcharged.
- The charging voltage is always higher than the discharging voltage at a given state of charge. This voltage difference, which depends on current density, represents a loss.
- The life of a battery in terms of number of charge-discharge cycles generally varies with depth of discharge in that deep discharges shorten cycling life.
- Battery price is affected by the cost of materials as well as the technology of manufacturing. The price of lead has varied by a factor of two during a year.

Some limits in the technology of storing energy in batteries are:

- Silver-zinc batteries can store over 100 watt-hours/lb, but their life is limited to around fifty charge-discharge cycles.
- An automobile starting-lighting-ignition battery, available for around \$50/kWh, is the lowest cost secondary battery.
- A sealed nickel-cadmium spacecraft battery with 5 percent depth of discharge has survived 30,000 charge-discharge cycles in a ground test.
- An energy density of 109 watt hours/kilogram is expected from sodium-sulfur batteries.

Battery capacity is commonly measured in ampere hours at a given temperature and discharge rate. Often the capacity of a battery is quoted as a "C/x" rate where C is the ampere hour capacity of the battery, and x is in hours. If a constant current of 50 amperes will completely discharge a 500 ampere hour cell or battery in ten hours, then this current is called a "$C/10$" rate.

Since all of the energy in a battery cannot be extracted quickly, the capacity of a battery for short discharges will be less than for long ones. A five-hour discharge ($C/5$) is a common rate for defining battery capacity. Battery capacity is usually affected by temperature, with the capacity dropping as temperature is lowered. Also pertinent to a battery capacity designation is the end-of-discharge voltage that can be accepted.

Sometimes a lead-acid battery is said to reach its end-of-life point when its

TABLE 7-1. Storage Batteries Developed for Specific Applications

Application	Duty	Electrochemistry	Key characteristics
Auto starting-lighting-ignition	Low depth of discharge, high current surge	Lead acid	Low cost, thin plates, short life
Diesel starting	High current surge (e.g. 3000A, 30 seconds)	Lead acid	High-current; thick grids; copper inserts in terminals to carry current
Lift trucks, golf carts	Deep discharge, high current surges	Lead acid	Over 1000 charge discharge cycles
Solar energy storage	Deep discharges, moderate current	Lead acid	Low currents permit high efficiency
Telephone and power utility standby	Very few deep discharges, long life	Lead acid	Reliability is important
Submarine	High capacity deep discharges	Lead acid	Electrolyte circulation prevents stratification
Power utility peak sharing	High capacity, deep discharges	Lead acid	Electrolyte circulation prevents stratification
Portable tools and appliances	Deep discharges, simple charging	Nickel cadmium	Light weight, high current
Spacecraft	Depth of discharges consistent with long life	Nickel cadmium Nickel hydrogen	High reliability, low weight
RAM memory sustaining	Low current	Nickel cadmium	Small size

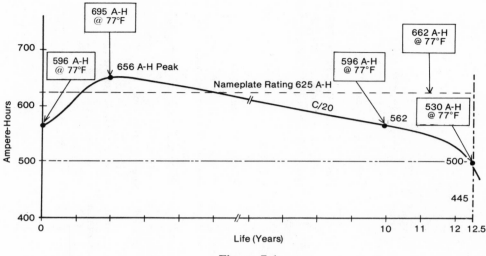

Figure 7-1

available ampere hour output is 80 percent of rated capacity. A battery designed for cycling service may improve in capacity during the first few hundred cycles, so at times its actual capacity can be more than its rated capacity (Figure 7–1). Depth of discharge is expressed as a percent of the ampere-hour content of the battery that is extracted during one discharge. For example, a 60 percent depth of discharge takes 60 ampere hours out of a 100 ampere hour battery, leaving 40 ampere hours.

LEAD ACID BATTERIES

A lead-acid battery is unique in that both plates are at least partly lead, and when the battery is discharged both plates are covered with lead sulfate. During charge the negative plate is reduced to pure lead and the positive plates becomes lead-oxide (Figure 7–2). Note that during discharge for every pair of electrons circulating through the external circuit, one lead atom in the negative electrode and one in the positive electrode must change state. The Faraday constant (F) relates the number of coulombs (ampere seconds) of current to the weight of lead required:

$F = 9.649 \times 10^4$ coulombs or 26.80 ampere h/gram molecule

The required amount of lead explains why lead-acid batteries are so heavy. Typical energy density is ten watt hours/lb (twenty-two watt h/kg). In contrast, a battery using lithium as an electrode can deliver over 200 watt-hour/lb. Lithium has an atomic weight of 6.9, compared with 207 for lead.

Lead Acid Batteries

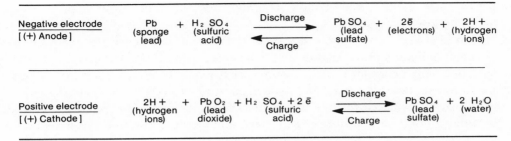

Figure 7-2

Charging current restores the discharged condition of a lead acid battery by converting the positive plate to lead dioxide and the negative plate to pure lead. After full charge is reached, an ordinary lead acid battery at around 2.5 volts per cell becomes an electrolyzer, breaking down the water into oxygen at the positive plate and hydrogen at the negative plate. The electrolyzed water must be replaced during battery maintenance.

In one form of a no-maintenance battery the negative plates are designed to have more capacity than the positive plates. The positive plate becomes fully charged first, and generates oxygen that diffuses to the negative plate. There oxygen combines with the pure lead and sulfuric acid, in effect discharging the negative plate so that it cannot become fully charged and liberate hydrogen. The battery with the overdesigned negative plates emits very little gas during overcharge, so it can be sealed, forming a maintenance-free battery.

The sulfuric acid used in lead acid batteries is diluted to achieve a specific gravity of around 1.250 in a fully charged battery. In some starved-electrolyte batteries the electrolyte has a specific gravity of 1.310 at full charge, dropping to 1.150 at the end of discharge. The state of charge of a lead acid battery can be inferred easily from the electrolyte density, a feature not available in nickel cadmium and other cells with alkali electrolyte.

Long life: available, but at a price. The key to a practical battery is exposing as much lead surface as possible to the sulfuric acid for the electrochemical reactions. Manufacturers have various ways of making a spongy lead paste that has lots of surface area but is so weak that it has to be supported by a lead grid to form a plate. Also, the grid functions as a current collector. Grid thickness and the alloy used affect the life of the battery. Pure lead resists corrosion, but it is too weak structurally for most batteries. Alloying the grid lead with calcium or antimony gives the grid strength, but the alloying affects battery performance and life. For example, antimony alloying in the negative plate promotes hydrogen formation.

A lead acid battery can be designed for low production cost. Its grids would have the minimum cross-section area, consistent with allowable voltage drop at specified output current. Lowest cost materials would be used for separators and the strongest alloys would be used for lead parts. On the other hand, batteries with pure lead plates lasted for over thirty years in standby telephone service. These batteries are big and costly for their ampere-hour capacity. With materials and technology now available, long battery life in charge-discharge service can be obtained at reasonable cost. Joseph Szymborski identifies these failure mechanisms:[1]

- Shedding from the positive plate reduces the effective active material in the electrode, hence ampere-hour capacity declines. Unless the shedding material is restrained it accumulates in the bottom of the battery case where it eventually short-circuits the plates, ending the useful life of the cell.
- Separator cracking or deterioration allows shedded positive-plate material to form a short circuit.
- Corrosion of the positive grid reduces its cross section, thus increasing resistance. Finally the grid can no longer support the active material.
- Sulfating of the plates occurs when the battery is stored in a discharged condition. The sulfated surface eventually becomes so complete that it is a barrier to the diffusion of ions, thus forming an insulating layer for the charging current.

Design features that lengthen the life of sealed lead acid batteries include:

- Wrapping plates with small-pore fiberglass separators, and packing the plates tightly to prevent sloughing off of active material from the plates. This also prevents mossing of the negative plate.
- An alternative approach, used in Europe, has the active material packed in sleeves formed from perforated polyvinyl chloride with braided fiberglass reinforcing. The current is conducted by a lead spine in the center of the sleeve. Absence of grid intersections is claimed to avoid a corrosion-prone zone.[2]
- Design the positive plate grid with enough thickness to provide sufficient strength and conductivity after corrosion during its lifetime has taken place.

[1]Joseph Szymborski, "Development of a Totally Maintenance-Free Lead-Acid Battery for Photovoltaic and Other Renewable Energy Source Systems," *Proceedings of the 17th Intersociety Energy Conversion Engineering Conference,* IEEE (1982), pp. 630–634.

[2]Mark Eggers, "A Long-Life Deep Cycle, Tubular Lead Acid Battery," *Proceedings of the 19th Intersociety Energy Conversion Engineering Conference,* ANS (1984), pp. 868–874.

Lead Acid Batteries

The improvement in battery life that can be achieved by avoiding antimony alloy in the positive grid has been demonstrated in a long-term development program at C & D Power Systems. J. S. Enochs and his colleagues reported achieving over 2,000 charge–discharge cycles with 80 percent depth of discharge with lead-calcium-tin grids.[3]

The remaining consideration in long-life batteries is positive grid corrosion. If the grid corrosion rate is 0.005 inches (0.125 mm) per year, and if at the end of fifteen years the battery needs 0.022 inches of grid radius, then 0.097 inches must be provided initially. An alternative approach, investigated by the U.S. Navy, is to use pure lead in the form of a composite with titanium or alumina as an internal reinforcing. The alumina didn't work out well. Titanium, costing around $40/lb, does not corrode in sulfuric acid and satisfactorily stiffens pure-lead positive grids.

Scandia Laboratories tested advanced GNB cells under a regime that simulated a solar power plant in which the battery was cycled between 80 percent and 20 percent of full charge. This is a severe service, not only because of the discharge to 20 percent residual charge, but also because a battery has better life if it is discharged from a fully charged condition. The first batteries lasted for over 1,500 charge–discharge cycles, corresponding to four years of solar power service.

Cost of lead acid batteries. Lowest in cost of conventional lead–acid batteries is the starting-lighting-ignition battery, prices being quoted in newspapers. The battery supplies a current of a few hundred amperes, typically between 9 and 11 volts, for a few seconds to start the engine. As temperature drops, the engine oil becomes more viscous, increasing friction, while at the same time the internal resistance of the battery rises. Also, the capacity of a battery degrades as it gets colder (Figure 7–3). This cold-starting situation is the worst-case design consideration, and batteries are designed and rated accordingly. An engine-coupled generator normally recharges the battery and maintains the charge by supplying 14.3 volts at its terminals. A battery is said to be on "float charge" when maintained at a terminal voltage slightly higher than its open-circuit voltage.

The normal discharge from a starting-lighting-ignition battery is shallow, with less than 10 percent of the battery capacity being extracted. Attempts to use this type of battery for serious energy storage requiring deep discharges generally have failed. We once tested a variety of makes of these batteries with 70 percent discharges. Most of them failed in less than 100 charge–discharge cycles. At

---

[3] J. S. Enochs and colleagues, "Nonantimonial Lead-Acid Batteries for Cycling Applications," *Proceedings of the 19th Intersociety Energy Conversion Engineering Conference,* ANS (1984), pp. 850–856.

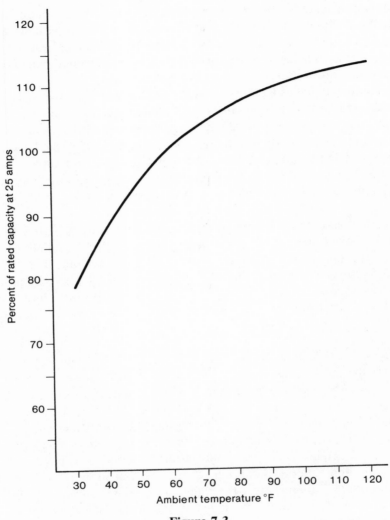

Figure 7-3

NASA Lewis Research Center batteries of this type that had been modified for electric vehicle service failed in 61 to 233 charge–discharge cycles. The SAE J227a Schedule D driving cycle had been simulated.[4]

[4]John G. Ewashinka, "Results of Electric-Vehicle Propulsion System Performance on Three Lead-Acid Battery Systems," *Proceedings of the 19th Intersociety Energy Conversion Engineering Conference,* ANS (1984), pp. 727–731.

Lead Acid Batteries

Standby battery with twenty-year life. A product of GNB Batteries is a 3,700 ampere hour lead-acid cell battery that is about 20 inches high, 16 inches long, and 13 inches wide. It weighs 750 lb when filled with electrolyte, and costs around $1,000, a price that is affected by the price of lead. It is designed to supply standby power for telephone and utility power plants and sub-stations. This service requires occasional deep discharges, but most of the time the battery is on float charge at around 2.17 volts/cell. Long life is achieved by making the grid so thick that after twenty years of corrosion, the grid will still have sufficient cross section to carry the full current. The cells are normally sold in sets of twenty-four cells connected in series.

At $1,000/cell the battery cost per kWh of stored energy (CB) then is:

$$Cb = \frac{\$1000 \times \text{Cell} \times 1{,}000 \text{ volt amperes}}{\text{cell} \times 3{,}700 \text{ ampere hours} \times 2 \text{ volts} \times \text{kWh}} = \$135/\text{kWh}$$

Added to this would be the cost of installation and the price of chargers and converters for the stand-by power unit. Note that the battery is designed for a low float charge current and occasional deep discharge. It would be designed otherwise if long life in charge–discharge cycling is required.

Battery for charge–discharge cycling service. Another product of GNB Batteries is a 450 ampere hour lead-acid "Absolyte" battery that is designed specifically for charge–discharge cycling as required for fork-lift trucks and stand-alone solar power plants. The electrolyte within the cell is all absorbed in the fiberglass separators, and the negative plate has greater capacity than the positive one has, so that free hydrogen will not form. The cells are sealed, making them maintenance-free.

A 450 ampere hour, twelve-cell battery costs $3,112, weighs 897 lb, and is 12.8 by 23.3 by 30.6 inches in size. In one installation we had forty-eight cells in series to produce 96 volts dc, which we converted to 120 volts 60 Hz ac and other voltages. We float charged the batteries at 2.25 volts/cell to get the longest possible life. A higher or lower voltage could increase the corrosion rate of the grid.

The cost of storing energy in our battery (Cb) is:

$$Cb = \frac{\$3112 \times 4 \times 1{,}000 \text{ volt-amperes}}{450 \text{ ampere h} \times 96 \text{ volts} \times \text{kWh}} = \$288/\text{kWh}$$

The battery was designed for 1,500 charge-discharge cycles.

Nickel-cadmium batteries. A nickel-cadmium cell is commonly made in two forms, a round unit containing rolled electrodes and separator, and a prismatic unit in which plates and separators are stacked. The electrolyte is water that contains 25 to 35 percent dissolved potassium hydroxide. The cells can be sealed, but most cells intended for terrestrial use have pressure relief devices.

Energy features of nickel-cadmium batteries include:

- A nickel-cadmium battery will store up to twice the energy that can be stored in a lead-acid battery having the same weight, but at higher cost.
- A pocket-plate nickel-cadmium battery with concentrated electrolyte will have output at $-40°F$, a temperature at which a lead-acid battery is not usable.
- Battery lives of over 30,000 charge–discharge cycles have been achieved with low depths of discharge.
- The variable charge voltage required for nickel-cadmium batteries makes them inconvenient for service where they float on the line that supplies power when the main power source fails. The float charge voltage would have to be compensated for battery temperature.

The concentration of potassium hydroxide in nickel-cadmium cells does not change appreciably with the state of charge. Thus the state-of-charge can't be determined by simply measuring electrolyte specific gravity. Voltage is not a good indication either, because the discharge has a long flat region in which the voltage is nearly constant (Figure 7–4). Voltage variations from changes in

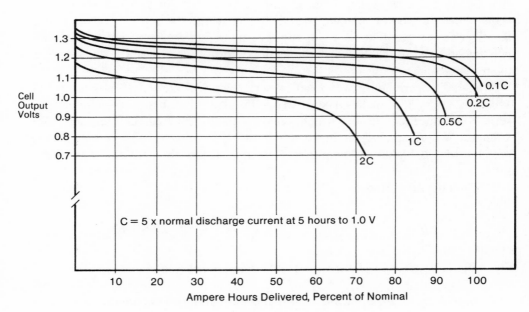

Figure 7-4

temperature are usually greater than variations from state of charge. One way to measure state of charge is to discharge the cell or battery completely and measure the energy delivered. Another way, used in spacecraft, is to integrate current going in and current going out of the battery to keep track of its energy content.

A useful source of performance data for nickel cadmium batteries is the NASA Applications Manual.[5]

Nickel-cadmium battery design. The typical nickel-cadmium battery has a positive electrode consisting of a sintered nickel substrate coated with nickel hydroxide. Absorbing electrons from the external circuit during charge converts the surface to NiOOH. Each electron converts one molecule.

The negative electrode also has a nickel substrate, but it is covered with cadmium metal when charged. Discharging converts the surface to cadmium hydroxide, releasing two electrons from each cadmium atom converted. The chemical equation is:

$$Cd + 2\ NiOOH + 2\ H_2O \rightleftharpoons Cd(OH)_2 + 2\ Ni(OH)_2$$
$$\text{(Charged)} \qquad\qquad KOH \qquad \text{(Discharged)}$$

The following example illustrates the amount of energy that can be stored in the active materials. Energy Research Corporation designed for the U.S. Bureau of Mines a nickel-cadmium battery that replaces the lead-acid battery used since the 1930s for powering miners' cap lamps. The specifications for the nickel-cadmium and lead-acid batteries are compared in Table 7–2.

William H. Lewis and Elio Ferreira chose roll-bonded rather than sintered-plate electrodes for this application.[6] In the roll-bonded structure, the electrodes are fabricated by forming a conductive mix of nickel hydroxide or cadmium oxide, graphite, and teflon into a bonded structure by rolling in a calendaring mill. The steps in manufacturing the electrodes are continuous and automated. This structure weighs less than conventional plates, but gives fewer charge–discharge cycles.

Faradaic calculations show that it takes 3.65 grams (g) of nickel hydrate $[Ni(OH)_2 \times \frac{1}{3} H_2O]$ for one ampere hour of capacity. The designers found experimentally that using a 10 percent overcharge, the utilization of nickel hydrate is 80 percent and that of cadmium oxide is 60 percent of theoretical. Slight

---

[5]Willard R. Scott and Douglas W. Rusta, "Sealed-Cell Nickel-Cadmium Battery Applications Manual," NASA Scientific and Technical Information Branch, Publication 1052, 1979.

[6]William H. Lewis and Elio Ferreira, "New Developments in Personal Lighting Systems for Miners," *Information Circular 8938,* U.S. Department of the Interior, Bureau of Mines.

TABLE 7-2. Comparative Battery Specifications

Specification	Nickel-cadmium (prototype)	Lead-acid (typical)
Weight lb	2.42	4.66
Voltage (average) V	3.5	3.7
Capacity A × h	15	12
Energy W × h	52	44
Energy density W × h/lb	21.4	9.5
Size in	1.77 × 3.92 × 6.65	1.64 × 4.90 × 6.65
Volume in^3	46.1	53.4
Energy density W × h/in^3	1.13	0.82
Charging cycle life cycles	>500	≤400
Cost	$50[1]	$30

[1]Cost is a function of production volume. The projected cost figure is based on a production volume of 20,000 units/year.

variations can be obtained as the number of electrodes, amount of graphite, and the discharge current density are varied. Based on the amount of active material (nickel hydroxide) on the positive electrode of 11.3 grams, the nominal capacity (Cn) of the cell in ampere hours (Ahr) can be calculated as follows:

$$Cn = \frac{11.3 \text{ g} \times 6 \text{ plates} \times \text{Ahr} \times 0.8 \text{ utilization}}{\text{plate} \times 3.65 \text{ g}} = 15 \text{ Ahr}$$

The separator between plates is composed of one 5-mil Pellon (nonwoven polyamide) bag heat-sealed around the positive electrode, followed by a layer of U-wrapped Celgard K 306 microporous polypropylene film. The separators are important in vented nickel-cadmium cells because they prevent cadmium shorts from penetrating to the positive plate during extended cycling.

Nickel-cadmium performance at low temperature. Nickel-cadmium batteries can be used at temperatures that are too low for lead-acid batteries. However, the charge and discharge voltages vary with temperature, so a battery operating at variable temperatures requires a charger that senses battery temperature and biases charge voltage appropriately. Also, the loads must accept the voltage that varies with temperature as the battery is discharged. The charge and discharge voltages for a General Electric 1.5 ampere hour cell are mapped with temperature as a parameter in Figure 7–5.

Charge–discharge cycle life of a nickel-cadmium battery. The manufacturer rates a battery on the ampere hours that it can deliver before reaching a

Lead Acid Batteries

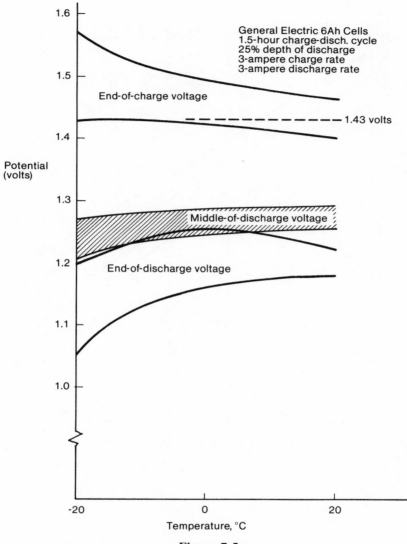

Figure 7-5

cutoff voltage, often 1.0 volt, at a rate calculated to exhaust the battery in a specified time. A common rating discharges the battery in five hours. When delivering its rated content the battery's storage capacity can be 25 ampere hours/lb. However, the battery charge–discharge cycle life would be short if it is discharged fully each time.

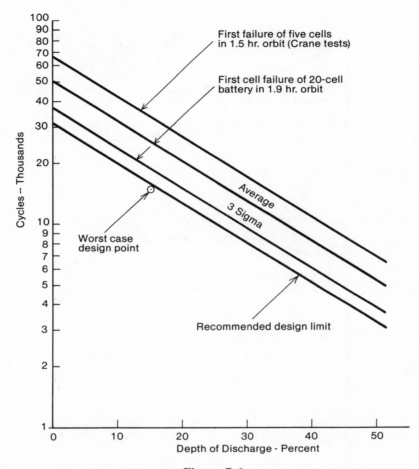

Figure 7-6

Sid Gross, a spacecraft battery specialist, uses curves, such as in Figure 7-6, in selecting battery size. For example, if he needs a battery for a low-altitude spacecraft that orbits the earth once every ninety minutes for three years, the service life (L) for the battery would be:

$$L = \frac{3 \text{ years} \times 365 \text{ days} \times 24 \text{ h} \times 60 \text{ minutes} \times \text{cycle}}{\text{year} \times \text{day} \times \text{h} \times 90 \text{ minutes}}$$
$$= 17{,}520 \text{ cycles}$$

From Figure 7–6, he finds that 14 percent depth of discharge is consistent with the required life. Using a battery rated 14 watt hours/lb gives him an actual performance (P) of:

$$P = 14 \text{ watt h/lb} \times 0.14 = 1.96 \text{ watt h/lb}$$

Such is the price for long life.

Data on the charge–discharge cycle life of nickel-cadmium batteries is costly to acquire. The U.S. Navy's Naval Weapon Support Center at Crane, Indiana has been testing nickel-cadmium cells for space use for decades. Each test requires ten to fifty cells because at each temperature there must be a statistically significant number of samples. Sometimes the life is long, for instance, 10 to 15 years, and the technology that produced the successful cells is lost during the test time.

Reliability of nickel cadmium batteries. An example illustrates the reliability of nickel-cadmium batteries for spacecraft. Traditionally redundant batteries have been used to avoid loss of spacecraft functioning if a cell fails. A lightweight mechanization of redundancy is to add an extra cell or cells in series in the battery so that if one cell fails shorted, the battery will still produce minimum voltage. Some batteries have diodes to bypass cells that fail in the open-circuit mode.

Thomas Dawson of the U.S. Air Force Space Divsion in planning future FLTSATCOM satellites wondered if this redundancy is necessary. He interviewed project engineers responsible for six spacecraft, and found that during 107.2 satellite years of operation there had been no open-circuit cell failures.[7] One satellite had experienced four cell shorts in thirty-three satellite years. Two others had experienced no battery cell failures in fifty-three years of satellite operation, with the longest satellite life exceeding eight years.

Major Dawson concluded that rather than installing bypass diodes, the money and weight could better go into extra battery weight that reduces the stress on the batteries and increases their life. Eliminating the battery cell bypass electronics also increased reliability of the battery assemblies from 0.9691 to 0.998.

Nickel battery cost. Nickel cadmium batteries cost more than lead acid batteries. For example, nickel costs around $3/lb, and cadmium, a by-product of zinc refining, costs around $1.50/lb. Lead is around 25 cents/lb. In seeking a

---

[7]Thomas Dawson, "Elimination of Battery Cell Bypass Electronics on FLTSAT-COM," *Proceedings of the 19th Intersociety Energy Conversion Engineering Conference,* ANS (1984), pp. 72–77.

lower cost battery for electric automobiles, M. Klein and A. Charkey built nickel cadmium batteries using the technique that had been developed for the miner's lamp battery.[8]

Klein's nickel cadmium battery powered a Saab with a curb weight of 2,440 lb. The range was 75 miles. With an improved battery, the car would weigh 2,180 pounds and have a range of 100 miles. However, Klein sees the electric car becoming competitive with an internal-combustion engine driven car making 27.5 miles/gallon only when the price of gasoline reaches $2.45 a gallon.

Gas–metal and other new batteries. The nickel-hydrogen battery is a successful gas–metal battery, now being used in COMSAT synchronous-orbit satellites. The nickel hydrogen cell is like a nickel cadmium cell, except that the cadmium electrode is replaced by a platinum electrode that consumes hydrogen gas during discharge and evolves hydrogen when being charged. The hydrogen pressure can reach 600 lb/in^2 when the cell is fully charged. Hughes Aircraft Company, using a cell design developed for the U.S. Air Force, has tested 25 ampere-hour nickel-hydrogen cells at 80 percent depth of discharge for 7,000 charge–discharge cycles. The observed capacity at that point reported by E. Levy was in the range of 23 to 30 ampere hours.[9] A nickel-cadmium cell at 80 percent depth of discharge would last only around 1,000 cycles.

An advanced design nickel-cadmium battery might deliver 25 watt hours/lb at 100 percent depth of discharge. A nickel-hydrogen battery can deliver 30 watt hours/lb. Recent cells have a life goal of 30,000 charge–discharge cycles.[10] Although nickel-hydrogen batteries are good for spacecraft, they are not necessarily useful for terrestrial uses. The platinum catalyst suggests high cost, even if the pressure vessel need not be chem milled to keep weight down.

Redox flow battery stores energy in liquids. The redox flow battery, developed at NASA Lewis Research Center, stores energy in electrolytes, from which the energy is extracted with carbon electrodes. The electrolytes can be stored in tanks, which can be made as large as needed for the energy to be stored.

[8]M. Klein and A. Charkey, "Nickel–Cadmium Battery System for Electric Vehicles," *Proceedings of the 19th Intersociety Energy Conversion Engineering Conference,* ANS (1984), pp. 719–725.

[9]E. Levy, Jr., "Life, Engineering, and Acceptance/Qualification Test Data on Air Force Design Nickel Hydrogen Batteries," *Proceedings of the 19th Intersociety Energy Conversion Engineering Conference,* ANS (1984), pp. 85–88.

[10]Martin J. Milden and Charles C. Badcock, "Overview of Nickel-Hydrogen for Space Satellites," *Proceedings of the 19th Intersociety Energy Conversion Engineering Conference,* ANS (1984), pp. 191–195.

K. Nozaki and his colleagues developed improvements and built a 1-kilowatt 30–cell unit which they tested. With computer simulations based on test data they predict an 80 percent in-and-out efficiency for a 1 megawatt unit that they will build next.[11] In selecting electrode materials they had screened 100 varieties of carbon before adopting carbon knit from cellulose. No catalyst was required. The positive electrolyte was iron chloride dissolved in hydrochloric acid. The negative electrolyte was chromium chloride, also dissolved in hydrochloric acid.

H. L. Steele and L. Wein of the Jet Propulsion Laboratory list the redox battery as the potentially lowest cost mechanism for storing energy in electrochemicals.[12] They estimate that the initial cost of a redox battery will be $22/kWh stored, plus $4.15 for balance of plant. Table 7–3, from their paper, shows the alternatives.

Rechargeable zinc batteries. Zinc is a high-energy, low-cost electrode in primary batteries. An example is the common dry cell. Zinc has also been a successful electrode in silver-zinc primary (used once) batteries that deliver over 100 watt h/lb. These batteries can be designed to deliver their energy in a few minutes, making them useful for missiles, launch vehicles, and torpedoes. Such high-discharge-rate batteries are often constructed so that the electrolyte is not introduced until just before use.

Development of rechargeable silver zinc and nickel zinc batteries has been only partly successful. During charge the zinc is plated out of solution, and it tends to form dendrites through pinholes in the material separating the plates. The pinholes seem to provide the lowest resistance path that attracts dendrite growth. Some silver zinc batteries have achieved over 100 charge–discharge cycles.

Nickel zinc batteries have survived 700 cycles, but not consistently. Applications such as load leveling, vehicle traction, and solar energy storage require thousands of charge discharge cycles.

Sodium sulfur batteries for 100 percent coulombic efficiency. Sodium-sulfur batteries have been developed for electric cars. Their advantage is high power density, in the order of 110 watt hours/kilogram. A disadvantage is that the operating temperature is 350°C, which requires an oven.

Douglas M. Allen decided to test these batteries for possible use in space-

---

[11]K. Nozaki and colleagues, "Performance of ETL New 1 kW Redox Flow Cell System," *Proceedings of the 19th Intersociety Energy Conversion Engineering Conference,* ANS (1984), pp. 844–849.

[12]H. L. Steele and L. Wein, "Comparison of Electrochemical and Thermal Storage for Hybrid Solar Power Plants," *Paper 81-WA/Sol-27,* ASME Winter Annual Meeting, 1981.

TABLE 7-3. Cost of Batteries Replacement and Balance of Battery System

Type	Round-trip thruput eff %	Initial battery costs $/kW_e$	Initial battery costs $/kW_eh$	Total replacement cost $/kW_eh$	Number of replacements	Balance of pwr cond. syst.[b] $/kW_e$	Balance of pwr cond. syst.[b] $/kW_eh$
Fe-Cr Redox	69	$132[a]	$22	−1.3	0	101.75	4.15
ZnBr	69.3	—	48	20.7[c]	1	91.9	9.13
NaS	74	—	39.6	55.5[d]	1	95.12	7.52
ZnCl$_2$	66.5	59[a]	27	47.7	2	105.2	5.1
NiH$_2$	60	—	65	−1.82	0	101.2	34.5
LiM-FeS	78	—	54	84.2	3	95.12	10.2
NiFe	58.75	—	54	119[e]	4	101.2	34.5
FeAir	46	—	32	197[e]	9	101.2	34.5
NiZn	83	—	108	293[e]	4	101.2	34.5
Adv Lead Acid	75.5	—	123	27	2	101.2	34.5

[a] Electrolyte stored externally in tanks and pumped to power conversion section during operation.
[b] Initial cost, no replacement needed.
[c] Varies from 20.7 to 157 (3 manufacturers), the lowest value was used.
[d] Varies from 27.9 to 87 (3 manufacturers), the average value was used.
[e] High replacement cost due to 3, 9, and 4 replacements.

craft, where an excellent vacuum insulation is available for the oven. The test results were dramatic. By August 4, 1984, the number of charge discharge cycles at 80 percent depth of discharge had reached 2,400, with no discernible loss of performance.[13] Furthermore, the cells had 100 percent coulombic efficiency. This means that every ampere hour that went into the cells was available for extraction. The charge voltage was higher than the discharge voltage, so the overall efficiency was 80 percent. Open-circuit voltage of these cells is 2.07 volts.

MEGAJOULES IN SUPERCONDUCTING MAGNETIC STORAGE

A 30 megajoule superconducting magnetic storage unit was built to stabilize 0.35 Hz oscillations on the Pacific Intertie transmission lines. The equipment performed satisfactorily, but while it was being developed an alternative method of damping the oscillations proved effective and was adopted. Essential features of superconducting storage are shown in Figure 7–7.

In a subsequent study, R. J. Loyd and colleagues at Bechtel Group, Inc. analyzed designs for useful energy storage with this technology. Loyd's first design analysis showed that a 5,000 megawatt-hour superconducting energy storage would cost too much, that is, around $1,300/kWh capacity.[14] Batteries cost around $500/kWh of stored energy, and the Helms pumped hydro storage cost around $700/kW of power generation capacity.

Bechtel was contracted to study an alternative design that could have a cost comparable with pumped hydro. The 1,000-meter diameter coil, operating at 4.2 degrees Kelvin, would be built underground because of the radial stress, around 50,000 lb/in^2 from the 3 tesla field. The coil in the trench would be about twenty-five meters deep and one meter wide. The coil would take about ten years to construct. Sufficient helium for cooling can be extracted from natural gas.

Superconducting storage has an in-and-out efficiency of approximately 96 percent. Batteries and pumped hydro are around 70 percent. Energy stored in a superconductor is available instantly. Pumped hydro energy is available only as fast as the turbine gates can be opened.

---

[13]Douglas M. Allen, "Sodium-Sulfur Satellite Batteries: Cell Test Results and Development Plans," *Proceedings of the 19th Intersociety Energy Conversion Engineering Conference*, ANS (1984), pp. 163–168.

[14]R. J. Loyd and colleagues, "A Cost Study of Superconducting Magnetic Energy Storage for Large Scale Electric Utility Load Leveling," *Proceedings of the 19th Intersociety Energy Conversion Engineering Conference*, ANS (1984), pp. 1144–1149.

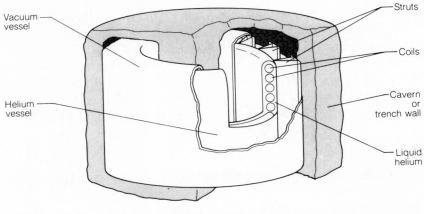

Figure 7-7

FLYWHEELS FOR SHORT- AND LONG-TERM STORAGE

A flywheel would be useful for storing the braking energy of a bus or train. This stored energy would be extracted later to help the vehicle accelerate to operating speed. One study showed that train frequency in New York subways was limited during the summer by the heat released from the braking of trains. Two cars equipped with flywheels showed in six months a net energy saving of 33 percent over conventional cars. Installation of flywheels on trains to absorb braking energy seemed economically sound, but was never implemented.

The energy that can be stored in a flywheel is limited by the strength of the material from which it is made. M. Olszewski reported that 80 watt hours/kilogram have been achieved in rim type flywheels using S-glass and graphite fiber.[15]

Sidney Gross found that flywheels can be useful for storing energy in spacecraft.[16] Availability of high strength carbon fibers permits high speeds and much stored energy in lightweight wheels. Low loss of strength from fatigue results in a long life in charge–discharge cycling service. Also conveniently

---

[15]M. Olszewski, "Flywheel Performance: Current State of the Art," *Proceedings of the 19th Intersociety Energy Conversion Engineering Conference*, ANS (1984), pp. 1150–1155.

[16]Sidney Gross, "Flywheel Energy Storage System for Spacecraft," *Proceedings of the 19th Intersociety Energy Conversion Engineering Conference*, ANS (1984), pp. 615–620.

Flywheels for Short- and Long-Term Storage

available in spacecraft is a vacuum environment where the spinning flywheels do not have an air-friction loss.

Particularly important to Gross's application was the long life with deep discharges available with flywheels. A 6 percent derating was all that was needed for a fifteen-year life. Nickel cadmium batteries, the usual spacecraft energy storage element, have long life only if not deeply discharged. With flywheels depth of discharge affects only the size of the generator, which must be rated for the speed range required to extract energy from the wheel. Gross proposed to extract 75 percent of the energy out of the flywheel by slowing it down to 50 percent of its peak speed. A 75 percent depth of discharge has no effect on the life of a flywheel. In contrast, a nickel cadmium cell could not be discharged more than 35 percent if more than a few thousand cycles are required.

Gross calculated a weight of 25 kg/kWh for a modern flywheel. Equivalent cycling performance was 70 kg/kWh for nickel hydrogen batteries and 130 kg/kWh for nickel cadmium batteries. Gross used a motor and generator efficiency of 98.5 percent, and 99.7 percent for bearings. His in-and-out efficency was 92.8 percent. He noted that the generator could be designed to produce regulated voltage, whereas with batteries auxiliary regulators are required.

Long-time storage of energy in a flywheel on the surface of the earth requires vacuum containment. Flywheels storing energy for periods like days have been developed, but are not in production.

Comparison of energy storage mechanisms. A key requirement in stand-by service is instant power availability. Gas turbines running from stored compressed air and diesel engines need a finite time for starting. Even a pumped hydro plant would require several seconds for the gates to open and the turbine to come up to speed. The Helms pumped hydro project, described in the chapter on hydro power, cost some $700/kW. Superconducting energy storage, costing around $1,000/kWh, could deliver instant power like batteries can. So could a flywheel that is running continuously.

The Electric Power Research Institute in its *EPRI Journal* published the comparison of energy storage mechanisms reproduced in Table 7–4.[17]

---

[17]Robert B. Schainker, "Superconducting Magnetic Energy Storage," *EPRI Journal,* May, 1984, p. 60.

TABLE 7-4. Capital Cost Comparison: Energy Storage Technologies (1982 $)

Technology	Capital cost[1]			Efficiency	
	Power related ($/kW)	Energy related ($/kWh)	Total: 5-h storage ($/kW)	Round trip[2] (%)	Overall[3] (%)
SMES (Superconducting)					
State of the art	194	231	1349	91	31
+ 15 years' R&D	158	143	873	91	31
Pumped hydro	600	10	650	70	24
Underground pumped hydro	575	30	725	70	24
Compressed air	490	10	540	—[4]	30
Thermal (oil)	600	45	825	70	24
Batteries					
Lead-acid	100	140	800	75	26
Advanced	100	100	600	70	24
Hydrogen-air	750	12	810	50	17
Hydrogen-halogen	425	35	600	67	23
Flywheel	120	340	1820	85	29

[1] Capital cost is based on "overnight construction" and comprises two components; before they are added for a total, the energy-related component must be multiplied by the storage capacity (hours) to convert to $/kW.
[2] Round-trip efficiency refers to the energy storage technology only (ac in, ac out).
[3] Overall efficiency encompasses the entire energy system—from primary fuel through a 10,000 heat rate baseload power generator to energy storage technology, including any supplemental fuel used in the storage facility.
[4] Compressed-air energy storage requires 0.72 kWh (electricity) + 4000 Btu (oil or gas) to produce 1 kWh.

8

Electrochemical Conversion Avoids Carnot Limitations

Energy released as heat of reaction is converted into mechnical energy with engines that are limited to Carnot cycle efficiency. Chemical energy can also be converted directly into electricity in electrochemical reactions that are not limited in efficiency to the Carnot cycle. For example, a dry cell at room temperature delivers electric power until the zinc case is completely corroded away. Muscles operating at 98°F (37°C) convert the energy of foods directly into mechnical energy with an efficiency between 25 and 50 percent. Electric eels convert food energy into high voltage pulses delivered into the water.

Batteries that recover the smelting energy of metals are common. Fuel cells that recover energy from gaseous and liquid fuels are in varying states of development. Processes for direct conversion of chemical energy to mechanical energy are not very well understood.

Energy features of electrochemical energy production include:

- Primary batteries are a convenient source of power even though power from them costs more than power from fuel burning engines.
- Fuel cells and batteries produce direct current that must be converted to alternating current for many uses.
- Virtually all the losses in a fuel cell are recoverable in heated water, which can be used for processes such as washing clothes in laundries.

Fuel cell power plants running on natural gas and petroleum products are hard pressed to achieve efficiencies of better than 60 percent. Gas turbine combined cycle plants achieve better than 50 percent at lower cost. Neither can operate directly on low cost coal as fuel.

PRIMARY BATTERIES: OLD AND NEW

Primary batteries are used once and discarded. In secondary or storage batteries, which are treated in Chapter 7, the energy can be repeatedly extracted and replaced.

Energy features of primary batteries are:

- Primary batteries are convenient sources of small amounts of power for portable flashlights, radios, watches, and calculators.
- Continuous power for fixed equipment requiring more than a few watts can be generally provided at lower cost from public utility power lines, or fuel consuming generator sets.
- Primary batteries are useful for short time loads such as in missiles, torpedoes, mines, and rocket boosters.
- Decades of time can separate the invention of a good battery concept from its commercial success. Many, like aluminum and magnesium dry cells, fail to make the grade commercially.
- In most batteries the output voltage drops as the battery discharges.

Limits of primary batteries include:

- Refining from ores the materials used in making a battery requires energy, usually in the form of heat. A battery can deliver no more energy than has been invested into its components.
- The highest theoretical energy denisty from a lithium-fluorine battery is around 2,800 watt hours/lb. Other electrochemical couples have lower limits.

An electrochemical cell consists of two plates separated by an electrolyte. The negative electrode, which contributes electrons to an external circuit is also called an "anode" during discharge. The reaction at the negative electrode is called "oxidation." The positive electrode or cathode absorbs electrons from the external load. The reaction at the cathode is called "reduction." A memory help is "red cat"—*red*uction occurs at the *cat*hode.

Two or more cells in series are always called a "battery." A single cell is also called a "battery."

The energy density of a battery is measured in watt hours/lb or watt hours/kilogram. The output of a battery is usually specified in ampere hours (Ah)

Primary Batteries: Old and New

of current delivered before the terminal voltage falls to a designated value. Characteristics of batteries pertinent to selecting one for a given application are:

Characteristic	Comment
Peak current	The energy content of a battery with high internal resistance cannot be extracted quickly.
Temperature sensitivity	Voltage under load can drop severely when a battery is at sub-zero temperature because of reduced electrolyte conductivity.
Self discharge	The shelf life of batteries varies widely and is affected by temperature.

Zinc-air cell illustrates electrochemistry. A common anode for primary cells is zinc, a plentiful metal that costs around 50 cents/lb. It provides a convenient 1.2 to 1.7 volts per cell when oxidized to zinc oxide or zinc chloride.

The zinc-oxygen reaction is useful because only the anode, electrolyte, and a non-reacting carbon cathode are required. This is the reaction in the "air cell" B battery used in early vacuum tube radios. It is also used in the zinc-air battery made by McGraw Edison for railroad signalling, ice floe beacons, radio repeaters, and off-shore buoys. The buoys are hard to service and the need to recharge a storage battery adds to the complexity of the task. The zinc-air battery can be discarded after use. At a buoy where a lead acid storage battery of given weight has to be recharged every six months, the zinc-air battery of the same weight will deliver power for five years.

The zinc air battery uses a porous carbon electrode for conveying oxygen from the air into the electrolyte (Figure 8–1). Here oxygen reacts with water, receiving electrons from the external circuit, to form hydroxyl ions:

$$O_2 + H_2 + 2e^- \longrightarrow 2\ OH^-$$

The hydroxyl ions diffuse through the electrolyte to the zinc anode for the oxidation reaction:

$$Zn + 4\ OH \longrightarrow ZnO_2^{--} + 2\ H_2O + 2\ e^-$$

The released electrons circulate through the external circuit to the carbon cathode, delivering energy in the form of current times voltage, which is watts. The zinc consumption is related to delivered current by the term "faraday." One gram mole of zinc is its atomic weight, 65.38, expressed in grams. One gram mole of an element has 6.0225×10^{23} atoms, and one electron is worth $1.602 \times$

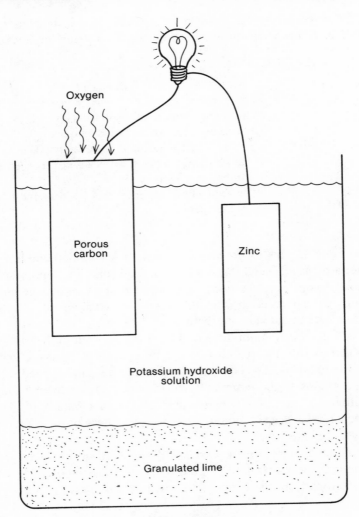

Figure 8-1

10^{-19} coulombs. With these constants we calculate the value of one gram of zinc consumed in a battery (Vz), noting that each atom contributes two electrons to the reaction.

$$Vz = \frac{2 \times 6.02252 \times 10^{23} \text{ electrons} \times 1.602 \times 10^{-19} \text{ coulombs} \times \text{mole}}{\text{mole} \times \text{electron} \times 65.38 \text{ grams}}$$
$$= 2951.57 \text{ coulombs/gram}$$

Primary Batteries: Old and New

TABLE 8-1. Constants for the Zinc-Air Primary Battery

Per 1,000 amp hours (1.25 kWh)		
Zinc	1219 g	2.69 lb
Oxygen	298 g	.66 lb (7.9 ft^3)
Lime	1382 g	3.05 lb

One ampere hour is 3,600 ampere seconds, so the gram of zinc gives:

$$Vz = \frac{2951.57 \text{ coulombs/gram}}{3600 \text{ coulombs/Ah}} = 0.82 \text{ ampere hours/gram}$$

This corresponds to 1.22 grams/ampere hour.

At the negative electrode four OH^- ions are consumed for every two generated at the positive electrode. In the McGraw Edison zinc-air cell the zinc oxide formed at the negative electrode is absorbed in a bed of lime in the bottom of the cell. The reaction is:

$$Ca(OH)_2 + ZnO_2 \longrightarrow Ca\,Zn\,O_2 + 2(OH-)$$

This reaction releases the two extra OH^- ions needed at the negative electrode. The amount of lime needed is shown in Table 8–1.

One of McGraw Edison's zinc-air units, designated Type TR, weighs 42.5 lb, occupies 0.5 cubic ft, and delivers 94 watt hours/lb and 8 kWh/cubic ft. It is rated 3,300 ampere hours and its nominal voltage is 1.25 volts. The cost of the cell is around $53, making the energy cost (Ce):

$$Ce = \frac{\$53 \times 1,000 \text{ volt-ampere-hours}}{3300 \text{ ampere hours} \times 1.2 \text{ volts} \times \text{kWh}} = \$13.38/\text{kWh}$$

Alternative costs for electric power per kWh are:

U.S. public utility electric rates	$0.04 to 0.12
Diesel generated power in Alaska	$0.16

The best applications for the McGraw Edison zinc-air cells are those in which the battery is exhausted at a low rate, like months. At high rates the air cannot penetrate the carbon positive electrode fast enough to prevent OH ion starvation in the electrolyte. The TR cell, for example, is rated at one ampere discharge (Figure 8–2). The zinc air cell has been replacing silver-oxide/zinc

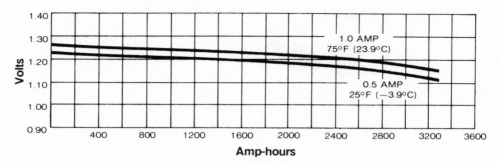

Figure 8-2

cells for hearing aids. These button-type air cells are small, light, and inexpensive.

A zinc cell producing a higher voltage is the ordinary dry cell, which also has a carbon cathode and zinc anode. Here the electrolyte releases chlorine that reacts with the zinc anode, producing around 1.5 volts.

Batteries with higher performance. The zinc air primary battery is hard to beat for low-cost electrochemical energy. However, low weight is advantageous for applications like rockets and missiles that boost spacecraft into orbit. Here low cost isn't as important as the need to release the battery's energy content in a few minutes of time.

High performance is obtained when the reactants available at the electrode surface support high current density. The zinc-air cell fails in this respect because the oxygen reaching the cathode must diffuse through the carbon. Also contributing to high performance is close spacing of plates. A dry cell needs generous space for the electrolyte between the carbon rod in the middle and the zinc anode around the periphery. On the other hand, a high-current silver-zinc battery that can melt nails placed across its terminals has electrodes arranged in plates separated by a thin layer of electrolyte absorbed in a separator. Making the electrolyte thin contributes to low internal resistance in the cell.

Aluminum looks good for an anode because it is light in weight, releases three electrons when oxidized, and is manufactured in huge quantity. Aluminum could deliver 2.94 ampere hours/gram, compared to 0.82 for zinc. An aluminum-air battery could run a car for 1500 miles, after which the aluminum oxide would be removed and new plates would be installed. The aluminum cost would correspond to around $3/gallon for gasoline. The catch is that the aluminum continues to be consumed even when no current is discharged from the battery. The research is directed toward finding one or more alloying elements that would inhibit self-discharge.

Primary Batteries: Old and New

Another contributor to high energy density is an energetic reaction. An electron with a potential of 3 volts is twice as good as one with 1.5 volts. Very few of the possible energetic chemical reactions can be used in batteries. Some are so energetic that an inert battery structure is best. For example, silver zinc batteries are often assembled and stored dry, with provisions for injecting the electrolyte into the cells when power is needed.

Lithium batteries. Lithium, a reactive lightweight metal, has been successfully packaged into abatteries that deliver over 200 watt hours/lb (440 watt hours/kg). The U.S. Air Force has sponsored development of large units for its applications, and a spin-off has been long-life batteries for watches, emergency radios, and flashlights.

The energy content of a lithium battery cannot be extracted quickly. For example, an Altus AL 1700–500 cell will deliver 500 Ah at nearly constant 3.5 volts if discharged for 500 hours at one ampere. If the current is increased to 6.25 amperes the cell gives only 300 Ah with rapidly dropping voltage (Figure 8–3). Performance of some other Altos cells is summarized in Table 8–2.

The lithium battery has about one-fourth the energy density of TNT. Accidents in the early development of lithium batteries motivated research leading to fail-safe devices. For example, Altos has tested its lithium thionyl chloride cells in every conceivable accident scenario, including shock, vibration, puncture, crush, fire, reverse voltage, and forced charging. No catastrophies resulted.

For the light weight of lithium batteries a price must be paid. For example,

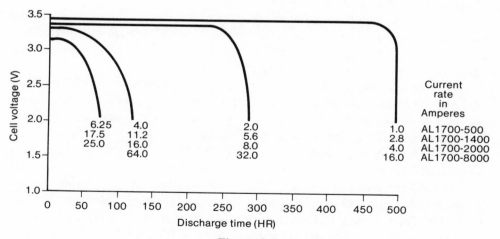

Figure 8-3

Specifications, Altos Cells

Model	Capacity at 250 hr Rate, 20°C	Energy Density WH/lb Wh/kg	Maximum Continuous Current	Dimensions Diameter	Height	Weight
AL1700-500	500A-H	125/275	7A	17/43.2 cm	0.5/1.27 cm	16 lb/7.27 kg
AL1700-1400	1400 A-H	180/396	16A	17/43.2 cm	1.38/3.5 cm	30 lb/13.6 kg
AL1700-2000	2000 A-H	200/440	25A	17/43.2 cm	2.0/5.1 cm	39 lb/17.7 kg
AL1700-8000	8000 A-H	215/473	100A	17/43.2 cm	7.35/18.7 cm	125 lb/56.8 kg

an AL 1700–2000 battery good for 200 watt hours/lb costs from $2,500 to $5,000, depending on quantity and configuration. At $2,500, the price/kWh is $320. A zinc air cell costs about $15/kWh.

Gas-metal battery. The zinc-air battery produces 1 watt hour for every gram of zinc consumed. The reaction of zinc with chlorine and bromine releases more energy, hence gives higher voltage. Such a battery is also called a hybrid fuel cell because one electrode is gas and the other is metal. We once evaluated such batteries for an application needing many megawatt hours of stored energy. These electrochemical couples had already been developed for possible secondary batteries in electrical automobiles.

Marti Klein of Energy Research Corporation developed for us concepts of a zinc/bromine battery. One of his concepts has a zinc slab electrode that is consumed as the battery delivers energy (Figure 8–4).

With a zinc slab battery the electrolyte space widens as the zinc is reacted and carried away in solution. This increases the internal resistance of the battery and hence its internal losses. An alternative suggested by Klein is to deliver the zinc to the battery in the form of a slurry.

Bromine is easy to store, being a liquid at temperatures up to 138°F at normal atmospheric pressure. The zinc bromine cells would self-discharge, so the battery would have to be used within a few weeks after it is activated.

Klein's mass balance for a battery that delivers 1 kilowatt for a year is in Figure 8–5. A year has 8,760 hours. At 50 cents/lb, the one-year supply of zinc would cost $8,000.

A zinc chlorine gas primary battery was adapted for us from an electric automobile development by Energy Development Associates in Madison Heights, Michigan. Their battery had the zinc in a comb-shaped cross section. Carbon electrodes in the spaces between the comb teeth were constructed with chlorine passages. The energy content of the battery was limited to its zinc content. The same principle is used in the firm's 500 kWh load-leveling battery for utility use.

Cells That Make Power from Fuels

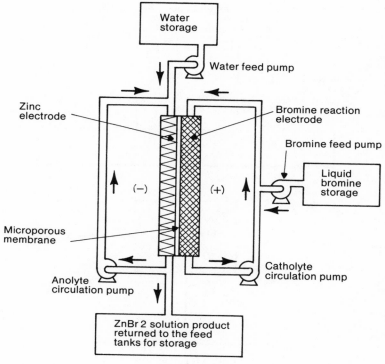

Figure 8-4

CELLS THAT MAKE POWER FROM FUELS

A fuel cell is a box with catalyzed electrodes, gas passages, coolant passages, and electrolyte that convert the chemical energy of incoming gases, usually hydrogen and oxygen, to electric power. Usually the fuel cell is more efficient and compact if the reaction takes place under a pressure of several atmospheres and the temperature is above the boiling point of water.

The reactants that enter the fuel cell are fuel and oxidizer. The fuel is usually hydrogen, coming from a tank, hydride bed, or a reformer that converts methanol or methane into hydrogen. The oxidizer is usually pure oxygen or air.

Energy features of fuel cells include:

- Conversion of fuel and oxidizer into electric power is not a heat-engine process, so Carnot efficiency limits do not apply.

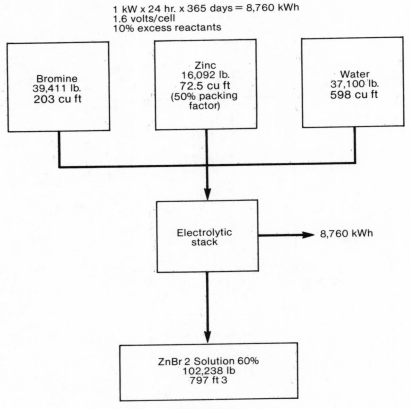

Figure 8-5

- Fuel cells have been successful power sources for manned spacecraft where power is needed during boost and re-entry mission phases as well as during cruise. High purity hydrogen is needed for the fuel in spacecraft fuel cells.
- Fuel cells have supplied commercial and utility power, but the economic advantages of fuel cells over alternative power sources have not been proven.
- Platinum-type catalysts are generally required in fuel cells.
- Hydrogen is hard to store, so most terrestrial fuel cells are designed to use methane or liquid fuels. The required reformers add to the complexity and losses in the power plant.

An inherent energy loss of at least 17 percent occurs when hydrogen and oxygen are electrochemically converted into water. Also, the latent energy of vaporization of the product water is released as heat.

The key performance parameters of fuel cells are voltage and current density. The cost of a fuel cell is related to electrode area, so designers look for high current density. Every fuel cell reaction has a theoretical voltage and an achievable voltage. The difference represents losses, and in a sense purchased fuel that produces no benefits. The fuel cell designer's task is to find the best compromise of costly catalyst, clever cell configuration, operating temperature, and sophistication of auxiliaries.

The performance of a single cell is usually plotted with voltages a function of amperes/square ft, or of milliamperes/square cm. Efficiency is the ratio of power output of the cell to the higher heating value of the fuel used to produce this power. A dc to ac inverter is required if the fuel cell plant supplies power to a utility. Inverter losses must be added to fuel cell losses when comparing fuel cells with power sources having ac generators.

Spacecraft fuel cells. The fuel cells used in manned spacecraft had sintered nickel electrodes and a potassium hydroxide solution for the electrolyte. This type of cell is costly to make, but more important, any carbon compounds getting into the cell degrade it quickly. Therefore the hydrogen must be a costly high-purity gas. No manufacturer is seriously considering offering spacecraft type fuel cells for terrestrial use.

Phosphoric acid fuel cell. A fuel cell configuration that has performance confirmed by test has been developed by the United Technologies Corp. at South Windsor, Connecticut. Sponsors of the development are the Electric Power Research Institute, U.S. Department of Energy, and several public utilities. The prototype power plant, rated 4.8 megawatts electrical output, has carbon electrodes catalyzed with platinum, and a phosphoric acid electrolyte (Figure 8-6 from Ref. 1). The fuel is hydrogen, obtained by reforming methane, methanol, or naptha. The oxidizer is oxygen from the air. The acid electrolyte can tolerate carbon dioxide in the air. Circulating water cools the cells.

The first 4.8 MW fuel cell power plant was installed by Consolidated Edison at a station on Manhattan Island, New York. It turned out to be a development laboratory that in eight years identified numerous unexpected difficulties and failure modes.[2] Important examples were (a) imposition of refinery-construction building codes because naptha was being stored and reformed into hydrogen, (b) breaking of heat exchangers when trapped water froze, and (c) degradation of cell electrodes and separators during storage. The sponsors spent $22.5 million on the development. The project was eventually abandoned.

[1]*EPRI Journal,* April 1976, p. 14.
[2]Gadi Kaplan, "New York's Fuel-Cell Power Plant: On the Verge of Success," *IEEE Spectrum,* December, 1983, p. 60–65.

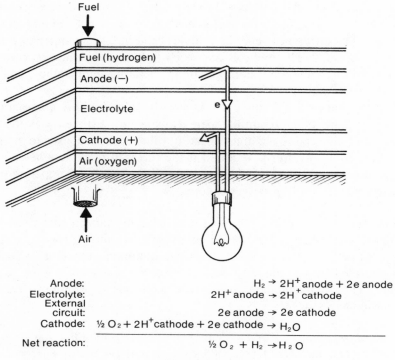

Figure 8-6

In the second 4.8 MW power plant, built for Tokyo Electric in Japan, United Technologies used newer fuel cell construction. The plant went into operation with few start-up difficulties, and demonstrated a heat rate of 9,600 Btu/kWh, using natural gas as a fuel. United Technologies plans to sell fuel cell plants to utilities in 11 MW increments. These units would have 0.93 square meter cells, arranged in twenty-cell stacks.

40 kw natural gas fuel cell. Another fuel cell development at United Technologies, sponsored by the Gas Research Institute, is a 40 kW unit that runs on natural gas and produces heat as well as electricity. For example, a unit installed in a Portland, Oregon laundry provides hot water for the plant's operation. Efficiency of electricity generation of these units is around 40 percent, but if the 150,000 Btu of heat energy each hour in the product hot water is considered then the efficiency becomes 80 percent. The hot water could have been produced by burning the natural gas in a water heater, so the electricity comes as a bonus. No attendant is required for this plant.

TABLE 8-3. Key Operating Characteristics of the Onsite Fuel-Cell Field-Test Power Plant

Parameter	Design value
Electric Power Output	
Steady State	0-40 kW
Transient	56 kW for 5 sec
Energy Form	3 phase, 60 Hz, 120/208 volts ac
Electric Generation Efficiency	40% at half load to full load
Thermal Output	
Rate	150,000 Btu/hr
Temperature	160°F to 275°F
Fuel Input	Pipeline gas at 4-14 inches of water column pressure
Features	Unattended, automatic load-following operation; low emissions and noise; all-weather operation
Fuel Efficiency	Overall: 80% at full load

A total of forty-nine of these 40 kW units are being built. Several run on contaminated methane gas gathered from garbage in landfills. Characteristics are summarized in Table 8–3. The unit is 9 ft long, 8 ft wide, and 6.5 ft high, and weighs 8,000 lb. It contains a reformer that converts the methane gas into hydrogen and carbon dioxide.

Fuel cell performance. Fuel cell performance is measured in heat rate, expressed as the higher heating value of the fuel consumed in generating 1 kilowatt hour of electricity. A 100 percent efficient fuel cell would need only 3,412 Btu of fuel/kWh. The United Technology 4.8 MW plant at Tokyo Electric had a fuel rate of 9,600 Btu/kWh. The best U.S. steam plant in 1983 had fuel rate of 8,987 Btu/kWh.

The electrochemical performance of a fuel cell is plotted in volts as a function of current density, commonly in amperes/square ft. The energy released when hydrogen and oxygen combine in combustion corresponds to 1.48 volts output of a fuel cell if water is the product, and 1.23 volts if it is steam. Neither voltage can be achieved because of the irreversibility of the electrode processes, activation polarization, and concentration or activity gradient in the electrodes.

An objective of fuel cell developers is to get the highest practical voltage from the cell. A. J. Appleby, of the Electric Power Research Institute, in a paper presented to a fuel cell session at Orlando Florida, developed a value for this voltage.[3] He assumed that the annual fixed cost of a fuel cell is $200/kW year, and that the plant would have a 5,000 h/year availability. If the fuel, natural gas, costs $7/million Btu, then a 40 millivolt improvement gives the same overall electricity cost change as a 7.5 percent decrease in capital cost of the power plant. He suggested these routes to improved efficiency:

Heat rate Btu/kWh	Cell voltage, volts	Cell temperature °C	Pressure, atmospheres	Feature
9600	0.73	205	8	Acetylene black based carbon supports platinum or platinum alloy cathode catalyst.
8400				120 psia reformer used.
7600	0.92			Titanium carbide support for catalyst.
7200		240		Higher temperature.
6800				Internal reforming of methane.

Cost of fuel cell power plants. Postulated costs of fuel cell power plants have ranged from $600 to $1,000 per kW. Meaningful costs will become available when plants are produced in a reasonable quantity.

Alternatives to fuel cells. The fuel cells being developed for public utilities use natural gas, methanol, and petroleum products for fuels. These fuels cost 3 to 5 times as much as coal on an energy basis. Thus, the coal burning steam plant is a logical alternative to fuel cells. The coal plant that is only 38 percent efficient can beat a fuel cell plant that is 50 percent efficient from a fuel cost viewpoint. The advantage of a fuel cell plant is that it can be bought in small increments that can be installed near loads and operated automatically. Also, the heat losses at a coal plant are not generally available to the customer. The laundry with the 40 kW natural gas burning fuel cell gets the hot water from the losses.

The other alternative to fuel cells is the gas turbine. Gas turbines with steam bottoming cycles can exceed 50 percent in efficiency, and their cost is around $300 to $400 per kilowatt installed.

[3] A. J. Appleby, "Acid Fuel Cell Technology–An Overview," *Fuel Cell Seminar at Orlando Florida,* November, 1983.

9

Economics, The Final Decision in Comparisons

The Washington Public Power Supply System (WPPSS) is an entity managed by directors selected by the controlling boards of seventy-one publicly-owned electric power utilities in the State of Washington. WPPSS owned and operated a steam power plant and hydro plant. When building nuclear power plants was fashionable, WPPSS decided to build five. Ownership was distributed among Bonneville Power Administration, and privately-owned and publicly-owned utilities. WPPSS, a public entity, was able to sell tax-exempt bonds to finance the projects. To attain a high credit rating the publicly-owned utilities signed "hell or high water" contracts that guaranteed payment of interest and principal on bonds even if the plants were not built. When difficulties developed, these utilities, with approval by local courts, reneged on the guarantee, leaving investors who had invested in Plants 4 and 5 with $2.2 billion in worthless bonds.

Nuclear power plants can be made to work. Even WPPSS is generating power in its Plant 2. The WPPSS difficulties were economic. The main causes included:

- The five power plants were being built at the same time, so no learning-curve benefits accrued. Duke Power, with six successful nuclear plants generating one-half of its power, builds them one at a time, using the same management and engineering team.
- Inexperienced construction management resulted in unexpected price escalation. In 1984, Duke Power completed Catawba No. 1 for a little over $1,000/kilowatt. At about the same time WPPSS completed Plant 2 for $2,904/kilowatt. Cost of nuclear

power plants are particularly sensitive to inexperienced management because of continuously changing Nuclear Regulatory Commission (NRC) requirements and the thoroughness of NRC inspection.

- Small utilities bought more generation capacity than they could use, intending to sell this capacity at a profit as they had done with hydro capacity. As the cost of the nuclear plants rose, the output power was no longer marketable. The only way it could be used was to mingle it with low-cost hydro power and retail it to a utility's own customers.
- Power demand turned out to be elastic in that higher power price inspired customers to conserve energy.
- Power demand diminished because of a recession. Regional power forecasts had not considered recessions.
- As interest rates rose, WPPSS had to spend more of its new borrowings to pay interest on the old borrowings.

Another example of a technical success, but economic failure, is the nuclear ship, Savannah. Submarines and icebreakers have been successfully powered with nuclear plants. The use of nuclear power on the Savanah was expected to give low propulsion power cost and a long operating range without a need to refuel. However, in the operating cost of this ship, the fuel cost was small compared to the crew cost. The Savannah, an American ship, had to carry an American crew, and the affected unions established a crew size that made the ship non-competitive.

These examples show how economic factors can overwhelm technical ones in an energy project. Predicting future economic trends accurately is difficult. Those who have been successful predictors of future interest and inflation, either by skill or accident, have done well in the stock market. They are not likely to reveal their techniques, and engineers would not be likely to believe them anyway!

The importance of economic considerations varies among energy projects. For example, the decision of whether to insulate a structure better than is required by the building code could be made simply on the basis of the years of fuel saving needed to pay back the extra cost. At the other end of the scale is the evaluation of future power generation options by Peter D. Hindley and associates for Pacific Gas and Electric Co.[1] All costs of each option for its operating lifetime were evaluated, discounted in future costs, and reduced into terms of how much a resource alternative reduces revenue requirements for each dollar of capital spent.

---

[1]Peter D. Hindley and associates, "Economic Assessment of Alternative Generating Sources," *Proceedings of the 19th Intersociety Energy Conversion Engineering Conference*, AIChE (1984), pp. 1039–1044.

What Determines Price?

They then calculated sensitivities of their conclusions to differing economic and political environments.

WHAT DETERMINES PRICE?

The price that the user pays for fuels, energy, and manufactured equipment is established by one or more of the following methods:

Pricing technique	Examples
What the traffic will bear	Clothing, sports equipment, jewelry, furniture, art, oil well drilling rigs (Rigs cost $90,000 per day in 1981, $20,000 in 1984)
Supply and demand	Coal, lumber, copper
Monopoly	Diamonds
Cartel	Oil between 1971 and 1985
Low bid	Power plant construction
Price book	Transformers, circuit breakers, engines
Regulated	Electric power and gas rates
Negotiation	Mineral-rights leases, right-of-ways, real estate
Cost plus fixed fee	Engineering and construction management

Today's cost of an energy project can be estimated from established prices, suppliers' quotations, estimating guides, and history of similar projects. For example, the cost of a diesel engine plant can be synthesized from the following sources. A detailed checklist of needed equipment and structures appears in a book published by the Kohler Co.[2]

Item	Sources for cost data
Engines, generators, blowers pumps, tanks, and auxiliaries	Manufacturers' catalogs
Construction labor and materials	"National Construction Estimator," and "Richardson Construction Trend Reporter"
Repair services	Subcontractor
Spares and repair parts	Supplier quotation
First tank of oil	Diesel fuel supplier
Interest	Local bank or *Wall Street Journal*
Tank of oil 10 years later	?

[2] "Engineer's Guidebook to Power Systems," Kohler Co., Kohler, Wis.

The difficult cost to estimate is the cost of fuel oil ten years from the present. Corporations and banks have gone bankrupt because they wrongly predicted the future price of oil. The U.S. Government set up Synthetic Fuels Corp., which spent billions to design and build plants that could make gasoline out of coal. A subsequent oil surplus made the plants uneconomical. The price of commodities that are traded freely follows a supply and demand relationship.

SUPPLY AND DEMAND ECONOMICS

The laws of supply and demand are basic to modern economic theory. For example, people living in cities having low electricity rates consume more electrical energy than do people who have to pay high rates (Figure 9–1). The economist evaluates this relationship in terms of supply and demand. Regulated utilities have the little control over demand. Consumers as a group and their

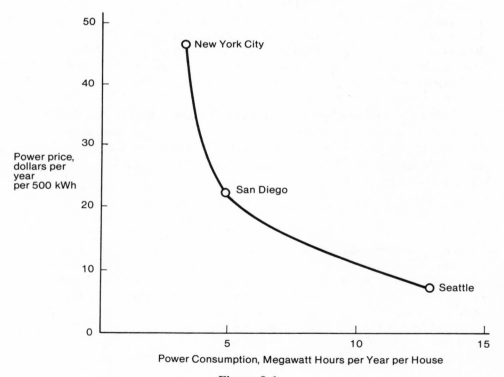

Figure 9-1

Supply and Demand Economics

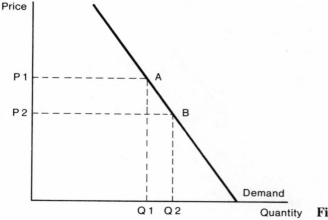

Figure 9-2

reaction to price establish demand. The law of demand states: the quantity of demand of a good varies inversely with the price.

Demand is usually represented by a line with a negative slope on a graph where price is plotted as a function of quantity (Figure 9–2). At price P1, the quantity sold is Q1. If we lower the price to P2, then we could sell the larger quantity Q2. The lower the price, the greater the demand.

The law of supply states that the quantity of supplies of a good usually varies directly with its price. For example, a seller would be willing to supply larger quantities at a higher price. The supply-side of the economy is generally slower to react than the demand-side. The plot of price as a function of quantity generally has a positive slope (Figure 9–3). At price P1, the sellers are willing to produce

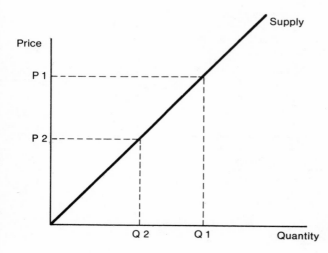

Figure 9-3

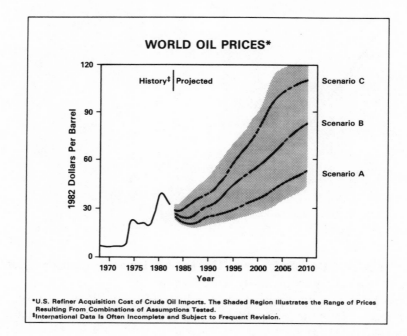
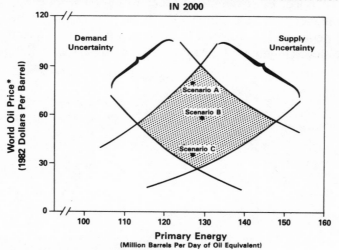

Figure 9-4

Q1. If the price is lowered to P2, then sellers are willing to produce only the smaller quantity Q2. The lower the price, the less sellers are willing to produce.

When supply and demand curves are plotted together, their intersection is called the equilibrium price (Figure 9-4). In theory this is the competitive price that makes the best use of our scarce resources. In practice this equilibrium price is seldom reached because sellers and buyers react to changes in the market place at different rates. It is much easier for a consumer to lower his consumption than it is for a utility to build a new power plant.

ELASTICITY RELATES PRICE AND QUANTITY

Elasticity is the slope of the curve resulting from plotting price versus demand on logarithmic scales. Elasticities are usually expressed in absolute values. The relationship between price and quantity is assumed to be causal.

$$\text{Elasticity} = \frac{\text{Percent change in quantity}}{\text{Percent change in price}}$$

Price-quantity relationships are perfectly elastic, relatively elastic, unit elastic, relatively inelastic, or perfectly inelastic (Figure 9-5).

Perfect elasticity is a useful textbook concept seldom found in the real world. The price will not change for any change in quantity, so the slope of the elasticity curve is zero. An example is power or gas from a utility with a regulated price. In the long run, however, the regulating authority would be forced to adjust the price to reflect what the equilibrium price might be. Spacecraft solar cells are another example. No one is going to increase production of spacecraft just because the price of solar cells dropped.

In relative elasticity, where the slope is greater than minus one, the percent change in quantity is larger than the percent change in price. Thus total revenue increases as the price is lowered. In Figure 9-2, total revenue is the area of the rectangle P1, A, Q1, O or P2, B, Q2, O.

Among many market pressures that affect relative elasticity is the time the market has to adjust to changes. The more time available, the more elastic it will become. The larger the percentage of the consumers' total budget, the more elastic the product will become. Availability of substitutes for a commodity makes the elasticity greater. There are many other pressures in the marketplace that affect elasticity.

Unit elasticity occurs when the slope of the elasticity curve is minus one. A percent change in price is matched by the same percent change in quantity, and total revenue to the seller doesn't change. Unit elasticity is the dividing point between elasticity and inelasticity.

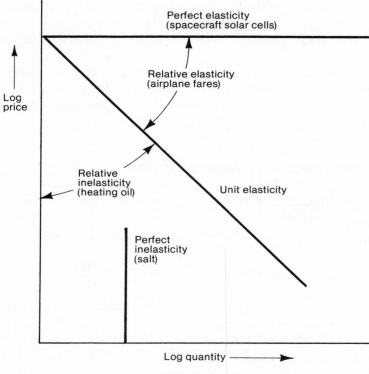

Figure 9-5

In relative inelasticity, where the slope is less than minus one, a smaller percentage change in quantity follows a given percent change in price. Total revenue decreases as the price is lowered. Note this is the exact opposite of relative elasticity. It follows that relative inelasticity would be caused by market pressures opposite from those that cause relative elasticity. For example, a short period of time for the market to adjust to change will tend to make the market more inelastic. The smaller the percentage of the consumer's total budget a commodity requires, the more inelastic the market for that commodity will become. The fewer substitutes that can be found for a commodity, the greater will be its inelasticity.

Perfect inelasticity occurs when the price–demand curve has infinite negative slope. A textbook example that comes close to some real situations is salt. Consumers need a certain amount of salt and are willing to pay almost any price for it. However, if you lower the price, they will not buy much more. Salt is unlike vacation trips, which increase in quantity if the price is reduced.

A study of the response of energy demand to higher prices by James L.

Interest and Future Expenses

Sweeney of Stanford University provides a good example of the elasticity of demand in energy.[3] He reports that energy demand increased exponentially from 1950 to 1973. From 1973 to 1982 the demand for oil and gas changed to an average annual decline of 1.4 percent. Also electrical load growth slowed to an annual rate of 2.1 percent. The cause for this change was the large increase in oil prices by the OPEC cartel.

Good substitutes contribute to a relatively elastic demand. Coal can be substituted for gas, wood heat for electrical heat, and capital investment for inefficient energy use.

Factors contributing to the elasticity of demand are argued about in journals on economics. The laws and rules of economics, unlike those of the natural sciences, are often changed.

INTEREST AND FUTURE EXPENSES

"Discounted fuel expense," "discounted replacement cost," and "present worth" are expressions that reduce the benefit of spending money today to save money in the future. For example, consider a battery that stores for night use electric energy generated by a solar array. Assume that the two choices are:

	Solar battery	Electric vehicle battery
Cost/kWh of storage capacity	$200	$400
Cost of a 10-kWh battery	$2,000	$4,000
Life, charge discharge cycles	1,000	2,000
Life, years	3	6

Without considering interest for the six-year period, the battery cost for the $4,000 electric vehicle battery is the same as for the $2,000 solar battery plus its $2,000 replacement. However, the money for the replacement solar battery could have been earning interest during the first three years (Figure 9–6). If $1,423 had been invested in U.S. Government bonds at 12 percent, and if the interest had been compounded annually, the money available for buying batteries (Mb) at the end of three years would have been:

$$Mb = \$1{,}423 \times 1.12 \times 1.12 \times 1.12$$
$$= \$1{,}423 \times (1.12)^3 = \$2{,}000$$
$$= \$1{,}423\,(1 + I)^t$$

---

[3] James L. Sweeney, "The Response of Energy Demand to Higher Prices: What Have We Learned?" *Proceedings of the 19th Intersociety Energy Conversion Engineering Conference,* ANS, (1984), pp. 1045–1050.

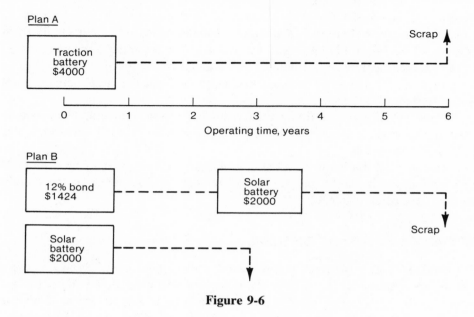

Figure 9-6

where I = interest rate in decimals
t = time of deposit in years

The $1,423 could have been calculated from the equation,

$$Vp = \frac{Vf}{(1 + I)^t}$$

where Vp is the present value of the future cost, Vf. The present value of a $2,000 replacement battery bought twelve years in the future, would be only $513 if the interest rate were 12 percent.

Inflation would increase the cost of the future battery. For example, at a 6 percent inflation rate the cost of the $2,000 battery ($Bi$), at the end of three years would be:

$$Bi = \$2,000 \times (1.06)^3 = \$2,382$$

An increment of inflation starts with a cost-of-living increase added to the cost of labor. The average labor content of all manufactured goods in the U.S. is around one-third so the average price will rise. However, elements such as construction have high labor content, whereas materials like cement and aluminum have lower labor content. One estimator uses the inflation predictions in

Learning Curves for Multiple Units

TABLE 9-1. Inflation Prediction Used by One Estimator

Year	DT&E	Labor	Material
82	1.000	1.000	1.000
83	1.070	1.107	1.104
84	1.125	1.228	1.226
85	1.178	1.365	1.369
86	1.231	1.509	1.506
87	1.286	1.657	1.636

Table 9–1, which starts from 1982. In his table DT&E means development, test, and engineering.

One life-cycle costing analyst uses a simple 10 percent discount rate to account for inflation and interest.

LEARNING CURVES FOR MULTIPLE UNITS

If two identical power plants are built one after the other, the second will cost less than the first. A third would cost less than the second. This relationship was quantified in airplane production in which the first airplane is so expensive that no airline could afford to buy it. The learning curve, based on previous production experience, is the tool that enables the manufacturer to set a price that is acceptable to the customers, yet will ultimately produce a profit. Not all learning curves are constructed in the same manner.

The learning curve that we use, assumes that each time the quantity of items produced is doubled, the last item of the lot costs a fixed percentage of the cost of the first item of the lot. The equation for the cost (y) of unit (x) then is:

$$y = A\, x^n$$

where A = the cost of the first unit

$$n = \frac{\log S}{\log 2}$$

where S = slope of the learning curve, expressed in decimals.

For example, if the first unit costs $100 million and the curve slope is 0.9, then the second unit should cost $90 million and the fourth should cost $81 million. From the equation, the cost of the fourth unit is indeed:

$$n = \frac{\log 0.9}{\log 2} = -0.152$$

$$y = \$100 \text{ million} \times 4^{-0.152} = \$81 \text{ million}$$

The cost of the 100th unit would be $49.6 million.

ECONOMIC OPPORTUNITIES FROM POWER GRIDS

A power grid is a group of electric utilities in a region, interconnected with high-voltage transmission lines. For example, the Western grid extends from British Columbia to the Mexican border, and from the Pacific Ocean to Wyoming. A grid can include publicly-owned as well as investor owned electric utilities. Connecting utilities into grids does not cause power rates to equalize.

At one time electric utilities were independent units that generated power and delivered it to customers. Each utility had a spinning reserve that was equivalent to the largest generator on the line, so that failure of any one generator would not collapse the voltage and frequency delivered to customers. Spinning reserve requirements in a grid correspond to the largest unit on the grid, reducing the total number of partly loaded or idle generators required.

Other benefits of grids are:

- Advantageous scheduled power exchanges can be arranged, particularly if power consumption peaks occur at different times in the region.
- A participating utility can shop for the best short-term power supply, or arrange a power sale to avoid load changes at steam plants.
- Unexpected overloads will not bog down a utility's generating plants.
- The output of non-firm solar and wind generators can be accommodated.

The following excerpt from Utah Power's 1983 annual report illustrates the benefits of grid connections to its customers: "Utah Power buys available surplus hydro energy from northwestern utilities to serve its own customers. The company then sells its own, more expensive coal-fired generation to southwestern utilities to displace their higher-cost oil and gas generation. These sales, called "surplus sales," resulted in $65 million in revenue in 1983."

The technical problems in assembling power grids have been solved. For example, the 500 kV ac and ± 500 kV dc lines provide the stiffness needed to keep the vast Western grid in synchronization. Stability questions can be resolved by computer modeling. A participating utility can instrument its interfaces and govern its generators to achieve specified power import or export. Protocols establish responses to underfrequency and overfrequency, and the necessary short-term exchanges are equalized at a convenient time.

Contractual arrangements within grids can be complex. For example, in the 1950s BC Hydro built storage dams that smoothed the yearly water flow in the

Columbia River. This enabled the downstream plants to increase their annual output. One-half of this extra output was given to BC Hydro as compensation for building the dams. However, BC hydro had no use for this "Canadian Entitlement," so it sold it to a group of Pacific Northwest utilities which in turn sold it to California utilities on a long-term basis. In a sense, the power flowed from the Columbia River hydro plants to the California customers, but the money traveled a different route.

In another sense the direction of power flow in a transmission line is not always clear. For example, on a section of one of the 500 kV ac lines connecting the Columbia River hydro plants with California is Portland General Electric's 108 MW Pelton hydro plant, which can be shipping power north at the same time that Bonneville Power Administration is wheeling power to California.

In the simplest power transfer agreement two participating utilities contract to exchange power on mutually favorable terms. For example, the City of Seattle at one time sent surplus hydro power during the summer to Sacramento Municipal District for air conditioning loads. In the winter the power flow reversed to help carry Seattle's peak heating load. Appropriate adjustments were made for the difference in cost of steam and hydro power. The owners of the transmission lines, among them Bonneville Power Administration and Pacific Gas and Electric Co., were paid fees for wheeling the power from one city to the other.

Non-firm power for grids. Solar arrays stop producing power at night and when clouds obstruct the sun. Windmills produce no power in calm air. These no-power periods cannot be scheduled. United States utilities are required by the PURPA act to accept wind and solar power, and compensate the producers with a price that is related to the cost of avoided generation. A small utility with 50 percent wind and solar power would have to keep alternative generation on the line at all times.

A large grid can readily accept solar and wind power because the variations in output of these plants is like the variations in load. However, the contractual arrangements become complex, because some entity has to accept and pay for this generated power. In the state of California one obvious entity is Pacific Gas and Electric, with a peak load of approximately 15,000 MW, and a willingness to accept non-firm power. On the other end of the scale is Washington's Orcas Island Public Utility District which has no generation and buys all of its power from Bonneville Power Administration. The contract negotiations could be complicated if an optimum wind power site were on Orcas Island.

10

Moving Energy in Wires, Pipes, and Ships

Energy or its equivalent can be moved on trains, on trucks, in pipes, in ships, through wires, or in microwave beams. Sometimes there is no question as to how the energy should be transported. If you are sawing cordwood and run out of chain-saw fuel, you will haul in a new supply with your car or truck. Likewise, there isn't much choice in how to have heating oil delivered to your home. For other situations the optimum method may not be obvious. For example, if you are planning a steam power plant, you cannot easily decide whether to build the plant at the mouth of the mine and transmit power, or build the plant at the load center and haul coal by train.

Energy moved in electric transmission lines is in a more valuable form because it can be converted to light and mechanical power without Carnot-cycle losses. If a train hauling coal to a power plant has 100 cars, then thirty-three of the cars can carry the energy that will be delivered as electricity from the plant, and sixty-seven cars can carry energy that will go up the cooling tower. On the other hand, of the energy flowing through a natural gas pipeline and destined for heating, 90 to 95 percent of the delivered gas can end up as heat where it is wanted.

ELECTRIC POWER TRANSMISSION

Electric power can be transmitted in conductors at high voltage for distances of over 2,000 miles to connect hydro power generating stations to loads in cities.

Electric Power Transmission

Transmission lines also allow placing fuel-burning steam plants where the fuel and cooling water are, away from populated areas. The visible transmission power line is a set of conductors, supported on poles or towers, stretched over a right-of-way. Other elements of the transmission line are the sub-stations at each end and at intermediate points, and controls that recognize, isolate, and bypass faults.

Energy features of electric power transmission include:

- Transmission lines can be designed to have low power loss, usually less than 10 percent, at full load. Most lines carry less than full load most of the time.
- Most transmission lines are part of a network, and the network stability considerations enter into the design of the line.
- The cost of constructing a 115 kV wood-pole transmission line across a desert is only around $20,000/mile. The cost of acquiring right-of-way elsewhere can be many times that cost.
- The capacity of existing transmission lines can usually be increased by installing more and bigger conductors, or by re-insulating and operating at a higher voltage. This avoids the difficulty of acquiring new right-of-way.

The technical limits of electric power transmission include:

- A line-to-line potential of 1,100 kilovolts (kV) seems to be a practical limit because higher-voltage transformers become so big that they can't be shipped from the factory. Field assembly of high-voltage transformers is not practical.
- The 1,100 kV conductors would be spaced 72 ft (22 m) apart, requiring a right-of-way 300 ft wide.
- 765 kV is the highest voltage in common use in the U.S.
- The longest transmission line, which is not part of a network, is a $\pm$ 533 kV dc line carrying 1,920 megawatts (MW) from Cabora Bassa water power station to Apollo near Johannesburg, South Africa, a distance of 1,400 km (870 miles).
- The longest high-voltage dc line being planned would carry 10,000 MW from hydro plants in the Amazon valley of Brazil to the industrialized San Paulo area, a distance of 2,000 miles (3,219 km).

Transmission line potential is commonly measured in kilovolts line-to-line, one kilovolt (kV) being 1,000 volts. The capacity of transmission lines is commonly measured in megavolt-amperes (MVA). The capacity in MVA is the capacity in megawatts if the power factor is 1.00. Otherwise MW is MVA times the power factor. The power factor of transmission lines is usually adjusted with tap-changing inductors to control voltage at the receiving end of the line.

The following are the important electrical characteristics of transmission lines.

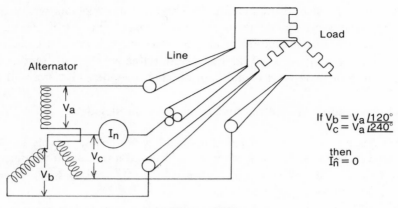

Figure 10-1

- Most electric power is generated as 60 Hz alternating current (ac) in North America, and 50 Hz ac in Europe. Alternating current transformers have been a low-cost convenient way of changing voltages.
- Most transmission lines carry three-phase power. The power could also be carried in three single-phase lines. The three independent circuits could share a common return circuit (Figure 10–1). However, if the current waves of the three circuits were spaced 120 electrical degrees apart, an ammeter in the return circuit would read zero. Thus the return conductor could be eliminated, and for a given voltage drop in a given conductor size the three-phase circuit requires only one-half the conductor weight required for the three corresponding single-phase circuits.
- Any current-carrying conductor at other than cryogenic temperature will have voltage drop and power loss. An alternating-current carrying conductor, if spaced apart from its return path, will have inductance that also causes voltage drop or rise, depending on the power factor of the current being carried.

If the load supplied by the transmission line is pure resistance, then the current in the line will be in phase with its driving voltage. The load is more likely to be inductive since more than 60 percent of the power generated in the U.S. is consumed in motor drives. The current fed to inductive loads lags the voltage, and the product of the current is called volt-amperes, kilovolt amperes (kVA) or megavolt amperes (MVA). This product (P) is related to real power, usually measured in watts, kilowatts (kW) or megawatts (MW) by the power factor (PF):

$$P = V \times A \times PF = V \times A \times \cos\theta$$

where θ = the angle between the voltage and current phasors in electrical degrees.

Electric Power Transmission

Transmission line optimization. A transmission line is optimized by comparing the cost of alternative voltages and type of power transmitted, ac or dc. The power carried by the line relates to the product of the current and voltage, which means that for a given power higher voltage requires less current. Loss is related to the resistance of the wire times the current squared, so for a given conductor cross section a higher voltage reduces current and losses.

As voltage is raised, the cost of insulators, transformers, and circuit breakers goes up, and the conductor must have a larger diameter for corona-free operation. At voltages of 345 kV line-to-line and above, limiting the corona becomes an important design consideration. Corona-free operation is achieved with large-diameter conductors and by bundling up to four appropriately-separated conductors in each phase.

In long transmission lines the cost of conductors, insulators, and towers predominate. In short ones, the cost of the receiving-end and sending-end substations becomes significant. In high voltage dc lines the cost of the converter stations at each end becomes important.

The components of a power transmission line are not expensive. For example, a 115-kV transmission line on wood poles in the desert costs only $20,000/mile. In populated areas, the cost of acquiring a right-of-way becomes important. One mile of overhead transmission line over exclusive right-of-way requires as much as twenty acres of land. As a result, we find interesting variations in line construction. For example, a 220 kV transmission line is normally supported on towers over a dedicated right-of-way. However, we now see 220-kV conductors supported on post insulators on tall poles along a public road.

The benefits of higher voltage are shown in Figure 10-2, from Ref. 1. A single 1,100 kV transmission line would carry as much power as five 500 kV lines. A 500 kV line will carry around 1,000 MW, the output of a large nuclear plant.

Underground transmission costs 5 to 15 times as much as an overhead line on the same route.

Transmission line voltage. The Institute of Electrical and Electronics Engineers has designated standard transmission line voltages. Line-to-line voltage is usually quoted. Line-to-ground voltage is line-to-line voltage divided by $\sqrt{3}$. Voltages of 35 kV and less, once used for transmission, today are considered distribution voltages. The common standard transmission voltages and distances for which they are being used are:

---

[1] "Transmission in Energy Researches," *EPRI Journal,* April 1974.

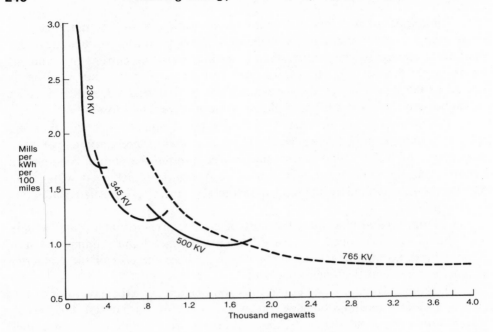

Figure 10-2

Standard power transmission voltages[2]	
Voltage, kilovolts	Typical distance, miles (km)
46	25 to 100 (40 to 160)
69	
115	
138	50 to 150 (80 to 240)
161	
230	
345	400 (650)
500	500 to 1500 (800 to 2400)
765	
1100	

Figure 10–3 shows typical dimensions of towers for transmission lines.

[2]Standard C 84, American National Standards Institute, Table 8.31.

Example of High-Availability Transmission Network   247

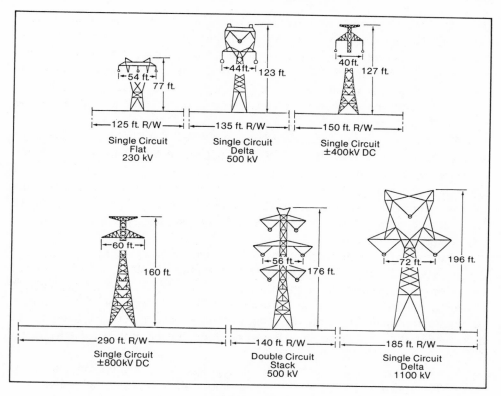

Figure 10-3

EXAMPLE OF HIGH-AVAILABILITY TRANSMISSION NETWORK

Once we made a preliminary design of a power transmission network for delivering power to 4,600 MX missile shelters in Nevada and Utah.[3] This MX missile basing concept was subsequently abandoned because alternative basing concepts turned out to be better. However, the steps in achieving an optimum power transmission network illustrate the use of reliability, stability, and cost comparisons.

The missile shelters were arranged in clusters of twenty-three, with shelters

[3] H. Oman and C. F. Bannon, "High-Availability Power for MX," *Proceedings of the 1982 IEEE Power Engineering Society Transmission and Distribution Conference*, 1982.

one mile apart. The basic requirement was to provide 0.9999 power availability at each site, starting with 0.99 availability at public utility interfaces. We attained the availability objectives by having dual feeders supplying power to each of 200 distribution centers from twelve transmission sub-stations arranged in a loop. This gave us 0.999 power availability at the distribution centers. At each distribution center a stand-by power plant with a spare engine brought the availability to the required 0.9999.

Several areas in the U.S. were being evaluated for MX deployment. Our studies were confined to the Eastern-Nevada and Western-Utah region. As shown in Figure 10-4, the only useful line in the area was a 230 kV link between Sigurd sub-station in Utah and Fort Churchill power plant east of Reno, Nevada. This line, owned in Utah by Utah Power Co., and in Nevada by Sierra Pacific Power Co., was fully committed to its 150 MW stability unit. The estimated demand for MX power was 190 MW.

Our transmission network analysis was done primarily by Engineering and Design Associates at Tigard, Oregon, where Stanley D. Reed and Michael W. Unger were the principal investigators. Their approach was to postulate plausible networks, test them with computer-based load flow analyses, and then compare the costs and reliabilities of the ones that worked. The final networks were tested in the model of the Western States Power Grid maintained at Provo, Utah, by the Western Systems Coordinating Council.

An important requirement was to have at least two independent sources of power. The transmission network had to supply all loads, even after loss of a utility-interfacing sub-station, or after any segment of the transmission network failed.

One of the best transmission networks is shown in Figure 10-4. The adopted transmission voltage was 138 kV line-to-line. Lower voltages failed to maintain required sub-station voltage under worst case outage conditions. Higher voltages required more expensive components and larger conductors to limit formation of corona. The significant conclusions from the analysis were:

- Interface switching stations, where power is received from supplying utilities, cost between $0.6 million and $3 million each. Transmission line costs about $100,000/mile. Therefore the lowest cost approach was to add switching stations and use existing transmission lines whenever constructing thirty miles of new transmission line could be avoided.
- Most networks failed in the load-flow analysis because they could not maintain voltage at all sub-stations after a worst case single outage. The usual worst case outage was in a line segment whose absence caused power to be fed from one end of the network to the other, instead of merely to the normally open point in the middle.
- Transmission losses in the best network were only 2 MW. These losses increased to 6 MW when the worst case segment of the network was open-circuited.

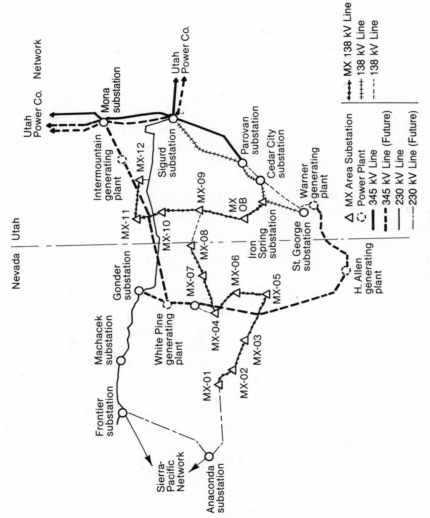

Figure 10-4

We were able to get a power availability of 0.99993 at a typical shelter with our best transmission network when we used stand-by generators at our distribution centers and looped laterals in the distribution network. The failure rates of most of the components (Table 10–1) were based on records of the Bonneville Power Administration, a federal agency in the Pacific Northwest that has 12,600 circuit miles of high voltage lines and 347 substations. Outages of power transformers were assumed to occur at a rate of one every 100 years. Appropriate spares were assumed so that a failed transformer could be replaced within twenty-four hours.

What's the voltage of that line? The energy systems engineer can impress family and friends while riding on a highway by describing the visible transmission lines in terms of voltage, power source, and loads. Great accuracy is not needed because experts who design power lines are sometimes confused about the voltage of unfamiliar ones. Speaking with an air of confidence is important, though.

The technique is simple—count the number of disks on the suspension insulator. Each disk is good for about 12 to 15 kV. For example, a 69 kV line would have around five disks, a 115 kV line around ten disks, and a 230 kV line around twenty disks. A 500 kV line has so many disks that they are not easy to count. The 765 kV lines are so rare that you will be right 99 percent of the time by ignoring them.

You get extra respect if you correctly identify a dc line.

DC power transmission. High-voltage ac power transmission lines have three conductors, one for each phase. When driving west of Death Valley, east of Reno, Nevada, or in Eastern Oregon you may be surprised to see a two-wire transmission line. It carries 2,000 amperes at $+500$ kV and -500 kV with respect to ground potential. The lines run for a distance, of 876 miles (1,410 km) from Celilo, Oregon, near The Dalles, to Sylmar, California. Its rating is 2,000 megawatts with both conductors in service. With one conductor out of service it can carry 1,000 MW, with return being through the ground. The ground for the return circuit is completed at Celilo in a string of 1067 cast iron pigs, buried in a coke-filled trench 3400 feet in diameter.

At each end of the line is a converter station that can make dc out of ac, or ac out of dc. The main advantages of dc transmission are:

- The highest possible amount of power is transmitted for a given conductor quantity.
- Transmission voltage is peak voltage. In ac the peak voltage, which sizes the insulator strings, is 1.414 times the root mean square (rms) voltage. The rms voltage corresponds to the power being carried.

TABLE 10-1. Component Reliability Data

System component	Forced outages per unit per year		Scheduled outages per unit per year	Duration of outage hours		Line number
	Normal weather outages	Adverse weather outages		Forced outage	Scheduled outage	
Bus, 230kV	0.024	0.024	0.00	6.0	0.0	1
Disconnect Switch, 230kV	0.000	0.000	0.10	0.0	2.0	2
Circuit Breaker, 230kV	0.018	0.018	1.00	8.0	1.5	3
Transmission Circuit, 230 kV	0.007/mi	2.00/mi	0.00/mi	5.0	0.0	4
Circuit Switcher, 230kV	0.018	0.018	1.00	8.0	4.0	5
Autotransformer, 230-138kV	0.010	0.010	0.10	—	7.0	6
Bus, 138kV	0.030	0.030	0.00	6.0	0.0	7
Circuit Breaker, 138kV	0.008	0.008	1.00	8.0	1.5	8
Disconnect Switch, 138kV	0.000	0.000	0.10	0.0	2.0	9
Circuit Switcher, 138kV	0.008	0.008	1.00	8.0	0.0	10
Subtransmission Circuit, 138kV	0.019/mi	2.00/mi	0.00/mi	5.0	0.0	11
Transformer, 138–25kV	0.010	0.010	0.10	—	0.0	12
Regulator, 25kV	0.050	0.050	0.10	4.0	0.0	13
Bus, 25kV	0.030	0.030	0.00	0.0	0.0	14
Disconnect, 25kV	0.000	0.000	0.10	0.0	1.0	15
Recloser, 25kV	0.020	0.020	0.00	4.0	0.0	16
Miscellaneous/230kV Terminal	0.310	0.310	0.00	1.0	0.0	17
Miscellaneous/138kV Terminal	0.170	0.170	0.00	1.0	0.0	18

- Stability is not a problem because the tie between load and source is not synchronous. In fact, power pulses transmitted in the dc line from Celilo to Sylmar are used to damp out subsynchronous oscillations in the 500 kV ac lines connecting the Northwest with California.

Disadvantages of dc power are:

- The converter stations are costly, and hence dc power transmission is not cost effective for short distances. However, there are short links, such as at Stegall, Nebraska. Their function is to transfer limited power between the regional grids, which are not necessarily in synchronism.
- There could be no branch circuits until high-voltage dc circuit breakers are developed. All faults have been cleared by interrupting the ac power at both ends of the line.

Brown Boveri has developed a high voltage dc circuit breaker and tested it on the 1,000 kV Celilo-to-Sylmar dc transmission line, so branch circuits may become practical.

HAULING ENERGY IN PIPES AND WITH VEHICLES

Coal is normally hauled in barges, ships, and dedicated "unit trains." It can also be pumped through a pipeline in the form of a water slurry. Natural gas and petroleum can be hauled in pipelines and ships. Natural gas hauled in ships is a cryogenic liquid. In general, expanding or modifying existing energy transportation means is less difficult and less costly than developing new ones.

Hauling energy is important because huge quantities of a relatively inexpensive commodity are moved. The cost of hauling can double the cost of the energy.

Features of energy hauling include:

- A supertanker operating in the open ocean is the most efficient hauler of energy.
- Hauling crude oil by supertanker costs around 2.5 cents/100 miles/ton of 6.7 barrels, or 0.37 cents/barrel/100 miles.
- Hauling coal by unit train costs around 1.6 cents/ton/mile.
- The best way of hauling energy shifts as the rate of energy flow diminishes. For example, crude oil is generally shipped over land in pipelines. However, tank trucks are used to empty the stock tanks from marginal wells that produce around twelve barrels a day.

Limits in the transportation of energy include:

Hauling Energy in Pipes and with Vehicles

- Maneuverability and other considerations seem to limit supertankers to around 500,000 tons.
- Coal trains are generally limited to around 100 cars, each carrying 100 tons (91 metric tons).
- A pipeline proposed for carrying Alaska natural gas to the contiguous 48 states is 48 inches in diameter and its pressure would be 1680 psig (11.6 Mpa).

The nomenclature and the units used in energy transportation are unique to each form of energy. Coal is sold by the ton, so the carrying capacity of coal-carrying railroad cars is measured in pounds or tons, and the cost of transportation is quoted in dollars per mile per ton of 2,000 lb (907.2 kg). The oil-carrying capacity of tankers is measured in barrels of 42 gallons (159 liters), or deadweight tons. A deadweight ton corresponds to around seven barrels of crude oil. For example, the ESSO "American Spirit" has a deadweight capacity of 265,000 tons of 2,000 lb, and carries 2 million barrels of petroleum. Pipeline capacity is quoted in barrels of oil per day. Natural gas is measured in units of cubic feet at standard conditions of sea-level atmospheric pressure (14.73 psia or 101.325 kPa) and 59°F (15°C). Natural gas pipeline capacity is quoted in standard cubic feet per second (sfc/sec).

The cost of moving different forms of energy can be compared by converting to units of cost/million Btu/mile. The energy of natural gas is close to 1,000 Btu/sfc. Crude oil yields around 5.8 million Btu/barrel. A ton of coal has around 22 million Btu. A Btu contains 1055.06 joules or about one kilojoule.

SENDING OIL THROUGH PIPELINES

An oil pipeline is a long tube through which the owner pumps crude oil or refined products such as gasoline or jet-engine fuel. For occasional oil deliveries, constructing a pipeline does not pay because interest and taxes have to be paid even though the pipeline isn't delivering oil. For example, oil from a well that produces around twelve barrels a day would be pumped into a stock tank from which a tanker truck collects the oil and delivers it to a pipeline terminal. On the other hand, a pipeline between the barge terminal at Renton, Washington, and the Seattle Tacoma International Airport replaced a small fleet of tank trucks.

Trans Alaska pipeline ships 1.6 million barrels per day. The 800-mile Trans Alaska pipeline is a 48-in (1.22-m) diameter pipe that delivers to the ice-free port of Valdez up to 1.75 million barrels of oil per day from the Prudhoe Bay oil field on the Arctic Ocean. The alternative of hauling the crude oil by tanker was not practical because only an undependable two months of ice-free period is available at Prudhoe Bay. The oil coming out of the wells is hot, around 180°F (82°C), and it is delivered to the pipeline at 140°F (60°C). The pipe is covered with 3 in (7.6 cm) of insulation to keep the oil from congealing in the event of a shutdown of up to twenty-one days. The winter temperature can fall to −80°F (−62°C). A galvanized steel jacket protects the insulation.

In zones where there is no permafrost, the pipeline is buried. When constrained by soil, the 0.462 to 0.562 in (1.17 to 1.42 cm) thick pipe walls can absorb the stress from thermal expansion and contraction. The three mountain passes crossed by the pipeline are 4,800 ft, 3,500 ft, and 2,800 ft in elevation.

About one-half of the 800-mi (1,287-km) length of this pipeline is supported on frozen underground soil, called permafrost. If the pipe were buried here, the heat in the oil would have eventually thawed the permafrost and overstressed the pipeline in its subsequent sinking. In the adopted design, the insulated pipe is supported on steel H-frames. The pipe is welded into a flexible trapezoidal zig-zag so that it can expand and contract as its temperature varies. Teflon sliding surfaces on the H-frame cross-pieces permit sidewise movement up to 146 in. The vertical supports are steel pipe, anchored into the permafrost. To keep the permafrost from melting, heat pipes transfer heat from the permafrost to radiators in the arctic air above the pipe. Operation of the heat pipe depends on condensed fluid draining to the bottom, forming a one-way heat valve that doesn't transfer heat from the air to the permafrost during warm days.

At more than 800 river and stream crossings the pipe is bridged over the waterway or buried beneath it. Fourteen special bridges were built, including a major bridge over the Yukon River that carries a highway as well as the pipeline. A 360-mile (580-km) road was built from the Yukon River to Prudhoe Bay for hauling construction supplies and for pipeline maintenance.

Ten pumping stations push 1.65 million barrels of oil per day through the pipeline. At two of these stations a drag-reducing agent is injected to increase the flow from 1.5 to 1.7 million barrels per day. At five of the pumping stations refrigeration plants keep the foundation permafrost from melting. At each station two 18,200 hp gas turbines drive two 25,000-gallons-per-minute pumps. A third turbine and pump are on stand-by. Natural gas from Prudhoe Bay powers the four northern pumping stations. Three of the other stations have small refineries that extract turbine fuel from the petroleum being pumped.

Pipeline terminals. The pipeline terminal at Valdez has eighteen covered-roof tanks that can store 180 million barrels of oil. One floating and three fixed berths can handle tankers of 16,000 to 265,000 deadweight tons at loading rates up to 110,000 barrels an hour. Three tanks store oil containing ballast water pumped out of incoming tankers. The 1 percent oil in the ballast water is recovered, and the remaining water is treated and discharged into Prince William

Sending Oil Through Pipelines

Sound. The terminal is powered by three 12.5 MWe steam turbine-generators. Recovered vapor from the storage tanks and fuel oil are burned in the steam boilers. Using the recovered vapor for the boilers reduced fuel oil consumption by 125,000 gallons per month.

The pipeline is controlled from the Valdez terminal through a microwave communication link. All pumping stations are manned. A staff of 1,092 employees operates the pipeline. Of the staff, 43 percent were technicians, 20 percent professional, 24 percent supervision and management, and 13 percent clerical. At remote sites employees work twelve-hour shifts on a week-on, week-off cycle.[4]

Cost of shipping oil in pipelines. The annual operating cost of the Trans Alaska pipeline has declined, as cost-reducing improvements are made, from $230 million in 1980 to $206 million in 1984.[4] During 1983, the pipeline delivered 600 million barrels of oil, so the cost per barrel (Co) was:

$$Co = \frac{\$205 \text{ million}}{600 \text{ million barrels}} = \$0.34/\text{barrel}$$

The cost of building the pipeline was $8 billion. Assuming a 20 percent interest and debt service cost, this interprets into a capital cost (Cc) of:

$$Cc = \frac{\$8,000 \text{ million} \times 0.2}{600 \text{ million barrels}} = \$2.67/\text{barrel}$$

The pipeline is 800 miles long, so the crude oil is transported at a cost (Cm) of:

$$Cm = \frac{(\$0.34 + \$2.67)}{\text{barrel} \times 8 \times 100 \text{ mi}} = \$0.37/\text{barrel-100 miles}$$

The energy content of a barrel of oil is about 5.8 million Btu. The cost of moving a million Btu (Cb) then becomes:

$$Cb = \frac{\$0.37 \times \text{barrel}}{\text{barrel} \times 100 \text{ mi} \times 5.8 \text{ million Btu}} = \$0.064/\text{million Btu-100 miles}$$

Ju'aymah-Yanbu pipeline. During 1984, transit of the straits of Hormuz by tankers became hazardous because of a war between Iraq and Iran. Marine insurers raised their rates to war levels. The normal route for tankers carrying crude oil from Saudi Arabia to Japan and Europe was through these straits. The alternative route was a 750-mile (1,207-km) pipeline that carried crude oil from

---

[4]Alyeska Pipeline Service Company, "Year in Review," March–April, 1984. Anchorage, Alaska.

the oil terminal at Ju'aymah on the Persian Gulf to the port of Yanbu on the Red Sea. Saudi Arabia tripled the shipments through the pipeline from 0.6 to 1.8 million barrels per day during the war emergency, and charged 25 cents/barrel. This charge, on a dollars/barrel/100 mi (Cm) becomes:

$$Cm = \frac{\$0.25 \times 100 \text{ mi}}{\text{barrel} \times 750 \text{ mi} \times 100 \text{ mi}} = \$0.033/\text{bbl-100 mi}$$

The cost per million Btu over 100 miles (Cb) then is:

$$Cb = \frac{\$0.033 \times \text{barrel}}{\text{barrel} \times 100 \text{ mi} \times 5.8 \text{ million Btu}} = \$0.0057/\text{million Btu-100 mi}$$

The 25 cents/barrel charge made by Saudi Arabia may not correspond to the true cost of building and operating the pipeline. This pipeline hauling cost may have been subsidized to avoid the loss of oil sales during the war. However, the cost of building and operating a buried pipeline in the desert is nowhere near the cost of building and operating the Trans Alaska pipeline.

OIL TANKERS FOR EFFICIENT TRANSPORTATION OF ENERGY

An oil tanker is a ship that carries up to 500,000 tons of crude oil in her hull, and is equipped with pumps that discharge the cargo to shore. The largest ships of this type, called "supertankers," haul crude oil from producing regions, such as the Persian Gulf, to consuming regions, such as Europe and Japan.

Tankers are important because they haul oil economically across oceans where pipelines cannot be built. A tanker requires no right-of-way that would have to be maintained and taxed when not needed, so small petroleum consumers are better served by tanker or barge than by pipeline or railroad tank car. In the free markets of Amsterdam and New York City, refined products are stored in tankers awaiting sales. Tankers are sometimes anchored at loading facilities to store petroleum until a transporting tanker arrives.

Energy features of oil tankers include:

- Direct cost of a supertanker can be as low as $0.0037/barrel/100 mi (161 km). This cost is developed in the chapter on hauling freight.
- The value of a supertanker can be less than the value of the product it carries. For example, a $20 million tanker might carry cargo worth $40 million.

Limitations in the use of tankers include:

- The 68-ft depth of the Malacca strait limits supertankers carrying oil from the Persian Gulf to Japan to around 500,000 deadweight tons.

Oil Tankers for Efficient Transportation of Energy

- Tankers sailing through the Suez Canal are limited to around 70,000 deadweight tons. The lowest cost way of delivering oil from the Mideast to European ports is in supertankers that sail around the Cape of Good Hope.
- Access by tankers to good port sites is often limited by local governments because of non-economic considerations.

The cargo carrying capacity of a tanker is generally given in deadweight tons, the ton being 2,000 lb (907.2 kg) in American practice. The specific gravity of crude oil is in the range of 0.7 to 0.9. Assuming Iran crude weighs 7 lb/gallon, the crude oil in a ton (Vt) would be:

$$Vt = \frac{2{,}000 \text{ lb} \times \text{gallon} \times \text{barrel}}{\text{ton} \times 7 \text{ lb} \times 42 \text{ gallons}} = 6.8 \text{ barrels}$$

TANKER COST

In May, 1984, when the Iran–Iraq war increased in intensity and several tankers were damaged by aircraft attacks, Lloyds raised the cost of insurance on tankers going to Kharg Island, Iran, to 7.5 percent of the value of the vessel and 3 percent of the value of the cargo. This cost covered a seven-day voyage. To Saudi Arabia, which was not in the intense war zone, the insurance rate was 0.25 percent of the vessel and 0.05 percent of the cargo. The value of most supertankers for insurance purposes is $20 million, and the value of the cargo is $40 million.[5]

At that time the tanker insurance to the rest of the world was 0.1 percent of the vessel and 0.0275 percent of the value of the cargo, plus 2.5 percent for storms. For a tanker charter costing $2 million the insurance would be $250,000 or one-eighth of the charter cost. To Kharg Island the insurance cost would rise to one-third of the charter cost.

The cost of moving petroleum products by tanker is small compared to the value of the products. For example, at 0.4 cents/barrel/100 mi, hauling a gallon of gasoline 3,000 mi across the Atlantic (Ca) would be:

$$Ca = \frac{3{,}000 \text{ mi} \times 0.4 \text{ cents} \times 1 \text{ barrel}}{100 \text{ mi} \times 42 \text{ gallons}} = 0.28 \text{ cents}$$

As a result, the tankers loaded with gasoline in the free market regulate the price of gasoline in New York harbor. A few cents difference in price can send tankers carrying gasoline across the Atlantic.

[5] *Wall Street Journal* (New York City), May 31, 1984, Sec. 1, p. 12.

NATURAL GAS PIPELINES AND SHIPS

Natural gas is important because it supplies 25 percent of the energy consumed in the U.S. and half of the energy used in heating buildings. A pipeline is a practical way of transporting gas from a gas field to places where it is used, provided a land route is available. Underwater pipelines are growing in length. North Sea natural gas is piped to England and Norway, and pipelines under the Mediterranean Sea are being considered for replacing natural gas tankers that carry gas from Algiers to Europe.

Pipes that carry natural gas. A thousand miles can separate a gas field from the city where the gas is used. In the U.S., pipeline companies are the agencies that buy gas from producers, pump it to using areas, and sell it to distributors.

As the gas resources in the contiguous U.S. diminish, gas imported from Canada and Mexico is filling the gap. Canada's 1984 contribution to the U.S. gas supply was less than 5 percent of the 17 trillion cubic ft consumed. Imports from Mexico were 300 million cubic ft/day, or 0.1 trillion cubic ft/year.

Gas could also be imported from Prudhoe Bay, Alaska, through a 2,200-mile pipeline, mostly through Canada. By the year 2000, the Alaska pipeline could contribute 1.4 trillion cubic ft/year to the nation's gas supply, according to a prediction by the American Gas Association. Alternative routes for this pipeline have been evaluated, but the diameter and pressure have not been resolved.

The Canadian interests favored a 56-inch diameter pipeline operating at 1,080 pounds per square inch (psig), and carrying 2.4 billion cubic ft per day. The cost estimate in 1978 was $4.4 billion. The United States Federal Energy Regulatory Commission preferred a 48-inch pipeline operating at 1,680 psig, which could deliver 3.6 billion cubic ft/day, and would have saved a 350 trillion Btu loss in the energy required to run the pumps during the pipeline's thirty-year life. The gross benefit would have been around $1 billion.

Gas transmission through pipelines is described in Ref. 6.

Natural gas across oceans by ship. Natural gas, being some 80 percent methane, can be liquified at $-263°F$ ($-164°C$) and shipped in insulated tanks across oceans. Shipments to Europe from Algeria, and to Japan from Alaska and Indonesia, supply base load with continuous movement of ships. Natural gas tankers also deliver gas which supplements local gas supplies for cold-weather peak loads.

[6]Douglas M. Considine, *Energy Technology Handbook*, (McGraw Hill Book Co.: New York), 1977, pp. 2–60 to 2–74.

Natural Gas Pipelines and Ships

Liquid natural gas imports to the U.S. will not be significant as long as gas can be imported by pipeline from Canada and Mexico. For example, the American Gas Association predicts that the U.S. natural gas consumption will grow from 1982's 17 trillion to 25 trillion cubic ft/year by the year 2000. Of the 25 trillion, only 0.2 trillion cubic ft would be imported as liquid natural gas.

El Paso's Algerian project. El Paso Natural Gas Company's Transatlantic liquid-natural-gas route was financially feasible when few new gas wells were being drilled in the U.S. to avoid price ceilings. The spot price of natural gas climbed to around $9/1,000 cubic ft. The firm negotiated a long-term contract with an Algerian agency, established sales contracts with U.S. distributors, obtained federal licenses, built ships and terminals, and shipped base load gas across the Atlantic. At that time Canada exported natural gas to the U.S. for $6/1,000 cubic ft. By 1984, the Canadian gas had dropped to $4.40, and the Algerian agency wanted $8.50. El Paso sold its tanker fleet and abandoned the operation.

11

Moving People with Less Energy

The moving of people and hauling of cargo are important because these activities account for more than 26 percent of the energy consumed in the U.S. More importantly, virtually all transportation is fueled with petroleum, an expendable resource having its greatest reserves in the Mideast. Around 6 million barrels of petroleum is consumed by the U.S. each day for highway transportation, and around 5 to 8 million barrels of petroleum is imported into the country every day. In the past, a steep change in the price of petroleum has produced a crisis with long-lasting effects. For example, the 1978 rise in the price of petroleum caused buyers to suddenly switch to smaller and more efficient imported cars, almost bankrupting the Chrysler Corporation.

Energy features of moving people include:

- The inefficient automobile is the key to moving people in America. A change from automobiles to alternatives will be costly and slow to implement.
- Fuel cost is not as important as crew and passenger time in moving groups of people.
- Nothing can match the bicycle from the standpoint of low-energy and low-cost means of moving people.

A measure of the efficiency of moving people is passenger miles per gallon. Some of the limits in efficient methods of moving people include:

- A vehicle with a $\frac{1}{4}$ horsepower diesel engine achieves 2,400 miles per gallon, carrying a lightweight person.

Moving People with Less Energy

- A bicycle can carry the rider over 1,000 miles for the food energy equivalent of one gallon of gasoline.
- Fully loaded trains are delivering 600 passenger miles per gallon.
- The best that a fully loaded ship can deliver has been around 135 passenger miles per gallon.
- A fully loaded 757 airplane can deliver 85 passenger miles per gallon on an optimum route.
- No fuel is consumed by sailplanes, sailing ships, or solar powered airplanes and cars.
- A 747 airplane flew 10,259 miles non-stop.

Fuel consumption of engines and earth-surface vehicles is commonly measured in gallons (3.785 liters). Fuel for jet airplanes is quoted in pounds when being loaded. A gallon jet fuel weighs 6.7 lb. Airplane fuel consumption is quoted in seat miles per gallon, which is passenger miles per gallon if the airplane is fully loaded. The propulsion power of electric trains is measured in kilowatts input to the locomotive. Diesel locomotives are rated in engine horsepower and drawbar pull in tons or pounds. Fuel consumption in marine diesels is quoted in gallons or grams per horsepower hour or liters per kWh.

Take the train, save fuel. A passenger train is a string of cars, rolling on steel rails, with propulsion power generated by an on-board engine or collected from a trolley or third rail. A steel wheel riding on a steel rail has a coefficient of friction of only around 0.3 percent. This makes the train the most efficient practical vehicle for moving people. As a result, costs other than fuel predominate in passenger train operations.

Passenger trains and buses can be so fuel efficient that they compete with each other on considerations other than fuel consumption. For example, in countries like Germany, France, and Japan an in-place well-maintained network of high-speed train routes provides service that cannot be matched by buses or even airplanes for travel between nearby cities. In contrast, in the U.S. lack of passenger-train density has caused railroads to reduce their track maintenance to the point where passenger trains can't maintain the 120 miles per hour speeds that they achieved in the 1930 to 1950 period.

Crew cost and fuel. The cost of passenger and crew time, rather than cost of fuel, are the important elements when comparing alternatives to airplane transportation for distance travel. For example, during a world fair in New Orleans, AMTRAK scheduled a twenty-six-hour train ride between New York and New Orleans. The train featured slumber coaches, Pullman cars, and a diner. Assuming that the train crew consisted of an engineman, fireman, conductor, two brakemen, two kitchen workers, and a porter for every two cars, then total crew time (Tc) for the eleven-man crew during a one-way trip would be:

$$Tc = 11 \text{ persons} \times 26 \text{ h} = 285 \text{ crew h}$$

This time, divided among say 150 passengers, amounts to 1.9 hours per passenger, for which the passenger must pay salary, overhead, and profit. One-half of a round trip fare for the trip was $142. Assuming $35 per hour for labor, overhead, and profit, the crew cost per passenger would be around $67 for the trip. In contrast, a 757 airplane could carry the 150 passengers to New Orleans in three hours, and the five-person crew cost (Cc) would correspond to:

$$Cc = \frac{5 \text{ persons} \times 3 \text{ h} \times \$50}{150 \text{ passengers} \times \text{h}} = \$5.00$$

If the train gets 400 passenger miles per gallon of locomotive fuel, the fuel cost to each passenger would be around $2.50 if diesel fuel cost $1 per gallon. The 757 airplanes bought by Florida Airlines deliver around 85 seat miles per gallon. Thus in a New York to New Orleans ticket, the passenger of a fully loaded airplane pays $12 for his share of jet fuel, assuming that the jet fuel also costs $1 per gallon.

Other costs, such as owning and maintaining a right-of-way, can contribute to the train passenger's fare, particularly if traffic density is low. The airplane needs no dedicated right-of-way.

Energy features of passenger trains include:

- A diesel powered coach train can easily deliver over 400 passenger miles/gallon.
- Operation of a commuter train, where peak travel occurs twice a day, is not compatible with efficient use of crews who work in eight-hour shifts.
- High train speeds are best achieved on a roadbed and right-of-way that are specifically designed for high speed travel.
- Around 170 mi/h (274 km/h) seems to be the limit for conventional trains. At higher speed, the current collection from the trolley and wheel adhesion become problems.

Propulsion power for trains. The force needed to pull a train on a straight and level track (Rl) at 10 mi/h is only 5 lb/ton for 35-ton cars. An equation developed by Davis for passenger trains[1] is:

---

[1] W. J. Davis, "The Tractive Resistance of Electric Locomotives and Cars," *General Electric Review,* 29 (October, 1926).

Moving People with Less Energy

$$R1 = 1.3 + \frac{29}{W} + 0.03\,V + \frac{0.00034\,A\,V^2}{Wn}$$

where $R1$ = drawbar pull required to keep train moving, pounds/ton
W = weight per axle, tons
V = train velocity in miles per hour
n = number of axles
A = cross section area of the lead car or locomotive, square ft

The 0.0034 constant for trailing cars includes the effect of an aerodynamic drag coefficient of 0.938. Streamlining the train, which reduces the drag coefficient to 0.3, is important at high speeds. For example, at 160 mi/h, a streamlined train with a 0.315 drag coefficient requires 2,371 lb of force to overcome air resistance. Note that the locomotive of a multicar train, which overcomes the full value of air resistance, has a 0.0024 coefficient.

For climbing hills, an additional 20 lb of drawbar force per ton of train is required for each 1 percent of grade. An additional force of 1 lb/ton/degree of curvature is required when cars go around corners because the wheels and axle are one piece, and the outer wheel must travel a greater distance than the inner wheel. Also, there are friction losses when the wheel flanges strike the rails. Energy is also consumed in accelerating the train when starting or increasing speed. This accelerating energy as well as energy released in downhill braking can be recovered in electric locomotives having regenerative braking.

The power required to pull a train can be computed from the velocity and train resistance, 1 horsepower being 33,000 ft lb/minute. For example, a 35-ton car pulled at 80 mph has a train resistance of 15 lb/ton. An early Union Pacific streamlined train incorporating lightweight structure weighed 85 tons. The train resistance would have been:

$$R1 = 85 \text{ tons} \times 15 \text{ lb/ton} = 1{,}275 \text{ lb}$$

The traction power expended is:

$$P = \frac{1{,}275 \text{ lb} \times 80 \text{ mi} \times h \times 5{,}280 \text{ ft} \times \text{horsepower}}{h \times 60 \text{ minutes} \times \text{mi} \times 33{,}000 \text{ ft-lb}}$$
$$= 272 \text{ horsepower}$$

The power car of the train had a 600 hp diesel engine, and the train could travel 110 miles per hour (mph). We can assume that the train requires full power when traveling for an hour at 110 mph. The fuel consumption of a modern Caterpillar diesel engine of about that size is 0.39 lb or 0.055 gallons/horsepower hour. The train, if equipped with this engine, would consume fuel at a rate (Fc):

$$Fc = \frac{600 \text{ hp} \times 0.055 \text{ gallons}}{\text{hp-h}} = 33 \text{ gallons/hour}$$

If the train carried 150 passengers, its fuel performance (Fr) would have been:

$$Fr = \frac{110 \text{ mi} \times \text{h} \times 150 \text{ passengers}}{\text{h} \times 33 \text{ gallons}}$$
$$= 500 \text{ passenger mi/gallon of diesel fuel}$$

Virtually all of the famous-name passenger trains in the U.S. were abandoned when passengers chose to fly in airplanes. Fuel efficiency was not a consideration at the time in the 1950s. Frequent and on-time train service is available in Europe and Japan where taxes drive gasoline and diesel fuel prices to twice what they are in America. If taxes convert Americans to train riders, then for intercity travel, the work done by the French railroads might be pertinent. The French engineers learned that high-speed trains require dedicated tracks, banked on curves for high speed.[2] The Paris-to-Lyons train travels on such tracks at 260 km/h (161 mi/h).

For commuter railroads, which haul workers in and out of central cores of cities, the economic key is subsidy. Randall Pozdena observed in 1977[3] that most rapid transit fares were within a few cents of 19 cents/passenger mile, providing little incentive for use of vehicles that do not provide door-to-door service. He observed that the San Franciso Bay Rapid Transit, built in 1960 for $2 billion, provides high-speed station-to-station transportation along 75 miles of track. However, the subsidy amounted to $4 per passenger.

PASSENGER SHIPS

The moving of people across oceans in passenger-carrying ocean liners reached its climax in the 1930s and 1940s when ships up to 1,020 ft (311 m) long were built for transatlantic service. These liners could not compete with airplanes, so most of them have been converted into cruise ships. Other passenger carrying ships are ferries and hydrofoils. Ferries, cruise ships, and ocean liners are displacement vessels that float because they displace a mass of water equal to their weight. In hydrofoils and planing hulls, the dynamic forces on the hull surfaces support the vessel as it moves at rated speed.

Energy features of passenger ships include:

- Passenger liners were designed for speed, and propulsion energy was not a significant element in the cost of operation.

[2] Jean Dupuy, "France's Superfast Train," *IEEE Spectrum*, July 1982, pp. 38–43.
[3] Randall Pozdena, "Making Rapid Transit Work," *Research Department of the Federal Reserve Bank of San Francisco*, December 7, 1979.

Passenger Ships

- Passenger ferries carry people with a fuel consumption of around 180 passenger miles per gallon of diesel fuel when fully loaded.
- The slowness of ships makes the cost of the time of crew and passengers the most important element in comparisons with airplane travel.

Energy-pertinent limits of passenger ships include:

- Propulsion power grows drastically as the characteristic hull speed of a displacement vessel is approached. The characteristic speed is proportional to the square root of the ship's length.
- At full speed the 2,000-passenger *Queen Mary* burned 6 barrels of Bunker C oil per mile. However, in the 1930s the fuel cost only around $2 a barrel. When fitted as a troop ship she carried 15,000 passengers, making her fuel consumption 59 passenger miles per gallon.
- The river steamer, *Hendrick Hudson,* built in 1906, was 380 ft long by 45 ft wide and operated with an 8-ft draft. She carried 5,000 passengers at 22 knots.

Passenger ships are generally described in length in feet and displacement in salt water, which weighs 64 lb/cubic ft. The contribution of passengers to the total displacement of the vessel is small. For example, the *Queen Mary* displaced 66,851 tons and carried 2,000 passengers. This corresponds to 33.4 tons per passenger. The total load of passengers weighing 160 lb each would contribute 0.2 percent to the displaced weight.

The speed of ships is generally measured in knots, one knot being one nautical mi/h, 1.15 statute mi/h, or 1.852 km/h. A nautical mile is 6,076 ft (1,852 m) or one minute of arc around the Earth.

Speed of passenger ships. Most of the power consumed in moving a ship goes into scrubbing the wetted surfaces of the ship with water and making waves. Thomas C. Gillmer relates these power losses into terms of pounds of resistance per ton of displacement as a function of speed-length ratio.[4] This relationship is plotted in Figure 11–1, where the numbers in the horizontal axis are

V = velocity of the ship, measured in knots (1 knot = 1.853 km/h)
L = length of the hull in feet

The speed–length ratio is important in ships because a value of 1.3, called "characteristic hull speed," is the practical limit for most displacement vessels.

---

[4]Thomas C. Gillmer, "Modern Ship Design" (Annapolis, Maryland, United States Naval Institute, 1970), p. 172.

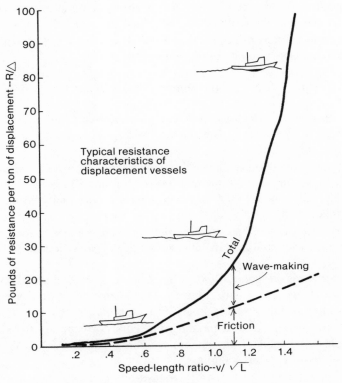

Figure 11-1

The friction of water scrubbing on the hull depends on the wetted area, the roughness of the surface, and whether the flow of water is laminar or turbulent. In passenger liners, this flow is generally turbulent. Mechanisms such as flexible hull surfaces, can promote laminar flow, but are not being used in passenger ships.

Perhaps the peak of the transatlantic liner technology was reached with the construction of the 66,348 ton ship *France,* which was 1,035 ft (315.5 m) in overall length. Waterline dimensions were 978.6 ft (298 m) long by 110.7 ft (33.7 m) beam by 55.7 ft (17 m) draft. She was launched in 1960, and carried 2,044 passengers and a crew of 1,044. Eight oilfired water tube boilers produced 910 psig (6.27 MPa) steam for the turbines. With 160,000 hp delivered to two propellers she achieved a speed of 34.14 knots. She was equipped with ship stabilizers for countering rolling in rough seas.

We can assume that 910 psig turbines, at that time, had a heat rate of 14,350 Btu/kWh. Then the fuel consumption of *France* would have been 6.8 barrels/mile, or 7.2 passenger mi/gallon. At the service speed, this improves to 8.7 passenger mi/gallon.

The passenger liner, *United States,* when built for a 1952 maiden voyage, cost about $75 million. A 747 airplane could be purchased in 1985 for about the same price. In one year, the *United States* would generate revenue from 487 million passenger miles of travel. The 747 would generate over a billion passenger miles.

FERRIES FOR LESS LUXURIOUS TRAVEL

A ferry is a ship that carries passengers, and often vehicles, across waterways. In Hong Kong, a ferry crosses the waterway between Hong Kong Island and Victoria in about fifteen minutes. A Puget Sound ferry takes all day to sail from Anacortes, Washington to Sidney in British Columbia. The Finnjet makes the Helsinki-Travemunde trip in twenty-two hours. Ferries crossing rivers have been replaced by bridges. Most of the remaining ferry routes are too long for bridges, and too short to attract airplane competition.

Ferries are important because they can provide low-cost fuel-efficient transportation in places where alternate transportation is not possible. A ferry, unlike a passenger liner, is configured to carry many people. Should the cost of fuel double from its around $1/gallon, the ferry fleets could become suppliers of recreational travel for residents and visitors in cities on waterways. A constraint is that ferries are designed to operate efficiently from terminals that have buffers and other structures for quick docking. Many ferries can run equally well in either direction and can load and unload from either end.

Most ferries carry commuters, so reliable operation is essential. Reliability is usually achieved with integrated redundancy in the ferry, rather than with a stand-by vessel.

Energy features of ferries include:

- Multiple diesel engines are commonly used for propulsion, rather than steam turbines. The *Finnjet,* an exception, uses gas turbines.
- Diesel electric drive, although more costly than clutched or direct drive diesel engines, facilitates desirable maneuverability and permits quick automatic response to equipment failures.
- Maintainability is as important as fuel consumption in ferry operation.
- Commuter ferries share with other commuter vehicles the twice-daily peak passenger loads, and light loads at other times.

Ferries crossing long distances over water become like ocean-going ships. For example, the Alaska ferries that cruise the inside passage between Seattle and

Alaska ports, like other ferries, carry people, cargo, and automobiles. They also have staterooms and dining rooms.

The limits applicable to ferries are:

- The gas-turbine powered *Finnjet* cruises at 30.5 knots with both turbines running. Its "get home" speed with one turbine is 24 knots. These speeds are not appropriate for ferries operating in crowded waters like the Puget Sound.
- The Staten Island ferry, *Barberi,* can carry 6,000 people. The Washington State Ferry, *MV Yakima,* carries 2,500 people and 160 cars.
- Diesel engine efficiency is around 30 percent for the types of engines used in ferries. A 10 percent reduction in fuel consumption is available for existing vessels with a new low-friction hull coating.

Carrying capacity is often designated as number of passengers and number of automobiles, rather than deadweight tons. Fuel consumption, measured in gallons per hour, varies with speed, load, and engine configuration.

FERRY POWER PLANTS: MOSTLY DIESEL ENGINES

Large marine diesel engines turning at 100 to 200 rpm are normally coupled directly to the propeller. The rotation of the propeller is reversed by stopping the engine and starting it in the reverse direction, or by using a propeller in which the pitch of the blades can be reversed. High-speed diesel engines rotating at 600 to 1,800 rpm, derived from railroad and truck engine design, occupy less space, weigh less, and cost less than the low-speed engines. They are not usually designed to be reversible, so if used without an electric drive these high-speed engines require reduction and reversing gears, and appropriate clutches.

A disadvantage of the clutched drive is that available engine power varies with speed, so only at the highest propeller rpm is full engine power available. A clutched drive could deliver full engine power at all ship speeds if the propeller pitch is variable. In the *Ward Leonard* electric drive (Figure 11–2) full engine power is available at all times, except as limited by the rating of the propulsion motor when it is running at low speed.

For ferry propulsion the *Ward Leonard* electric drive is the standard against which alternatives are compared. Disadvantages of the electric drive are the cost of electrical machinery and the 10 to 15 percent of the diesel engine output that ends in generator and motor losses. Some advantages are:

- Engines can run at full speed when full power is needed. At other loads the engines can be slowed to improve efficiency and lengthen life.
- Full engine power can be delivered to the propellers at any time to accelerate, decelerate, or reverse the ship.

Ferry Power Plants: Mostly Diesel Engines

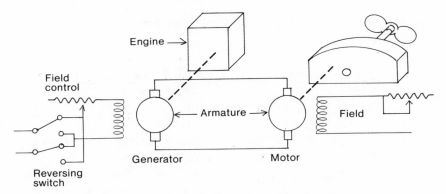

Figure 11-2

- Multiple engines can be arranged so that power will not be lost even if one engine or generator fails.
- In ferries that operate in two directions with the both bow and stern propellers, the propulsion engines and motors can be arranged so that no single failure can leave the vessel without power. This is especially important for ferries that use power for decelerating in approaching a dock.
- Where multiple engines are installed, an engine generator set can be electrically disconnected from service and overhauled while the ferry remains in operation.

Fuel consumed by ferries. At her 18 knot cruising speed, the 450-ft long Washington State ferry *Walla Walla* consumes around 230 gallons of diesel fuel per hour. If fully loaded she carries 2,000 people and 210 cars.[5] The combined rating of her four 6-cylinder General Motors 6-645 engines is 11,700 horsepower. The electric drive, which is 85 percent efficient, delivers 10,000 hp to her two propeller shafts.

The ship has two propellers, one at each end, so the ferry can operate in either direction. During cruise, the propulsion power is divided, 10 percent going to the forward propeller and 90 percent to the stern propeller. The propeller shafts are not coupled. Each shaft is driven at around 140 rpm during cruise.

Each engine drives an alternating current (ac) generator at 900 rpm. The ac power is converted to direct current (dc) with silicon controlled rectifiers. Generating ac avoids the commutator troubles and carbon dust from brushes, which had previously been experienced with big generators operating at the 900 rpm engine speed.

[5]Mary Stiles Kline and George A. Bayless, "Ferryboats: Legend on Puget Sound, Seattle," (Seattle, Wash.: Bayless Books, 1983.)

With a full load of passengers, the fuel performance (Pmg) of the *Walla Walla* would be:

$$Pmg = \frac{18 \text{ knots} \times 1.15 \text{ mi} \times 2000 \text{ passengers} \times h}{h \text{ knot} \times 230 \text{ gallons}}$$
$$= 180 \text{ passenger mi/gallon}$$

This fuel performance is around twice that of the best commercial airplanes, but far from the 500 passenger mi/gallon available with trains. Of course, the train can't cross water without bridges. Also, the ferry doesn't spend its working day cruising at 18 knots. Much of the time is spent in accelerating and decelerating and standing at the dock loading and unloading.

At 180 passenger mi/gallon, a passenger on a twenty-mile trip need pay only around $0.10 for his 0.1 gallon share of the fuel, if the ferry is carrying 2,000 people. However, the ferry may carry only a few hundred people during times other than rush hour, so the passenger's share of fuel might be $1 to $2.

Assuming that the *Walla Walla* has a crew of twelve who are paid $35 an hour in wages and fringe benefits, the crew cost would be $420/hour, compared with some $230/hour for fuel. One could ask, "Can we automate the ferry and reduce crew cost?"

Clark Dodge, Chief Engineer of the *Walla Walla,* explains that this has been done on the Issaquah class of Washington State Ferries, where the diesel engines are clutch-coupled to the propeller shaft, and speed is varied by computer-controlled variable-pitch propellers. The propulsion power plant is operated by remote control from the bridge. Dodge observes that minimum operating cost is the important criterion, and a skilled and dedicated engine room crew can do things that save money. For example, they can instantly recognize from engine sounds that something is wrong, and take the engine out of service before serious damage occurs. They can service on-schedule all filters, packings, injectors, and pumps so that 30,000 hours of operation between overhauls is being achieved. They can overhaul one of the engines while the ferry is operating.

Captain Ronald D. Hamrick, of the Alaska State Ferries, observed that contracted maintenance by shore service will appear to cost less than a full-time engine room crew during negotiations for such services. However, two years later when these services are needed quickly, they may not be available in time to prevent a delay in the scheduled departure of the ship. This results in costly inconvenience to travelers en route, plus cancellation of runs to catch up on schedule. Runs involve hundreds of miles of route with ferry terminals on islands that have no other transportation. His ferry, the *M/V Le Conte,* is the only one on the route, and there are no other ferries standing by to take its place if it goes out of service.

Ferry Power Plants: Mostly Diesel Engines

Table 11-1 Examples of recent ferries

Ferry	Capacity Passengers	Vehicles	Route	Length — Breadth — Draft	Notes
Baby Barberi	1280	0	Staten Island to New York	207 ft. 40 ft. 8 ft.	15 knots, 650 tons
Barberi	6000		Staten Island to New York		
Caribou			North Sidney, Nova Scotia to Basques, Newfoundland (200 miles)	165 m 25 m 8.4 m	28,000 hp, diesel electric drive, 22 knots
Klondike	250	8		72 ft. 29 ft. 6 ft.	26 knots, 1400 hp disel power, catamaran
Cheung Kong	700		Hong Kong to Macao	62.5 m 10.2 m	27.5 knots, 500 tons, crew of 200

Ferries for the future. For carrying people between well-designed terminals on waterways, a modern ferry is hard to beat for low cost and fuel efficiency. On long trips, like Seattle to Haines, Alaska, it is hard-pressed to compete with airplanes. In Europe, where the ferries haul people rather than cars, surface-piercing hydrofoils have been successful. In the Hong Kong to Macao route, boats with surface piercing hydrofoils compete with submerged-hydrofoil boats. This route does not require transporting automobiles since relatively few residents in either city have cars.

Meredith McRoberts notes that 80 percent of Americans live within twenty miles of a major body of water. He says that in more and more travel routes, a fast ferry can deliver people quicker than can cars and buses rolling on congested freeways.[6] An example that he cites is a $\frac{5}{8}$ mile, 3.5 minute ride on a 12-knot ferry between Weehawken, New Jersey and 38th Street in Manhattan. New ferries going into service described by McRoberts are summarized in Table 11-1.

[6]Meredith McRoberts, "Ferries on Second Leg of Round Trip," *Marine Engineering Log*, September, 1984, pp. 76-85.

HYDROFOILS FOR PASSENGERS

A passenger carrying hydrofoil is a cabin supported above water during travel by underwater wing-like hydrodynamic surfaces that generate an upward lift as they slide through water. While at dock or while maneuvering in a harbor, a hydrofoil descends and becomes a displacement ship. The hydrofoil's cruise speed is not limited by the power consumed in making waves, as is the cruise speed of displacement ships.

Energy features of hydrofoils include:

- A typical hydrofoil can cruise at 43 knots, roughly twice the speed of the fastest displacement ferries.
- Fuel consumption, measured in passenger mi/gallon, is roughly the same as for passenger ships.
- Hydrofoils cannot compete with airplanes on long trips, nor with buses or cars when paralleling land routes.

The limits of hydrofoils include:

- Above 45 knots cavitation begins to shorten the life of the foil surfaces.
- Thomas C. Gillmer estimates that 1,000 tons is the upper limit of hydrofoils, based on available lightweight power plants.[7]

In hydrofoil and other ship operations, the term *seas* means the height of wave crests above wave troughs.

Hydrofoil characteristics. After a hydrofoil vessel leaves a dock and enters clear water, it accelerates to become foil-borne. The foils are either surface-piercing or submerged (Figure 11–3). The hull that is supported on surface-piercing "V" foils automatically acquires an above-water elevation where the lift from the submerged foil equals exactly the weight of the vessel. However, when traveling through waves the hull rises and falls, making the passengers uncomfortable. Trips are cancelled when the seas become too high. Surface-piercing hydrofoils work well in protected waters and where trips can be cancelled without significant penalty.

The vessel with fully submerged hydrofoils is not affected by ordinary waves. However, it requires an active control that continuously measures the altitude of the hull above the average water surface, and adjusts foil lift as necessary to maintain this altitude constant. Such hydrofoil vessels cost more to

---

[7]Thomas C. Gillmer, "Modern Ship Design," *U.S. Naval Institute* (1970) p. 290.

Hydrofoils for Passengers

Figure 11-3

buy, but are more productive because they can operate in heavier seas than can surface-piercing vessels.

A hydrofoil vessel does not differ in power consumption from other displacement vessels until it reaches foilborne speed. Then its power consumption in engine horsepower (EHP) per ton drops, as shown in Figure 11–4, which was developed by Prof. Thomas C. Gillmer of the U.S. Naval Academy.[8] The foilborne hull is then limited in speed by cavitation problems rather than by wave-making considerations.

For energy performance comparisons, we can consider the Boeing jetfoil, a 250-passenger, two-deck submerged-foil vessel that is propelled by a water jet. Water-jet propulsion avoids the propeller cavitation that would otherwise be a problem at 43 knots foilborne speed. The water enters the foil strut at ram pressure, avoiding cavitation of the pump impellers. The pumps are driven by two 3,500 horsepower marine gas turbines, which are derivatives of aircraft jet engines. The gas turbine engines were selected over alternatives because of their light weight. With given foils a lighter power plant permits more payload. The hull is aluminum for the same reason.

The jetfoil is 90 ft long and has a 31 ft beam. When cruising at full power, 43 knots, the two engines consume 8 gallons of jet fuel per minute. This fuel consumption (Fc) corresponds to:

$$Fc = \frac{250 \text{ passengers} \times 43 \text{ knots} \times 1.15 \text{ mi} \times \text{minute} \times \text{h}}{\text{hour-knot} \times 8 \text{ gallons} \times 60 \text{ minutes}}$$
$$= 25.7 \text{ passenger mi/gallon}$$

This compares with over 80 passenger miles per gallon for a fully loaded 757 airplane. In addition to propulsion fuel, the vessel requires 10 gallons/h of fuel for the diesel engine that powers the auxiliaries. When maneuvering in a harbor the fuel rate is 40 percent of the full-power rate.

[8]Thomas C. Gillmer, "Modern Ship Design," U.S. Naval Institute (1970), p. 200.

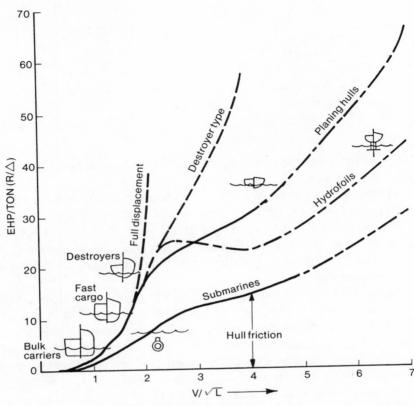

Figure 11-4

The Boeing jetfoil has been successfully used on the Hong Kong to Macao route and between islands in Japan. The Alaska Department of Transportation tested the jetfoil in Southeast Alaska where the runs between ports are long and service is also provided by conventional ferries. Here the speed, twice that of conventional ferries, is important. Also, the jetfoil, because of its 5.5 ft (1.67 m) draft when the foils are retracted, could dock near the downtown zones of Alaskan cities, whereas conventional ferries dock at terminals in the outskirts of cities. Captain Hamrick observed that service for gas turbines would not be available in the small Alaskan towns served by the ferries, and in this respect the jet foils would be in a disadvantageous environment when compared with Hong Kong, Hawaii, and Japan.

A fleet of three jetfoils for Alaska would cost $18 to $20 million each. In the same year American Airlines bought a fleet of McDonnell Douglas DC-9 airplanes for $18 million each. Boeing subsequently stopped marketing commercial hydrofoils.

BUSES CARRY PEOPLE EFFICIENTLY WHEREVER ROADS GO

A bus is a rubber-tired, wheeled vehicle that travels on streets and roads, hauling people to single or multiple destinations. In American cities, buses have replaced street cars that hauled passengers along routes that had tracks. Buses have replaced trains on suburban and rural routes that generate only a few busloads of traffic each day. Development of the long-life truck engine has made the passenger bus so reliable that with good maintenance, on-route failures are rare enough to be of little concern to the operator.

Buses are important because they can haul people more efficiently than cars can, and in contrast to railroads, don't need an exclusive right-of-way. In a crowded downtown area a bus delivers the passengers and departs. The same passengers traveling by car would need parking space. A bus traveling on public roads is flexible with respect to routing. If one route doesn't generate worthwhile traffic, the route can be changed or the buses can be used elsewhere.

Energy features of buses include:

- The fuel consumption of a bus varies. The number of stops per mile, the amount of hill climbing, and the route speed all affect fuel consumption. A firm that carries passengers along a freeway from Bellevue, Washington, to the Seattle Tacoma airport gets 6 to 8 miles per gallon with a 47 passenger bus.
- Costs other than fuel dominate in the operation of a public-transportation network that provides regular passenger service.
- Most urban and suburban public transportation in the U.S. is subsidized.

Buses are rated in passenger-carrying capacity. Long distance buses have seats for all passengers. Metropolitan commuting buses have seats and provisions for stand-up passengers. Fuel consumption is measured in America in mi/gallon. Elsewhere it is measured in km/liter. One mi/gallon is 0.425 km/liter.

Energy performance of buses. If a forty-seven-passenger bus gets 7 mi/gallon, then its performance (Pb) is:

Pb = 47 passengers $\times$ 7 mi/gallon = 329 passenger mi/gallon

A Japanese study quotes the actual 1977 energy consumption of public buses as 142 kilocalories/passenger-kilometer. This can be interpreted into our terms as follows:

$$Pb = \frac{\text{passenger-km} \times \text{mi} \times \text{kilocalorie} \times 137{,}750 \text{ Btu}}{142 \text{ kilocalories} \times 1.609 \text{ km} \times 3.968 \text{ Btu} \times \text{gallon}}$$

$$= 152 \text{ passenger mi/gallon}$$

The 137,750 Btu/gallon is the energy content of typical diesel fuel.

Using the Japanese fuel consumption and a fuel cost of $1.00/gallon, the fuel cost to the a passenger (Cf) for a ten-mile trip is:

$$Cf = \frac{\$1.00 \times 10 \text{ mi} \times \text{gallon}}{\text{gallon} \times 152 \text{ passenger miles}} = \$0.065/\text{passenger}$$

Thus the fuel cost for a fully loaded bus is not a significant element in the price of the passenger's ticket. More important are the labor costs of driving partly filled buses at times other than rush hour. Maintaining an eight-hour working day for bus drivers requires two crews in a commuter bus operation, one for the morning traffic peak and another for the evening peak.

A seventy-passenger, articulated bus costs around $200,000. At 20 percent interest finance cost, plus depreciation in seven years, the daily cost of capital (Cc) would be:

$$Cc = \frac{(\$200{,}000 \times 0.2) + (\$200{,}000/7 \text{ years})}{365 \text{ days/year}} = \$188/\text{day}$$

At 25 mi/h, 12 h/day, the bus covers 300 mi/day. This is 62 cents/mi. This daily cost persists whether or not the bus is carrying passengers.

Randall Pozdena observed, in 1977, that only 8 percent of transit service was provided by privately owned vehicles.[9] The rest was provided by public agencies whose problem was how to pay people to ride the transit. The San Francisco Bay Area Rapid Transit subsidizes its passengers at $4/trip, according to Pozdena. He proposes alternative ways for passengers to subsidize the public transportation in a manner that encourages good service.

Pozdena suggested revival of the jitney, a small privately-owned passenger vehicle that follows somewhat less regular routes, but unlike taxis, picks up passengers until it is filled. At one time, 62,000 jitneys operated in the U.S., but pressure from street car operators eventually forced them out of operation. Pozdena observed that while providing needed transportation, jitneys also offer employment for local low-skill workers who often hold midday jobs in addition to driving jitneys during commuting hours.

[9]Randall Pozdena, "Making Transit Work," *Publication of the Research Department,* Federal Reserve Bank of San Francisco, December 7, 1979.

Trolley buses that don't pollute. A bus-like vehicle that is driven by an electric motor powered from a pair of overhead copper wires is called a "trolley bus." Two trolley wires are required for completing the electric circuit, so the vehicle has two trolley poles. Street cars and electric trains need only one trolley because the return path for the electric current is through the steel rails. The common form of trolley power is 600 volts dc, converted with rectifiers from ac supplied by the local public utility. The propulsion motor is a series wound type that has high starting torque, giving the bus fast acceleration. Wear of friction brake pads is reduced by dynamic braking, an arrangement in which the motor acts as a generator, absorbing the kinetic energy of the vehicle for dissipation in resistors.

The trolley bus is more efficient than a diesel bus because its motor has no idling losses and has good part-load efficiency. The power for the trolley bus can come from a coal burning or nuclear steam plant, or from hydro plants, rather than from diesel fuel. The trolley bus power can be generated far from the downtown area, whereas the diesel engine discharges its exhaust wherever it is.

Trolley buses have operating cost problems. A new trolley bus costs around $300,000. The network of trolley wires must be maintained by costly personnel, and trolley wires must be replaced when they wear out.

A dual mode bus can operate with electric power from trolleys and then continue beyond the end of the line with diesel power. That costs around $350,000.

Buses versus airplanes for distance travel. A bus cannot compete with airplanes for long distance passenger service because of costs other than fuel, such as the wages of the driver and the value of the passengers' time. Buses had been able to compete with long-distance trains with fares that were lower than train fares.

AIRPLANES THAT FLY PASSENGERS AND FREIGHT

An airplane is a streamlined vehicle that is supported on aerodynamically shaped wings as it is pushed through the air by fuel-burning engines. An airplane can carry up to 600 people as well as cargo. Propulsion is achieved with either reciprocating engines that turn propellers, or gas turbines that drive propellers or create thrust with jet action.

Airplanes are important because they have made passenger trains for intercity travel obsolete in the U.S. as well as passenger ship travel between continents. They carry an ever-expanding portion of high-value freight. Airplanes are able to compete with buses, trains, and ships even though they are less energy efficient.

Some energy systems engineering features of airplanes include:

- The most efficient airplanes achieve over 80 passenger miles/gallon when fully loaded.
- Some 45 to 60 percent of the direct operating cost of a jet airplane goes for fuel.
- The high speed of a jet airplane makes crew cost such a small portion of the passenger's ticket price that it can't be matched by any other vehicle except by a bicycle.
- The high speed and productivity of a jet airplane also makes inconsequential the first cost of features that reduce maintenance.
- An airway, like an ocean, when not in use does not require maintenance, nor is it taxed.
- Empty seats generate no passenger revenue. However, airlines are resourceful in finding cargo to haul in airplanes that are not loaded with passengers.

The limits for airplanes are hard to predict. Many have tried. The following observations were made in 1985:

- Incorporating into an airplane the laminar flow of air over the wings, composite structure, advanced aerodynamics, and flight control will reduce fuel consumption by around 40 percent.
- Use of ceramic blading in high-temperature zones of jet engines ought to reduce fuel consumption by another 20 percent.
- A 747 airplane flew passengers 10,259 miles non-stop. Further increases in non-stop range are not needed, because very few airports in the world are further apart than 10,259 miles.
- Airplanes larger than the 747 are not likely to be built. Few routes generate enough traffic to fill a fleet of 600-passenger 747s.
- Supersonic airplanes are not likely to carry a significant portion of air travelers, mainly because of the fuel cost.
- Higher flight altitudes offer opportunities for more seat miles per gallon. However, a new generation of airplanes would be required, and a new airplane costs from $1 to $2 billion to develop.

Airplane weights are generally quoted in lb. The key weight is the maximum gross take-off weight. This includes the airplane, its passengers, crew, cargo, and fuel when it starts down the runway to take off. Fuel consumption is measured in lb for a designated flight profile for an airplane. The air speed of an airplane is generally specified in knots (1.151 statute mi/h). The air speed differs from ground speed when the airplane is flying in a moving air mass.

Airplanes that Fly Passengers and Freight

Fuel consumption of jet airplanes. Early fuel-thirsty jet airplanes could compete with reciprocating engine and propeller airplanes because the jets were faster and burned low-cost kerosene instead of high octane gasoline. An important characteristic of jet engines is their ability to operate at high altitudes where low air density makes airplane speeds approaching the speed of sound practical. The subsequent success of jet airplanes supported funds for competitive development of efficient engines, particularly after the fuel prices started rising after 1973. For example, in 1979, Pratt and Whitney announced a $1 billion program for developing a third-generation, fuel-efficient PW 2037 engine for the Boeing 757 airplane, even though a Rolls Royce RB211 engine could be readily modified for the airplane. Later Pratt and Whitney announced other third generation engine developments, the PW 4000 for the 767 and 747 airplanes, and a new V2500 for a future 150-passenger airplane. The last engine, to be ready in 1988, is expected to cost $1.3 billion to develop.

Paralleling the development of engines has been the development of new airplanes, with early jet airplanes such as 707s, 727s, DC-8s and DC-10s going out of production. New airfoils have improved aerodynamic efficiency, use of composite materials and computer-aided design have reduced weight, and computer controlled flight management has improved fuel efficiency.

The results can be illustrated by computation of fuel consumption during the mission profile shown in Figure 11–5. Following takeoff and acceleration, the airplane climbs to the first cruise altitude. Burning of fuel lightens the airplane during the first part of the cruise. It then climbs to a higher altitude for more efficient flight during the rest of the cruise. The flight management computers in the 757 and 767 airplanes optimizes the rates of climb and the distances in each cruise mode, based on constraints and flight data entered by the pilot. The reserves, shown on the right part of the chart, are required for flying to an alternate airport if the destination airport is closed.

Fuel consumption of Boeing airplanes for this flight profile have been calculated in Table 11–2 by Maurice E. Spencer, Manager of Airplane Economics at Boeing. The abbreviations on the chart are explained by the following notes:

Airplane. The first three digits identify the airplane type. The -100B, -200, and -300 are improved or lengthened versions of the original airplane.

Engine. CF engines are made by General Electric. PW engines are made by Pratt and Whitney.

MTOGW. The abbreviation means maximum take-off gross weight, the highest weight that the airplane can have as it starts down the runway.

OEW. The operating weight empty is the airplane without fuel, cargo, or passengers.

Block Fuel. Block length is the distance between airports. Fuel burned is in lb, and

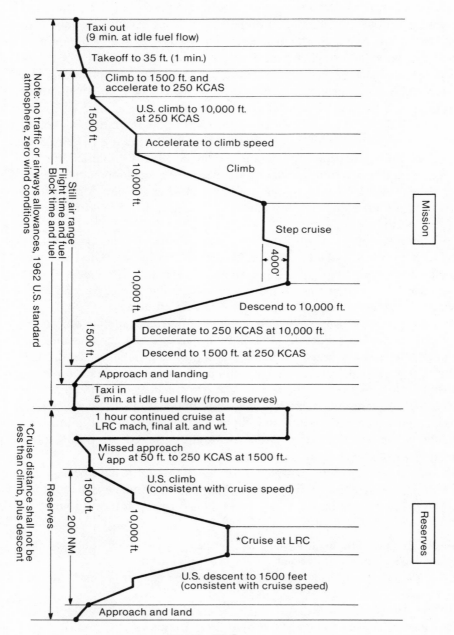

Figure 11-5

TABLE 11-2. Airplane Fuel Productivity Comparison
U.S. Domestic Mission/ATA Domestic Reserves

Airplane	Engine	MTOGW (lb)	OEW (lb)	Seats (38-40/34)	Block fuel–lb			ASM per lb of fuel		
					500 NM	1000 NM	2000 NM	500 NM	1000 NM	2000 NM
737-300	CFM56-3	135,000	70,010	118	7,040	12,585	24,500	9.64	10.78	11.08
747-100B	CF6-45A2	735,000	378,480	466	28,275	49,755	95,395	9.48	10.77	11.24
757-200	PW2037	220,000	126,100	186	9,250	16,450	31,730	11.56	13.00	13.48
767-200	PW-7R4D	300,000	176,870	211	12,170	21,330	40,525	9.97	11.38	11.98

the distance is in nautical mi. Chicago is 1,487 nautical mi from Seattle. San Francisco is 2,100 nautical mi from Washington, D.C.

ASM. Actual seat miles per lb of fuel is equal to the passenger miles per pound if the airplane is full. The values are in statute mi/lb. They can be converted to passenger miles per gallon by multiplying them by 6.8 lb/gallon for aviation kerosene.

The airplane data shown are based on a consistent set of ground rules so that they can be compared with each other. However, the data should not be compared with similar data from other sources. These data are a snapshot of airplane performance, and improvements come almost daily.

Cost of airplane flying. A consequence of the high speed and long life of jet airplanes is that the first cost of the airplane is not a very significant part of the cost of the passenger's ticket. For example, Air Florida, when it bought three 757s with engines and spare parts, announced that this contract had a value of about $200 million, with each airplane being worth $60 million. We might assume that the airframe has a life of around 40,000 flight hours and has 223 seats in an all-economy configuration. Then the cost to the airplane buyer of a seat mile (Csm) is:

$$Csm = \frac{\$60 \times 10^6 \times h \times \text{airplane}}{\text{airplane} \times 40{,}000 \text{ h} \times 500 \text{ mi} \times 223 \text{ passengers}}$$
$$= \$0.013$$

How much fuel cost in a ticket price? The fuel consumption of an airplane for a given flight scenario, such as shown in Figure 11–5, can be precisely predicted. However, the actual performance for a given flight can have many variables.

Useful airplane operating cost data derived from Civil Aeronautics Board (CAB) reports is published in the Air Transport World. For example, the 1983 operating costs of 747 airplanes are shown in Table 11-3.

Note that fuel, oil, and taxes accounted for 47 percent of the operating expense of American Airlines' 747 and 59 percent of the operating expense of Northwest Airlines' airplane. American Airlines' 747 cost $14.24/mile to fly, and if 300 passengers shared the cost, each one would have to contribute 4.74 cents, of which 2.24 cents went for fuel. At 83.46 cents a gallon, this represents 37 passenger mi/gallon. Passengers on a full plane would have made a smaller contribution to the fuel cost.

CAB data shows the 1983 airline yield for ticket sales as ranging from 10.41 to 13.45 cents/passenger mile. This includes costs other than direct flight costs and the cost of flying partly-filled airplanes.

Airplanes that Fly Passengers and Freight

TABLE 11-3. Cost per hour of flight, dollars

Cost element	American Airlines 747-100	Northwest Airlines 747-200
Crew, salaries, and expenses	$ 978.35	$ 738.71
Fuel, oil, and taxes	2781.73	3233.14
Insurance and other	27.28	28.27
Maintenance, with burden	773.72	793.14
Cash aircraft operating expense	4561.08	4793.26
Depreciation	1336.30	669.82
Total aircraft operating expense	5897.38	5463.08
Cost per revenue mile	14.24	12.88
Cost per available ton mile	17.00 cents	11.59 cents
Cost per gallon of fuel	83.46 cents	94.44 cents

A comparison with a fuel-efficient passenger railroad train is useful. The train can deliver 500 passenger miles/gallon, so if we assume that the railroad diesel fuel cost is the same as the cost of jet fuel for the 747 airplane, we have this comparison with Northwest Airlines' 747:

	150-passenger RR train	300 passengers in Northwest 747
Speed, mi/h	60	424.15
Fuel cost, cents/passenger mi	0.2	2.54
Crew cost, cents/passenger mi	4.10	0.58

The crew cost for the train assumed that the eight-member train crew is paid one-half as much as the eight-member airplane crew. The high speed of the jet airplane diminishes the crew cost of a passenger mile.

Reliability is designed into airplanes. To illustrate the importance of reliability we use round numbers. Assume that a $100 million 747-300 airplane flies at 500 mi/h and carries 500 people, each of whom pays 12 cents a mile for his or her ticket. The income (Ip) for this airplane is:

$$Ip = \frac{500 \text{ people} \times \$0.12 \times 500 \text{ mi}}{\text{mi/person} \times \text{h}} = \$30,000/\text{hour}$$

When this airplane fails to depart from the gate because of mechanical difficulties it generates a loss of $30,000/h.

The airline industry measures the probability that a revenue flight will not be delayed more than 15 minutes for mechanical reasons. In its first year of revenue service, the fleet of Boeing 757 twinjet airplanes achieved a cumulative dispatch reliability of 97.4 percent. The record for the last week of that year was 97.8 percent. The top mark, 99.7 percent was achieved by British Airways. The twenty-eight planes had completed 25,567 revenue dispatches and a total of 41,476 flight hours. Air Europe, a charter carrier, had been flying its 757s 10.4 hours/day.

Engine reliability has increased to the point where two-engine airplanes are flying over oceans. The key requirement is that at no time the airplane be more than 120 minutes from a suitable airport at single-engine operating speed. This is 800 miles for a Boeing 767.

The future for airplanes. New airplanes will deliver more passenger mi/gallon as long as incentive and development money are available. The incentive comes from the high seat–mile productivity of jet airplanes, which makes the airplane purchase price a small element in the operating cost of the airline. For example, a 767 airplane can generate nearly a billion dollars of revenue during its lifetime. The cost of the airplane is only 4 to 8 percent of its earning power. One Japan Air Lines 747 has spent 60,000 hours, one-half of the time since it was delivered, in the air.

In practice, new developments make airplanes obsolete long before they wear out. Some such developments that will appear in future airplanes include:

- Higher efficiency will come from engines with ceramics in the first row of turbine blades, and higher pressure ratios.
- Control of air moving over the wings will achieve laminar air flow. Sucking away the boundary layer is one approach. Richard Wagner and Michael C. Fisher predict that laminar flow control will result in a 21.7 percent reduction in fuel cost of a 400-passenger airplane.[10]
- Use of composites containing high-strength graphite in airplane structure, when combined with advanced aerodynamics, will reduce the fuel consumption by 40 percent, according to Robert L. James, Jr. and Dal V. Meddalon.[11]
- Use of lithium alloys in structure will further reduce airplane weight.

The development of a new airplane requires the services of a 1,000-person experienced engineering team and over $1 billion of capital.

[10]Richard Wagner and Michael C. Fisher, "Fresh Attack on Laminar Flow," *Aerospace America,* March, 1984.

[11]Robert L. James, Jr., and Dal V. Meddalon, "The Drive for Aircraft Energy Efficiency," *Aerospace America,* February, 1984.

HAULING FREIGHT: BY AIRPLANE OR SHIP?

At one time freight was hauled in passenger airplanes when insufficient passengers left load-carrying capacity. Avoiding the cost of interest on high-value items was an obvious advantage of air freight, but other features soon appeared:

- Quick access to distant warehouses avoids the need to have slow-moving products stored locally. A big cost element in the price of a slow-moving product is the interest and rental of warehouse space. The ultimate in warehousing efficiency is one warehouse that stocks and re-orders items to serve the whole world.
- For products such as money transaction papers, contract proposals, and valuable metals, the shipping cost is insignificant compared to the value of prompt delivery.
- Overnight transit of design data, drawings, and prototypes opens opportunity to use remote design and production facilities.

The estimated 1983 world airline freight traffic was 17.1 billion ton-kilometers, a 16 percent growth over 1982.[12] The estimated revenue was $4.8 billion, which corresponds to an average cost (Cf) of:

$$Cf = \frac{\$4.8 \times 10^9 \times 1.609 \text{ km}}{17.1 \times 10^9 \text{ ton km} \times \text{mi}} = \$0.452/\text{ton mi}$$

The biggest hauler of air freight was Japan Airlines, which hauled 2.257 billion ton kilometers for $609.7 million. Its average cost was $0.44/ton mi.

The growth of air freight has been explosive. For example, in the first ten months of 1983 Northwest Airlines' freight traffic grew to 790 million ton miles, an increase of 34 percent over the previous year. In the month of October, 1983, the airline flew 100 million cargo ton miles. Six of its thirty plane fleet of 747s were dedicated to handling cargo.

The lowest cost carrier of cargo is a ship. To compare air freight with marine freight we consider the bulk carrier, *Irena Dan,* which has a cargo carrying capacity of 72,371 long tons and cruises at 15.9 knots. The time (Ts) required for her to generate Northwest Airlines' October 100 million ton miles is:

$$Ts = \frac{100 \times 10^6 \text{ ton mi} \times \text{long ton} \times 2{,}000 \text{ lb} \times \text{h} \times \text{nautical mi}}{72{,}371 \text{ long tons} \times 2{,}240 \text{ lb} \times \text{ton} \times 15.9 \text{ nautical mi} \times 1.15 \text{ mi}}$$
$$= 67.5 \text{ h or } 2.8 \text{ days.}$$

In fuel conservation the *Irena Day* wins by a factor of 108. We assume that this ship has diesel engines delivering 16,000 shaft horsepower at cruise, and that the

---

[12] "Special Report: Air Freight Industry Expects Healthy Gains in 1984," *Air Transport World* (February, 1984), p. 26–27.

fuel rate is that of Sulzer engines, 133 grams/horsepower h. The fuel consumption for 100 million ton mi (Fs) would be:

$$Fs = \frac{133 \text{ g} \times \text{lb} \times \text{gallon} \times 67 \text{ h} \times 16{,}000 \text{ hp}}{\text{horsepower-hour} \times 453.6 \text{ grams} \times 7.1 \text{ lb}}$$
$$= 44{,}270 \text{ gallons}$$

The Northwest 747, if a model 747-100 B, has a takeoff gross weight of 735,000 lb and an operating empty weight of 378,480 lb, leaving 356,520 lb for fuel and cargo. Its fuel consumption is around 50 lb/nautical mi for flights in the 1,000 to 2,000 nautical mi range. Assuming an average flight of 2,000 nautical mi, this leaves 256,520 lb of carrying capacity for cargo. The fuel (Fa) consumed then becomes:

$$Fa = \frac{50 \text{ lb} \times \text{gallon} \times \text{nautical mi} \times 2{,}000 \text{ lb}}{\text{nautical mi} \times 7.1 \text{ lb} \times 1.15 \text{ mi} \times 256{,}520 \text{ lb} \times \text{ton}}$$
$$= 0.0477 \text{ gallons/ton mi}$$

Hauling 100 million ton miles then consumes 4.8 million gallons. This is 108 times the fuel consumption of the *Irena Day* for the same ton miles.

BICYCLES, SAILPLANES, AND SUNPOWER

People can be moved from one place to another without significant energy consumption. Sailplanes and sun powered vehicles require no fuel. The bicyclist may consume extra food to replenish the energy he expended in pedaling the bicycle. The dirigible needs no energy for overcoming gravity, and does not need a ground path, but energy is consumed in overcoming the resistance of air to motion.

Bicycles. A bicycle is a two-wheeled vehicle, propelled by the rider who pushes on cranks through pedals to turn a sprocket called a "chainwheel." A set of sprockets, called a "freewheel," is fastened to the rear wheel. The chainwheel and freewheel are coupled with a chain. Transferring the chain among various sized sprockets on the freewheel and on the chainwheel gives a range of gear ratios. The bicycle was developed between 1850 and 1900, and more recently multiple ratio transmissions and lightweight construction have revolutionized the capability of the bicycle. Energy systems engineering features of bicycles include:

- The rolling resistance of the wheels of a bicycle on smooth pavement with lubricated ball bearings and high pressure tires is negligible compared with wind resistance and the force needed to climb hills. At 20 mi/h 80 percent of the bicyclist's energy goes into displacing some 1,000 lb of air/minute.[13]

- The weight of apparatus accompanying the bicyclist can be under 20 lb. Only unicycles, skateboards, and skates can do better. In contrast, a passenger's share of hull and machinery on the *Queen Mary* is 33.4 tons.
- A bicyclist has to overcome gravity only when climbing a hill. A pedestrian or runner is continuously lifting his weight against gravity.
- A bicycle provides the most efficient possible way of using muscle power for travel. Furthermore, chemical energy is converted directly into mechanical energy in muscles, without a heat engine. Carnot-cycle efficiency limits do not apply.
- The cost of the bicycle, in cents/mile, can be insignificant.

The energy consumed in bicycling comes from the food that the bicyclist eats. Converting the food energy, measured in kilocalories per serving, into its petroleum equivalent gives around 1,000 passenger mi/gallon. This value has been exceeded only by small lightweight diesel-powered cars specifically built for breaking miles-per-gallon records.

Bicycles are much used in places where motor-vehicle fuel cost is high relative to income. The crowded bicycle lanes in Holland illustrate the popularity of bicycles in Europe. Bicycles carry freight as well as riders in the People's Republic of China. There it is common to see a farmer carrying produce to the market in barrel-sized panniers, and three-wheeled bicycles that haul furniture, household goods, and even steel beams.

The bicyclist needs no crew to accompany him on his trips. He need not pay for the cost of hauling empty seats that is paid by passengers on airplanes, buses, trains, and ships. The first cost of a bicycle, around $150 for a twenty-five-lb 10-speed unit, is 1 cent/mile for a 15,000 mi life. This is about the same as the first cost of an airplane when distributed over a lifetime of 30,000 hours and 250 passengers flying 500 mi/h. Bicycle maintenance consists of repairing punctured tires plus occasional repair of mechanical parts.

The best cyclists cover a lot of ground. In 1984, Pete Penseyres bicycled 3,047 miles across the U.S. in 9 days, 13 hours, and 13 minutes. A twenty-four-hour track record, set by Jim Elliott, in April, 1984 was 502.3 miles, with 2-minute miles in the last 30 minutes. Susan Notorangel broke the women's twenty-four-hour record with 401 miles in 24 hours.

The most important disadvantage of the bicycle for long trips is the cost of time. For an employee being paid $100/day plus 40 percent for fringe benefits, a

---

[13]Thomas A. Bass, "By Land and By Sea and Even By Air, Pedal Power Wins," *Smithsonian,* January, 1985, p. 96.

3,000-mile, one-day trip across the U.S. costs the employer $140 plus fare if the employee rides a jet airplane. A realistic cross-country time for a good cyclist would be tweny days at 150 mi/day. The wages and fringe benefits would cost $2,800, to which would have to be added food and lodging. Flying is obviously cheaper.

SAILPLANES FOR FAST TRANSPORTATION WITHOUT FUEL

A sailplane, also called "glider," is a winged vehicle that generally carries one or two persons and has no propulsion power. The sailplane is usually towed from the ground to a working altitude from which the pilot eventually glides back to the ground. Given skill and luck he can find upward air currents that raise him from time to time. Using new graphite-epoxy in spars, designers have achieved lift-to-drag ratios of over 56.

Energy features of sailplanes include:

- No ordinary non-energy-consuming passenger-carrying vehicle or mode can approach the speed achievable with a sailplane. Earth-orbiting spacecraft do better.
- A sailplane cannot offer economical and dependable scheduled transportation between two points.
- For much of the recreation travel by people the destination is not as important as the experience of traveling. A sailplane offers an exhilarating travel experience.
- As long as sailplanes are built in quantities of a few dozen to a few hundred at a time, they will be too costly for any significant hauling of people.
- Derivatives of sailplanes carrying small engines may become a practical mean of transportation.

The apparent limitations of sailplanes technology include:

- A lift-to-drag ratio of 56 will probably be exceeded.
- Catapulating of sailplanes into the air becomes practical with new lightweight high-strength ropes.
- High lift-to-drag ratios are difficult to achieve if more than two people are to be carried.
- The distance flown in one day will probably exceed the 1,022 miles flown by Tom Knauft in eleven hours.

The performance of a sailplane is described in terms of lift-to-drag (L/D) ratio. A wing with a $L/D = 50$ when flying through the air at a designated speed, will support a weight of 5,000 lb as long as it is pushed with a thrust of 100 lb. A sailplane with a L/D = 50 at a designated speed will glide horizontally 50 ft while dropping one foot.

SOLAR POWERED AIRPLANES AND CARS WORK

A crew drove a solar-powered car 2,566 mi (4130 km) across Australia in twenty days.[14] Solar-powered airplanes have, likewise, carried people. Propulsion power for both vehicles was generated with an array of silicon solar cells. Such vehicles can operate only when the sun is shining, unless they have batteries. Also, the array has to be oriented in the general direction of the sun. For example, the tail surface of the airplane was covered with solar cells to capture sunlight during banked turns.

The solar powered car weighed 125 kg (276 lb) and carried 720 solar cells. The car had a tubular steel frame and a fiberglass body. Bicycle wheels and racing tires assured a low road-friction loss. The solar array charged a battery when not supplying power to the motor. It cruised at 15 mi/h, although it had a top speed of 60 mi/h.

Cost is a factor that discourages solar powered vehicles. For example, assume that an 80 percent efficient, 10-horsepower motor is to drive the car through an 85 percent efficient drive train. The propulsion power (Pp) would be:

$$Pp = \frac{10 \text{ hp} \times 0.746 \text{ kW} \times 1,000 \text{ W}}{0.8 \times 0.85 \times \text{hp} \times \text{kW}} = 10,971 \text{ watts}$$

At $10/watt, the solar array would cost $109,710. Interest on the investment, at 15 percent per year, would be over $1,370 a month, which would buy a lot of alternative fuel.

ROUTES TO ENERGY EFFICIENT CARS

An automobile is a vehicle that carries one or more people, generally five or less, from one place to another, over a network of roads. Essential elements of moving people with automobiles are factories for making cars and repair parts, a network for distributing this hardware, garages in which cars are repaired, a fuel supply network, roads, and traffic control. Gasoline and diesel powered cars release air polluting chemical species in their exhaust; hydrogen and electric powered vehicles do not.

[14]Hans Tholstrup and Larry Perkins, "Across Australia By Sunpower," *National Geographic*, 164 No. 5, November, 1983, pp. 600–613.

Of the 15 million barrels per day of petroleum consumed in the U.S., 4.9 million barrels per day were used by private automobiles in 1983. The passenger miles per gallon of fuel consumed for automobiles varies from around 8 to 120. Thus, automobiles are a productive candidate for energy conservation.

Features of automobiles relative to energy conservation are:

- Cars can be designed to be practical and fuel-efficient. Over 50 statute mi/gallon is being obtained, and 100 mi/gallon is feasible.
- Cars built in a small production line, no matter how efficient, cannot compete on an operating-cost basis with cars built in quantities of over 100,000/year in a highly automated robot-equipped factory.
- No new car can approach the low operating cost of an old owner-maintained Volkswagen in the American economic environment.

The key to successful automobiles in the U.S. has been a continuously changing style that can be made appealing to buyers, and low manufacturing cost. The American automobile history is filled with nameplates of obsolete cars that were designed with other objectives. Energy-efficient examples include the Crosley, American Austin, 60-horsepower Ford V-8, and the Chrysler Airflow.

The low cost of cars, whether measured in terms of dollars per lb, or cost of machined parts, or hours of labor to assemble one vehicle, has been achieved by manufacturing automation and strenuous control of every element contributing to cost. For example, John C. DeLorean described how an $8 final inspection at General Motors had to be dropped because of its contribution to final cost.

An interesting energy-efficient exception to the styling requirement was the Volkswagen "Beetle." Dr. Ferdinand Porsche had about a year in which to design the car after being directed to meet these design objectives by Adolph Hitler:

Speed	100 km/hour (62 mph) on autobahns
Passengers	5
Weight	Under 1,000 kg
Cost	Under 1,000 marks
Fuel Efficiency	Better than 30 miles/gallon

To achieve low air drag, Dr. Porsche adopted an airflow design previously developed and abandoned by Chrysler. After testing many engines, including one with compression and expansion occurring in different cylinders, he selected a horizontally-opposed, 4-cylinder aircraft-engine design which he lightened by making the crankcase from magnesium. He designed the car with a rear engine. The first car was built in 1937. Forty-eight years later, Volkswagen "Beetles" were still being produced in Mexico and Brazil.

Dr. Porsche would never have heard the expression "life cycle cost," but he certainly understood the requirement. For example, he incorporated replaceable pistons and cylinders in his engine. The engine could be extracted from the car in less than a half-hour. Even the owner, following one of the many available instruction books, could completely restore his engine to new condition, using readily available parts. In the mid-1980s, a set of four new pistons and cylinders could be bought for around $50. Small firms would buy old Volkswagens, disassemble them, renew worn parts, and re-assemble the cars and sell them at a profit.

Automobiles, unlike power plants, are not usually selected by buyers on the basis of operating cost.

Routes to fuel efficiency of engines. Fuel consumption of automobiles can be improved by improving engine efficiency and by reducing the effort required to propel the car. We first examine engine efficiency.

The best industrial diesel engines have an efficiency of around 35 percent; that is, 35 percent of the energy in the consumed fuel appears at the output shaft. An air-cooled Volkswagen engine and transmission that we tested had 19 to 23 percent efficiency at nearly full throttle when climbing a hill.

Ways to improve the thermodynamic-cycle efficiency of an engine include:

- Use a more efficient cycle.
- Adopt higher pressure ratio in Otto cycle engines.
- Lower the temperature of the intake gas.
- Recover cycle losses with a bottoming cycle.

The steps to higher efficiency with better cycles goes from the Otto cycle to the Diesel cycle, and then to the Stirling cycle. Achievable efficiencies are discussed in the chapter on heat engines. Diesel engines are clearly more efficient than Otto-cycle gasoline engines. For example, the EPA economy rating for the 1984 Ford Escort with a diesel engine and a 5-speed manual transmission was 46 mi/gallon. The same car with a gasoline engine was rated 27 mi/gallon. Automobiles have been operated with Stirling engines,[15] but the expected fuel economy has not been achieved, and no Stirling powered engines designed for automobiles are in production. Higher temperatures are achieved in ceramic diesel engines, described in Chapter 2. R. Sekar, of Cummins, predicted 80 mi/gallon for a car powered with a ceramic diesel engine.

[15]D. G. Beremand, "DOE/NASA Automotive Stirling Engine Project Overview '83," *Proceedings of the 18th Intersociety Energy Conversion Engineering Conference*, AIChE (1983), pp. 681–688.

A bottoming cycle is used in turbo-charged diesel engines where both power output for a given cylinder displacement and efficiency are improved by a turbine that extracts heat energy from the exhaust stream. The turbine drives an axial compressor that adds input energy to the engine in the form of compressed air supplied to the engine intake. Engine efficiency can be further improved by cooling this compressed air with an air-to-water heat exchanger before it enters the intake valves. This cooling increases the temperature difference between the source and sink temperatures.

In an Otto-cycle gasoline engine a turbocharger can increase the power output of the engine at open-throttle operation. However, at other throttle positions, the turbo charging is ineffective because any pressure increase produced by the turbocharger is cancelled by the throttle.

Once, in testing alcohol–gasoline mixtures in an air-cooled Volkswagen, an associate and I discovered that the engine was more efficient and produced more miles per gallon when up to 15 percent of the fuel was alcohol. The increased mileage was especially surprising because methanol has only 56,500 Btu/gallon, while gasoline has around 123,500. At the time, we attributed the improved performance to the possible decomposition of alcohol in the engine. Subsequently we learned that methanol is just a better engine fuel, as pointed out in the chapter that discusses heat-engines and in Chapter 14 (page 347).

For this test we worked out a simple method of measuring the approximate efficiency of an automobile engine and transmission. For our test you need a long hill having a uniform grade (Figure 11–6). The hill we used was about two miles long and the car coasted down the hill at a constant speed of 50 mi/h. The speed when entering the test range was the same as the speed when leaving, so there was no change in kinetic energy. The potential energy of the car did change as it descended the hill, the amount being the weight of the car and its occupants times the difference in elevation. The elevation difference can be determined from maps or from an altimeter. We weighed the car at a truck weighing station . . . the attendant was not pleased with our planned experiments "with alcohol and cars."

The forces on the car going downhill at constant speed are the component of gravity parallel to the path of the car, and the windage and friction tending to slow it down (Figure 11–6). These forces can be calculated from the slope of the road and the weight of the car, although it is not necessary to do so.

To measure the energy going into the engine we brought the fuel line connecting the gasoline tank to the engine inside the car and intercepted the fuel flow path with a manifold and valves so that we could supply engine fuel either from the car tank or from one of the sample bottles. The bottles contained weighed quantities of methanol–gasoline mixtures. We drove the car up the test range at the same speed as it coasted downhill. At the instant of entering the test range we turned off the fuel tank and started extracting fuel from one of the

Reducing the Energy Required by the Car

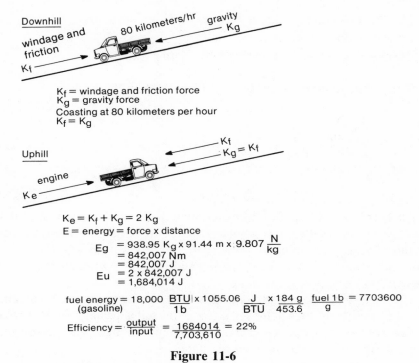

Figure 11-6

measured bottles. At the exit of the test range we turned the valves to feed fuel from the tank, and we set the partially emptied bottle aside for later weighing.

The energy supplied by the engine during the uphill run included the potential energy gained—the weight of the car times the difference in elevation. The engine also supplied the windage and friction energy, which was the same as supplied by gravity on the way down, since both runs were conducted with no significant wind. Thus the engine supplied twice the potential energy gain during the uphill climb. Weighing the consumed fuel, and accounting for the energy of its components, gave us the energy input to the engine. A sample calculation and some of our results are shown in Figure 11–6.

REDUCING THE ENERGY REQUIRED BY THE CAR

The energy in fuel consumed by an automobile engine goes mostly into these actions:

- Pushing air out of the way and through the engine.

- Squashing tires and turning shafts, gears, and bearing journals.
- Lifting the car and occupants over hills.
- Creating momentum for later dissipation during braking.

The power required to push a flat plate through the air is:

$$\text{Power} = C_d \, A \, V^3$$

where A = the frontal area
       V = velocity
       C_d = drag coefficient

The drag coefficient can be 1.35 for a parachute and 0.05 for an airfoil. Emmett J. Horton and his associates reported that a drag coefficient of 0.4 for a 1984 Ford Escort resulted in 6.5 horsepower being required to push the car at 50 mi/h through air. The rolling losses were 4.2 horsepower, giving a total road load of 10.7 horsepower.[16] A 400-lb experimental car with a drag coefficient of 0.15 achieved 200 mi/gallon of gasoline.

The energy that heats soft tires has to come from the engine. Switching from bias ply to radial ply tires achieves a few extra miles per gallon. Participants in economy contests generally over-pressurize their tires. In Japan tires are available that are designed to have low road resistance.

Lifting cars and occupants over hills need not be an efficiency penalty if the climbing energy can be recovered. Regenerating braking recovers this energy in electric battery cars. The heavily loaded engine is generally more efficient when climbing hills than it is when driving on level road. In a fuel injection engine the injectors are continuously closed if the throttle is closed and the engine is being dragged at above its idle speed. Thus the downhill miles are covered without the consumption of fuel. In a car with a free-wheeling transmission, the car coasts downhill, but the engine is consuming idling fuel in the meantime. With an ordinary transmission, the engine is driven at high speed while going downhill, with the cylinders sucking gasoline-enriched air through the idling passages in the carburetor.

As a car accelerates from a stopped condition some of the engine power goes into creating kinetic energy in the moving vehicle. This energy is dissipated in the brakes when the vehicle is stopped. In electric cars this energy can be recovered and stored in the battery. It could also be recovered into flywheels or hydraulic accumulators, but no such devices are in production.

The energy used in overcoming road friction, climbing hills, and starting

---

[16]Emmett J. Horton and associates, "Technological Trends in Automobiles," *Science*, V 225, August 10, 1984, p. 587.

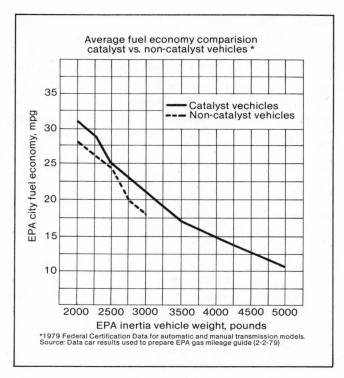

Figure 11-7

after a stop relates to the weight of the car and its occupants. Most of the weight is in the car. For example, two 165–lb persons add 8 percent to the total weight of a 4,000 lb car. The EPA gas mileage chart in 1979 related the observed EPA city mileages to the car weights (Figure 11–7).

Common to all of these approaches for increasing the efficiency of cars is the requirement that they be incorporated in mass-produced automobiles. The manufacture of special cars or the incorporation of efficiency features by rebuilding standard cars is generally so costly that the fuel consumption benefits don't pay for the extra cost of acquiring them.

ELECTRIC AUTOMOBILES

An electric automobile gets its propulsion energy from electrochemicals, carried in battery cases or in tanks for fuel cells. A hybrid electric car would carry an auxiliary engine for recharging batteries while on a trip. Electric automobiles used for short-distance travel include golf carts and wheelchairs. Battery powered electric vehicles have not been practical for cruise distances of over fifty miles between recharges. Other electric vehicles, such as trolley buses and street cars, collect propulsion power from a trolley or third rail, thus avoiding the need to carry batteries or fuel, but at the expense of being confined to electrified routes.

Electric cars are important because they are a practical substitute for automobiles powered with internal combustion engines when petroleum resources are depleted and the cost of non-petroleum fuels rises. The energy features that follow apply to vehicles that carry their own energy in batteries or fuel-cell reactant tanks.

- Battery weight, cost, and charge–discharge cycling life are keys to successful battery-powered electric vehicles.
- Starting-ignition automobile batteries are unsuitable for the deep discharges required in electric vehicle batteries.
- A starved-electrolyte lead-acid battery for which a 2,000-charge–discharge cycle life has been demonstrated would have satisfactory life, but its 10 watt-hours/lb energy density limits the cars it could power to a range of around fifty miles.
- A battery or fuel cell plus fuel that can store 20 kWh of energy, that weighs under 200 lb, and that will last 3,000 charge–discharge cycles, could make electric cars successful if the cost were reasonable.
- Climbing hills reduces the driving range of an electric car.
- If the electricity is generated by burning oil or natural gas, then powering vehicles with electricity doesn't significantly reduce consumption of oil or gas.
- Much of the propulsion energy carried by an electric vehicle during the first half of a trip goes into hauling the battery for the second half of the trip.
- Regenerative braking is practical with electrically powered cars. It is not practical in cars powered with internal-combustion engines.

The limits of electric powered automobiles are:

- Automobile fuels are taxed to collect money for building and maintaining roads. If electric cars displace gasoline and diesel powered cars, then they too would have to be taxed. Until then, electric and natural-gas powered cars enjoy a tax benefit.
- The 150-mile fuel for a 60 miles-per-gallon diesel powered car weighs eighteen lb. The energy delivered by the engine is 35 kWh.
- The same energy could be delivered with 3,888 lb of readily available starved-electrolyte lead-acid batteries. These batteries deliver around 9 watt-hours/lb.
- A rechargeable battery delivering 100 watt-hours/lb, if available, would weigh only 350 lb and could result in an electric car that would be competitive in performance with a diesel-engine powered car.

Batteries for electric automobiles. Once we were asked by Seattle City Light to evaluate the effect of having a significant population of electric cars recharging their batteries from the utility's network. Among other tasks, we tested starting-lighting-ignition car batteries for charge–discharge efficiency, which we found to be around 65 to 70 percent. We also discovered that these batteries, designed for engine starting and ignition service, could survive only 30 to 80 cycles of deep discharging when over 80 percent of the stored energy in the battery was extracted during each discharge. In tests at NASA Lewis Research Center, golf-cart batteries were down to 40 percent of capacity by the time they had undergone 250 charge–discharge cycles using a SAE J227a driving profile.[17] Any electric automobile using such short-life batteries would not be practical.

Lead-acid batteries specifically designed for traction service are available. They cost around $300 to $400/kWh of capacity. GNB Batteries, Inc. has developed a starved-electrolyte battery that has been tested for 2,000 deep discharge cycles. Design features of this and other batteries are described in the chapter on energy storage.

High-energy density and a potential for long life are possible in molten-salt batteries being developed specifically for electric vehicles. Douglas M. Allen had tested sodium-sulfur batteries at the U.S. Air Force Wright Aeronautical Laboratories. He found that the coulombic efficiency of the batteries was 100 percent, in that every ampere hour of charge was recovered on discharge.[18] Overall charge–discharge efficiency was around 80 percent. His paper, written when the tests were getting started does not record his announcement that by August, 1984 the batteries had accumulated 2,400 charge–discharge cycles with no evidence of degradation. In each discharge, 80 percent of the energy content of the cell was extracted. An important characteristic of the sodium sulfur battery is that it must be kept in a 350°C oven when it is in operation. However, it delivers 110 watt hours/kg.

Motor, control, and drive train. Availability of high-current field-effect transistors (FETs) and sophisticated integrated circuits makes possible efficient electric drive for automobiles when suitable batteries become available. Dozens of studies have evaluated drive configurations ranging from direct-drive, variable-frequency, alternating-current motors in every wheel, to a single dc motor coupled to a conventional transmission.

A. Borges de Oliveira, an engineer who has studied electric cars for many

---

[17]John G. Ewashinka, "Results of Electric-Vehicle Propulsion System Performance on Three Lead-Acid Battery Systems," *Proceedings of the 19th Intersociety Energy Conversion Engineering Conference,* ANS, 1984, pp. 727–735.

[18]Douglas M. Allen, "Sodium Sulfur Satellite Batteries: Cell Test Results and Development Plans," *Proceedings of the 19th Intersociety Energy Conversion Engineering Conference,* ANS, 1984, pp. 163–168.

years and has built and operated a hybrid vehicle, predicts that the drive train of a successful electric car will have these features:

Element	Type	Reason
Motor	DC shunt motor with samarium-cobalt field boost	Satisfactory speed-torque characteristic Simple implementation of regenerative braking
Transmission	Two-speed plus 20% field control	Good starting and low-speed acceleration
Armature current control	FET chopper	High efficiency plus safety from inductive damage
Vehicle power control	Digital computer	Fast response to driver commands Operates car in most efficient mode at all times Integrates current to display remaining battery energy Easy implementation of driver conveniences, such as cruise control Programs regenerative and friction braking in response to pedal pressure

Alternatives to electric cars. Methane, delivered as natural gas, like electricity, is not currently taxed as a fuel. It is compressed and carried in a vehicle in an appropriate tank. A dual-fuel carburetor that delivers methane to the manifold of an ordinary spark-ignition engine can be switched to gasoline if the methane supply is exhausted during a trip. The retail price of natural gas is around 60 cents for a 100,000 Btu therm, which corresponds to 74 cents a gallon for 123,500 Btu/gallon gasoline. The operating cost of a methane-powered car must include the cost of that car's share of the methane compressor as well as the tank for gaseous fuel and the special carburetor.

CONSERVING ENERGY IN TRANSPORTATION

If energy conservation becomes important in the U.S., then the key candidate is the 9.2 million barrels of petroleum that we burn each day in vehicles that move

Conserving Energy in Transportation

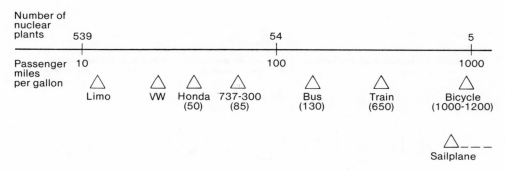

Figure 11-8

people and freight. In the year 2000, the U.S. is predicted to consume 8.2 million barrels of petroleum per day in transportation. With hydrogen and battery-powered vehicles, the same propulsion energy could be supplied to the driveshafts of our vehicles by 539 nuclear power plants 19.

Conservation steps would reduce this number of nuclear power plants, as shown in Figure 11-8, where the number of plants is related to the passenger miles per gallon achieved by vehicles.

---

[19]H. Oman, "Carbon Dioxide in Our Atmosphere—Slowing the Buildup." *Proceedings of the 21st Intersociety Energy Conversion Engineering Conference*, ACS, 1986.

12

Moving Freight with Ships, Trains, and Trucks

Survival of people living in cities depends on a continuous arrival of food, fuel, and goods. Most of this cargo moves on cargo ships, tankers, freight trains, and cargo trucks.

HAULING CARGO WITH SHIPS

A cargo ship is a steel hull, backed by framing to give it strength and rigidity, and shaped to carry the most possible cargo consistent with reasonable propulsion efficiency. Cargo ships are configured as tankers, bulk carriers, container ships, and miscellaneous-cargo carriers. Cargo ships are important from an energy viewpoint because they offer the most energy-efficient possible means of hauling cargo from one port to another. Trains, trucks, and airplanes can haul cargo faster, but not with less expenditure of energy.

Energy features of cargo ships include:

- Cruising speeds are generally kept well below the characteristic hull speed to optimize the cost of the fuel and propulsion power plant.
- Common diesel propulsion engines, available in ratings up to 62,000 horsepower, have efficiencies of 30 to 35 percent. New engines approach 50 percent efficiency. This cannot be matched by steam or gas turbines.
- No mode of hauling freight can approach the super tanker in ton-miles of cargo carried per gallon of fuel consumed.

Hauling Cargo with Ships

Many meanings of the word "ton" are used in ship design and operation. For example, the tonnage capacity of a cargo vessel, expressed in "tons," is the cargo volume in cubic feet, divided by 100. Displacement tonnage is the weight of salt water displaced by the vessel. In British documents this value is expressed in long tons, which contain 2,240 lb (1,016 kg). A metric ton is 2,205 lb. Displacement in long tons is related to water volume, salt water occupying 35 cubic ft/ton, and fresh water 36 cubic ft.

The most useful measure of a cargo ship's capacity is deadweight tons, which is simply the difference between the ship's displacement when loaded and when unloaded. Deadweight tonnage is what the ship can carry. It includes fuel, fresh water, stores, crew and their effects, passengers and their baggage, and cargo. Displacement of most tankers is measured in deadweight tons of 2,000 lb.

Table 12-1 shows displacements and dimensions of representative cargo-carrying ships.

Fuel used in hauling cargo. In Table 12-1, the biggest tanker is the *Esso Atlantic*, which can haul 611,200 cubic meters of oil. Oil is commonly measured in barrels throughout the world. The capacity (Ca) of the *Esso Atlantic* is:

$$Ca = \frac{611{,}200 \text{ cubic m} \times 264.2 \text{ gallons} \times \text{barrel of oil}}{\text{cubic meter} \times 42 \text{ gallons}}$$
$$= 3.845 \text{ million barrels of oil}$$

The U.S. uses around 15 million barrels of oil per day, so four of these tankers arriving each day could keep the country supplied with crude oil. Also, at \$30/barrel, the oil in the tanker when full is worth \$115 million dollars.

The steam turbine on the *Esso Atlantic* burns 198.4 grams of fuel/shaft horsepower hour. A useful value for comparing fuel consumption is ton miles per gallon, where a ton is 2,000 lb. The ton miles per hour (Tmh) produced by the vessel is:

$$Tmh = \frac{508{,}731 \text{ long tons} \times 2{,}240 \text{ lb} \times \text{ton} \times 16 \text{ knots} \times 1.15 \text{ mi/h}}{\text{long ton} \times 2{,}000 \text{ lb} \times \text{knot}}$$
$$= 10.48 \times 10^6 \text{ ton miles/hour}$$

The energy consumption (Ec) can be calculated from the 198.4 grams of fuel oil per shaft horsepower hour:

$$Ec = \frac{198.4 \text{ grams} \times 45{,}000 \text{ shaft hp} \times \text{lb} \times \text{gallon}}{\text{shaft-horsepower h} \times 453.6 \text{ grams} \times 7.1 \text{ lb}}$$
$$= 2772 \text{ gallons/h}$$

TABLE 12-1. Representative Cargo Ships

Ship	Dimensions, feet		Tonnage, deadweight	Shaft horsepower	Propulsion	Service speed, knots	Notes
	Length	Beam					
James Baines	216	36	2500	0	32,000 sq ft. sail	21 max	Carried 700 passengers plus 140 tons cargo, 14,000 miles in 63.5 days, London-Melbourne
Liberty	428	57	10,500	2500	Triple expansion steam, 220 psi	11 at 76 rpm	Ship designed in 1879 by Sunderland and Co. 2770 ships built, 1940–1945
Irena Dan	741	105.8	72,381 bulk	18,500		15.9	Bulbous bow, crew of 25, unmanned engine space
American Spirit	1,100	178	262,376 2.014 M barrels of oil	35,000	Steam Turbine	15	Cost $81.4 M 11,950 tons fuel for 20,000 mi
ESSO Atlantic	1,334	233	508,731 3.84 M barrels of oil	45,000	Steam Turbine	16	198.4 g/shaft-horsepower-hour. 5 stages feedwater heating, unmanned engine room.

Hauling Cargo with Ships

The measure of performance of (Pa) the *Esso Atlantic* then becomes:

$$Pa = \frac{10.48 \times 10^6 \text{ ton mi/h}}{2772 \text{ gallons/h}}$$
$$= 3782 \text{ ton mi/gallon}$$

A reference would be a diesel powered truck hauling 60,000 lb of cargo and getting about 5 mi/gallon of fuel. The truck would then deliver 150 ton miles/gallon.

An amusing extrapolation gives the cruise range of the *American Spirit* if the captain chooses to tap his cargo tanks. His cruising range with his 11,950 tons of fuel in the bunkers is given as 20,000 miles. With his 262,366 tons of cargo oil, the ship's range (Ra) is:

$$Ra = \frac{262,366 \text{ tons} \times 20,000 \text{ mi}}{11,950 \text{ tons}}$$
$$= 439,106 \text{ mi}$$

The 198.4 g/horsepower hour fuel consumption of the *Esso Atlantic* permits estimating the efficiency (ηc) of its power plant.

$$\eta c = \frac{\text{Output}}{\text{Input}} = \frac{1 \text{ hp h} \times 453.6 \text{ g} \times 7.1 \text{ lb} \times \text{gallon} \times 2547 \text{ Btu}}{198.4 \text{ grams} \times \text{lb} \times \text{gallon} \times 140,000 \text{ Btu} \times \text{hp h}}$$
$$= 0.298$$

The steam turbine power plant is a rather sophisticated one for the ship, having five stages of feedwater heating. For comparison, a MAN-B&W 45,000 hp K90MCE diesel engine has a fuel rate of 121 g/horsepower hour.

Cargo ship design. Cargo ships are designed by naval architects who consider many factors as well as energy efficiency.

The power required to push a displacement hull through the water is related to its actual speed with respect to characteristic hull speed (Vch):

$$Vch = 1.3 \sqrt{l}$$

where l = hull length in ft
Vch = characteristic speed in knots

Thomas C. Gillmer develops a general relationship for the propulsion power required for a ship once its length and characteristic hull speed are calculated. The plot of this relationship from his work[1] appears in Figure 11–1 in Chapter 11.

---

[1]Thomas C. Gillmer, "Modern Ship Design," *U.S. Naval Institute* (Annapolis) 1970, p. 200.

Super tankers tend to be very long, some over 1,000 ft (305 m), to give them a high hull-speed characteristic. Operating at substantially below this speed permits low engine horsepower per ton.

Propulsion power is affected by hull shape and smoothness. Ideal shapes are modified by requirements for maneuverability and stability in heavy seas. These factors are derived from model tests. Lower shipbuilding cost results from hulls that have simple curvature rather than compound curvature. Also, a very smooth surface, which has low water friction, can be achieved for a price during shipbuilding, but it deteriorates from corrosion and barnacle growth. Part of the propulsion power goes into making waves. An underwater protrusion on the bow of a vessel reduces wave-making power at cruise speed. This bow can be optimized for a single cruising speed, so it works well with tankers.

Selection of ship propulsion power. The choices for propulsion power for ships are steam turbines, diesel engines, and gas turbines. The selection of a propulsion power plant for a ship is no longer difficult. Lloyds' Register of Shipping shows that 96 percent of all merchant ships of over 100 gross tons are diesel powered.

Steam for steam turbines is produced in boilers that can burn the cheapest of fuels. However, the engine room of a ship is not a convenient place to build a high-efficiency steam plant. Furthermore, ship design is controlled to an extent by registration agencies such as the American Bureau of Shipping, Bureau Veritas, and Lloyds of London. Use of high-pressure steam has been slow to develop. Not many ships have steam plants with pressures higher than 910 lb/in^2. *Esso Atlantic*'s 29 percent efficiency is probably in the upper end of steam-plant efficiencies.

On the other hand, diesel engines with efficiencies of above 35 percent are common. The slow-speed marine engine does not require a reduction gear and boiler that are essential with steam engines. For example, the 865-foot container-carrying *Polycrusador* is propelled at 16.4 knots by a single 23,900-horsepower, 114-rpm diesel engine. It can carry a respectable 118,500 long tons of bulk cargo. Diesel engines adapt readily to remote control from the bridge, and unmanned engine rooms are common. The latest development in marine diesel engines is the Sulzer engine with an exhaust turbine coupled to the crankshaft. Its efficiency under optimum conditions exceeds 50 percent. The MAN B & W MCE engines are, likewise, in the 50 percent efficiency class.

Bunker C oil, the low cost fuel for steamships, was the residual left from refinery processes that extracted higher-value products from crude oil. New refineries can crack this residual fuel and reform it, narrowing the price difference between Bunker C and diesel fuel. For example, in February, 1985, the price of Bunker C ranged from $164 to $193 per metric ton ($20 to $24 per barrel),

Economics of Ship Operation

depending on the port.[2] Diesel fuel ranged from $224 to $315 per ton ($29 to $41 per barrel).

The direction of ship propulsion is clear. The list, "U.S. Flag Ocean Going Fleet of 2,000 Gross Tons or Over," shows about 80 percent of the vessels being turbine powered.[3] However, looking into the future, in the *World Orderbook for Ships, 1,000 Gross Tons and Above as of April 15, 1984,* a ship not powered by diesel engines is rare.[4]

ECONOMICS OF SHIP OPERATION

Many elements other than fuel enter into the cost of operating a ship. They include interest on investment, depreciation, insurance, time spent idle in ports, and fees for harbors, canals, and pilots. A ship also has a crew that must be paid, and propulsion machinery that burns fuel. A few examples will show that no other mode of transportation can approach an efficient cargo ship with respect to fuel and crew cost on a ton-mile basis.

A ship is not a particularly expensive capital investment on the basis of dollars per ton of cargo-carrying capacity. For example, U.S. Lines bought twelve ships, each having 57,714 long tons of capacity, for $47 million each. That is only $814/long ton.

MAN-B & W in its diesel engine catalog shows that the engine costs represent about 50 percent of a ship's operating expenses. Of the engine costs, 81 percent is fuel oil, 13 percent capital, 4 percent maintenance, and 2 percent lubricating oil.

The ultimate in energy efficient hauling of freight across water is the supertanker. To illustrate, our calculations showed that the *Esso Atlantic* consumed 2,770 gallons of Bunker C fuel oil per hour. This corresponds to sixty-six of the forty-two-gallon oil barrels. In 1984, when gasoline was sold at retail for about $1.10 per gallon, residual fuel No. 6 containing 1 percent sulfur was selling at $30 a barrel in the New York harbor. The hourly cost of fuel (Ch) for the *Esso Atlantic* was:

$$Ch = 66 \text{ barrels} \times \$30/\text{barrel} = \$1980$$

The hourly crew cost (Cc), at say $30 per hour for wages and fringe benefits for 20 workers, was:

---

[2] "Data Update," *Marine Engineering Log,* March, 1985, p. 13.
[3] *Marine Engineering Log,* June 15, 1984, pp. 93–118.
[4] *Marine Engineering Log,* June 15, 1984, pp. 51–79.

$$Cc = 20 \times \$30 = \$600$$

Crew and fuel would then cost $2580/hour, and on a ton-mile basis the direct cost of hauling crude oil (Co) would be:

$$Co = \frac{\$2580 \times h \times \text{nautical mi} \times 2{,}000 \text{ lb} \times \text{long ton}}{h \times 16 \text{ nautical mi} \times 1.15 \text{ mi} \times 508{,}731 \text{ long tons} \times \text{ton} \times 2{,}240 \text{ lb}}$$
$$= 0.025 \text{ cents/ton mi, or } 2.5 \text{ cents/100 ton mi}$$

No land vehicle can approach this low cost of hauling oil.

Modern bulk carriers. Ships that haul products like wheat, coal, and cement also achieve respectable performance. For example, the *Irena Dan* has a deadweight tonnage of 72,371 long tons, which corresponds to 81,055 of our 2,000 lb tons. If she had a Sulzer diesel engine that consumes 131 grams of fuel an hour/horsepower hour at 80 percent of the rated 18,500 horsepower, then each hour she would burn in fuel (Fh):

$$Fh = \frac{18{,}500 \text{ hp} \times 0.8 \times 131 \text{ g} \times \text{lb} \times \text{gallon}}{\text{hp} \times h \times 453.6 \text{ g} \times 7.1 \text{ lb}}$$
$$= 602 \text{ gallons/h}$$

The manufacturer claims that the engine will run on "low quality heavy fuels," and it has Nimonic alloy valves, presumably capable of withstanding corrosion from sulfur. Assuming that the engine will run on $30/barrel fuel, then the hourly fuel cost (Ch) is:

$$Ch = \frac{602 \text{ gallons} \times \text{barrel} \times \$30}{h \times 42 \text{ gallons} \times \text{barrel}} = \$430/h$$

If we assume that the twenty-five-person crew of this automated ship works eight-hour shifts, then the crew cost (Cc) becomes:

$$Cc = 9 \text{ crew members} \times \$30/h = \$270/h$$

For hauling freight, the combined crew and fuel cost (Cf) would be:

$$Cf = \frac{(\$270 + \$430) \times h \times \text{nautical mi}}{h \times 81{,}055 \text{ tons} \times 15.9 \text{ nautical mi} \times 1.15 \text{ mi}}$$
$$= 0.048 \text{ cents/ton mi}$$

The 128,000 deadweight ton bulk carrier *Jade Phoenix* illustrates the competition in hauling bulk cargo. In 1984, she embarked on a 12,000 mile voyage from Seattle to Egypt with 111,000 tons of soft white wheat. Two years earlier the average cost of hauling wheat to Egypt was $91.74 a ton. The *Jade Phoenix* carried it for $52.50 per ton. That's 0.44 cents/ton mi. The *Jade Phoenix* had

Economics of Ship Operation

been built as a liquified natural gas carrier in 1978, and converted in 1983 into a bulk carrier. She is equipped with grain blowers for unloading to a lighter in Egypt.

In contrast, we show later that a truck carrying a 60,000-lb load costs about $1.25/mile to operate. The cost of hauling freight (Ct) then becomes:

$$Ct = \frac{\$1.25 \times 2000 \text{ lb}}{\text{mi} \times 60,000 \text{ lb} \times \text{ton}} = 4.2 \text{ cents/ton mi}$$

Automation of engine control and navigation functions has enabled ship designers to reduce crew size. An example is *Arita*, a multipurpose cargo hauler designed by naval architects G. T. R. Campbell and built by IHI, a shipbuilder in Japan. Her deadweight tons is 16,975, and she is operated by a crew of fifteen. The 258,500 deadweight ton supertanker, *Tohkai Maru*, has a crew of sixteen. On the other hand, the cruise ship *Royal Princess*, which carries 1,260 passengers, has a crew of 500. Note, however, that her 9,500 hp main diesel engines have been fitted with economizers that heat water, presumably for the four swimming pools, two jacuzzis, and suana!

Moving people and freight without burning fuel. The *Preussen*, with a cargo capacity of 8,000 tons, was the largest sailing ship ever built. She was 438 ft (133 m) long, had a 54 ft beam, and 54,000 square ft of sail area. She made a run of 370 miles in 24 hours. Assuming a crew of 100 with twelve-hour shifts, and an overall speed one-half of her best speed, the crew cost (Cc) at $30/hour for wages and fringe benefits would be:

$$Cc = \frac{\$30 \times 50 \times 24 \text{ h} \times \text{long ton} \times 2000 \text{ lb}}{\text{h} \times 370 \text{ mi} \times 0.5 \times 8,000 \text{ long tons} \times 2240 \text{ lb} \times \text{ton}}$$
$$= 2.17 \text{ cents/ton mi}$$

No one is seriously interested in using sailing ships for hauling cargo.

Trains that haul freight. A freight train is a set of cargo-carrying cars that are coupled together and propelled on steel railroad track by one or more power cars called "locomotives." Freight trains are important because they can haul freight over land with low fuel consumption on a ton-mile-per-gallon basis.

Energy features of freight trains include:

- The low friction of a steel wheel rolling on a steel rail is hard to beat.
- The cost of energy is not the main cost of hauling freight on railroads.
- Trolley-powered and on-board-diesel-powered electric locomotives consume about the same amount of fuel energy. However, the energy for trolley power can be generated at lower cost in coal-burning power plants.

- Trains are not big consumers of fuel in the U.S. They account for less than 2 percent of the nation's fuel consumption.
- Trains haul about 40 percent of the intercity freight in the U.S.

The limits of hauling cargo with trains seem to be these:

- The allowable drawbar pull on standard couplings limits train lengths to about 100 fully loaded cars in the U.S.
- A unit train dedicated to hauling coal from a mine to a power plant can have 100 cars, each carrying 100 tons of coal, for a total of 10,000 tons. A train carrying mixed freight would carry fewer tons.

Freight train speeds in the U.S. are measured in miles per hour. The weights of cars, cargo, and locomotives are given in tons of 2000 lb (907.2 kg). The force required for pulling a single freight car at a given speed is called "resistance," and is measured in lb/ton of cargo plus car. Railroad grades are defined in percent. One percent means that in every 100 feet of track, the elevation changes by one foot. This corresponds to twenty lb of extra drawbar pull for each ton being pulled up each percent of grade.

Railroad locomotives in the U.S. are either diesel or electric. A diesel engine in a locomotive usually drives a generator that supplies power to traction motors that drive the axles. Diesel locomotives are rated in horsepower output of the engine and drawbar pull at various speeds. Electric locomotives are sometimes rated on their diesel-equivalent horsepower and at other times on kilowatts input from the trolley.

Energy consumed in hauling freight on railroads. The tractive resistance of freight cars has been quantified by W. J. Davis[5] as:

$$RL = 1.3 + \frac{29}{W} + 0.03\,V + \frac{0.0024\,A\,V^2}{W\,n}$$

where RL = rolling resistance of the lead car or locomotive, in lb
 W = weight/axle in tons
 A = cross section of the car in square ft
 V = speed in mi/h
 n = number of axles/car

The resistance for the following cars is:

---

[5] W. J. Davis, "The Tractive Resistance of Electric Locomotives and Cars," *General Electric Review*, October, 1926.

Economics of Ship Operation

$$RF = 1.3 + \frac{29}{W} + 0.045\,V + \frac{0.0005\,A\,V^2}{W\,n}$$

where RF = the rolling resistance of the following car, in lb.

In other investigations, the University of Illinois has published curves in which the resistance of a freight train having an average weight of 60 tons varies from 3.2 lb/ton at 10 mi/h to 6 lb/ton at 40 mi/h.

From Davis' equation we can calculate the force required to move a 60-ton box car at 60 mi/h. The car has 100 square ft frontal area and its weight distributed on four axles.

$$Rf = 1.3 + \frac{29}{15} + 0.045 \times 60 + \frac{0.0005 \times 100 \times 60 \times 60}{15 \times 4}$$
$$= 8.9 \text{ lb/ton}$$

Note that wind drag, the last term, is 3.0 pounds of the total.

The power required to pull this car on level track can be calculated from the definition of a horsepower, 33,000 ft lb/minute.

$$P = \frac{8.9 \text{ lb} \times 60 \text{ mi} \times 5{,}280 \text{ ft} \times \text{h} \times \text{minute} \times \text{hp}}{\text{ton} \times \text{h} \times \text{mi} \times 60 \text{ minutes} \times 33{,}000 \text{ ft lb}} \times 60 \text{ tons}$$
$$= 85.8 \text{ horsepower}$$

We assume that the 60-ton box car is $\frac{5}{6}$ cargo. The energy to haul one ton of cargo for the one minute required for traveling one mile (Em) becomes:

$$Em = 85.8 \text{ horsepower} \times 6/5 \times 1/60 \text{ h} \times 1/60 = 0.0286 \text{ hp-h}$$

A large diesel engine consumes fuel at a rate of around 0.35 lb/horsepower h. The fuel weighs 7.1 lb/gallon and contains 137,750 Btu/gallon in energy. Handbooks show that about 82 percent of the diesel engine power appears at the locomotive wheels. With these assumptions, the freight train's energy performance (Pe) becomes:

$$Pe = \frac{1 \text{ ton} \times 1 \text{ mi} \times 0.82 \times \text{hp-h} \times 7.1 \text{ lb}}{0.0286 \text{ hp-h} \times 0.35 \text{ lb} \times \text{gallon}}$$
$$= 581 \text{ ton mi/gallon}$$

Additional fuel is consumed in rounding curves and climbing mountains. For example, the energy in lifting a ton of freight over a 3,000 footpass (Em) is:

$$Em = \frac{3{,}000 \text{ ft} \times 2{,}000 \text{ lb} \times \text{minute} \times \text{horsepower} \times \text{h}}{\text{pass} \times \text{ton} \times 33{,}000 \text{ ft lb} \times 60 \text{ minutes}}$$
$$= 3.03 \text{ horsepower h/ton/pass}$$

Assuming that there is one 3,000-ft pass every 300 mi along the route, then the total energy (Et) consumed becomes:

$$Et = \frac{0.0286 \text{ hp-h}}{\text{ton-mi}} + \frac{3.03 \text{ hp-h} \times 1}{\text{ton-mi} \times 300}$$
$$= 0.0387 \text{ hp-h/ton mi}$$

The train's fuel performance then drops to 430 ton miles per gallon.

In comparison, we had calculated that the cargo ship, Irena Dan, if equipped with Sulzer diesel engines, would produce 2356 ton miles per gallon.

Cost of energy and labor. The cost of running a railroad includes fuel, labor, depreciation of rolling equipment, track and equipment maintenance, supplies, rents, and taxes. The train crew might be an engineman, fireman, conductor, and three others. Assuming that a crew member costs $40/hour for wages and fringe benefits, the cost/mi (Cm) for a 60 miles-per-hour train would be:

$$Cm = 6 \text{ persons} \times \$40/\text{person} \times 1/60 \text{ mi} = \$4/\text{mi}$$

For a train carrying 7,000 tons the cost of diesel fuel at $1/gallon (Cf) would be:

$$Cf = \frac{7,000 \text{ tons} \times \text{gallon} \times \$1.00}{430 \text{ ton mi} \times \text{gallon}} = \$16.27/\text{mi}$$

An Electric Power Research Institute report has an estimate of 0.27 cents/ton mi for maintenance of track that carries coal unit trains.[6] The Institute also analyzed the rates charged by railroads for hauling coal. The rates ranged from 1.56 cents/ton mi on a 1,650-mi route in the South to 3.59 cents on a 397-mi route in the East.[7]

TRUCKS FOR HIGHWAY FREIGHT

A truck consists of one or more freight carrying boxes, supported on many rubber tired wheels, driven by one person, and powered by an engine. Availability of reliable diesel engines made trucks so dependable that they don't need dedicated repair facilities en route. As a result, owner-drivers can compete with big transportation organizations.

---

[6]"Unit Train Coal Transportation Costs," *EPRI Journal,* September, 1983, p. 53.

[7]Edward Altouney, "Rail Routing and Costing System for Coal Transportation," *EPRI Journal,* 1985, pp. 54–55.

Trucks for Highway Freight

Energy features of trucks:

- Trucks cannot match the low fuel consumption of ships, barges, or pipelines for hauling cargo.
- Railroads cannot match the speed with which trucks can deliver goods within cities and between nearby cities.
- Fuel expense is about 15 percent of the cost of operating a freight carrying truck.
- The ceramic diesel engine will probably be the next improvement in fuel efficiency of trucks.

Size of trucks. Congress has authorized travel over interstate and some other highways by tractors pulling two trailers, each 48 ft long and 102 in wide. Gross vehicle weights up to 80,000 lb are allowed on these roadways. In comparison, a fully-loaded gondola car carries around 200,000 lb of coal, and a unit train might consist of 100 such coal cars.

Virtually all trucks that carry freight between cities have diesel engines.

Energy consumption of trucks. Large freight hauling trucks moving on interstate highways achieve around 6 mi/gallon.

The staff of the Congressional Budget Office has analyzed the energy expended in moving freight by truck and by other, alternative means.[8] Their data is summarized in Table 12-2. Note that their expression, "modal energy," includes the energy used in propulsion, maintenance, vehicle manufacturing, construction, and refinery losses. The Budget Office staff also accounts for movement of empty vehicles and following other than straight-line routes. They used 138,700 Btu/gallon for diesel fuel and 125,000 Btu/gallon for gasoline. In this way they were able to compare pipelines that don't have empty return journeys, with tankers that do, and airplanes that fly straight routes, with trucks that follow highways.

The report notes that for rail and barge transportation, propulsion consumes between 35 percent and 50 percent of the energy used. For inter-city trucks, propulsion accounts for about 60 percent of the energy, and for airlines about 90 percent. Not being able to follow a straight route requires 45 percent of the barge energy, 35 percent of the rail energy, and 20 percent of the inter-city truck energy. With few exceptions, none of the other components of energy use—vehicle manufacture, guideway construction, and maintenance—accounts for more than 10 percent of the total energy use.

In another study the U.S. Department of Energy contracted with Tech-

---

[8] "Energy Use in Freight Transportation," *Congressional Budget Office,* U.S. Congress, Washington, D.C., February 1982.

TABLE 12-2 Estimates of Typical Freight Energy Efficiency (In Btus per Ton-Mile of Cargo)

Mode	Modal energy
Rail—Overall	1,720
TOFC*	2,040
Unit coal train	890
Truck	
Average intercity	3,420
Barge—Overall	990
Upstream	1,280
Downstream	620
Air	
All-cargo plane	28,610
Belly freight	3,900
Oil Pipeline	500
Coal Slurry Pipeline	1,270

*Trailer on flat car.

nassociates, Inc., in Washington, D.C., to compare the energy consumed in hauling cargo between Chicago, Illinois, and Minneapolis, Minnesota. The train was the Milwaukee Road's *Sprint*, which carried trailers. Such a train is called "intermodal" by the railroad industry. The General Motors Model EMD SD-40 locomotive was equipped with a Trident 8 Type S, model 157 Neptune flowmeter. The truck, owned by CW Transport, Inc., was equipped with a Columbia System Fuel-O-Meter, model ERD-DT-50 for measuring fuel flow.

The results of the test are summarized in Table 12–3.

COST OF TRUCKS AND TRUCKING

The cost of operating trucks was surveyed for the Private Truck Council of America by A. T. Kearney, a management consulting firm. At the time (1983), the cost of operating a tractor was $1.29/mi.[9] It had risen 2 percent from $1.26/mi in 1982. The average tractor had traveled 97,000 miles that year. More

[9] "Fleet Costs Held in Check," *Trafic Management*, July, 1984, pp. 45–46.

Cost of Trucks and Trucking

TABLE 12-3. Comparative Results

The following data are *averages* for the 12 recorded intercity trips by each mode:

Item	Truck	Rail
Distance (mi)	420	412
Speed (mph)	42.6	38.2
Trailing Gross Weight (tons)	23.1	1,466
Fuel Consumed (gallons)	82.3	1,283.2[1]
Loaded/Empty Trailers	1/0	42/3
Trailer Miles Per Gallon	5.1	13.5
Trailer Gross Weight (tons)[2]	23.1	18.1
Gross Ton Miles Per Gallon	117.7	257.8
Trailer Revenue Weight (tons)	17.1	12.1
Revenue Ton Miles Per Gallon	86.9	172.9

[1] Includes linehaul and terminal fuel use.
[2] 45-foot highway trailers; 40-foot railroad trailers.

than 440 private fleets in twenty-five different industries participated in the survey.

The cost of running the big fleets averaged $1.37/mi, whereas operating smaller fleets cost only $1.25/mi. Leased rigs with non-union drivers cost only $1.11/mi to operate.

Fuel isn't a big element in the cost of operating a truck. At 6 mi/gallon, the fuel cost is only 16 cents/mi if fuel costs $1 a gallon.

Carrying trailers on freight trains. Intermodal transport, also called "piggyback," grew 20 percent between 1982 and 1983. In 1983, the railroads hauled 4.1 million intermodal trailers and containers. This cargo accounted for 12.4 percent of the total carloadings recorded by the American Association of Railroads. Only coal, which accounted for 27 percent of the carloadings, was greater. The Association claims that one train crew can move as much freight as 100 truck drivers on intermodal trains, using less than one-half of the fuel. The Association also observed that having cargo locked in trailers and boxes reduces pilferage.

13

Heating Buildings with Less Energy

In the U.S., around 9.5 quadrillion British thermal units (Btu) of the nation's total energy consumption goes into heating buildings. In 1983, space heating represented 12.9 percent of the nation's energy consumption. Cooling took only 4 percent. Heating of water consumed 6.4 percent. The energy that heats a building is typically released when fuel is burned in a furnace to heat water or air. Within twenty-four hours after this release, most of this energy has been carried off by infrared photons that are by then passing by the outer zones of the solar system on their way to deep space.

A building is a box from which in cold weather heat escapes to the outside world, mostly by conduction through walls, and in air that leaks through cracks and openings. A trivial amount escapes by radiation. Heating of buildings is an unusual opportunity for energy conservation because most old buildings were designed with little attention to energy cost. When natural gas sold for 20 cents/million Btu at the wellhead, and oil cost 16 cents/gallon, there was little incentive to insulate even the walls of a structure.

A building is heated by releasing thermal energy within the building at a rate that equals the heat escaping from the building when the air within the building is at its desired temperature, usually about 70°F in buildings occupied by people. Common heat sources are boilers or furnaces in which fuels are burned, heat pumps, electric resistance heaters, and fluids that bring heat from an outside source.

Heating Buildings with Less Energy

Energy features in heating buildings include:

- A new building can be designed to make the cost of heating almost trivial, compared to the interest on the money used to build the structure, by insulating walls, installing multiple glazing in windows, caulking all seams, and routing ventilating air through heat exchangers.
- Fuel resources are conserved when buildings are heated with heat pumps or power-plant losses, rather than by burning fuels or consuming electricity in resistance heaters.
- Heat source selection and degree of insulation are useful trades in the analysis of building heating.

The minimum added heat required to keep an occupied space comfortable is virtually zero. The smallest space that one person can occupy at $-20°F$ is an REI Aleutian sleeping bag, in which he is surrounded by 6 inches of goose down. Heat released by the body is sufficient to replace escaping heat.

On the other end of the scale is a State of Massachusetts office building in Boston, where the heat released by occupants and lights is stored in water tanks during the day, and recovered with heat pumps during the night. No other heat is required for the building.

Other limits in heat sources for buildings include:

- An electric resistance heater can be 100 percent efficient. All heat energy released by the unit goes into its occupied volume.
- Gas-furnace efficiency can approach 95 percent.
- Oil furnaces are generally limited to less than 90 percent efficiency because the flue temperature has to be above the boiling point of water. Otherwise, the condensation would erode the flue and stack.
- Heat pumps are Carnot-cycle limited. Coefficient of performance of over 3 is not common.
- Burning wood with an efficiency of greater than 50 percent requires apparatus better than an ordinary stove or fireplace.

Most buildings, particularly those that aren't densely populated, need heat in the winter. The amount of heat can be calculated from heat losses by conduction and infiltration, using degree-days data from the National Climatic Data Center.[1] The heating plant has to be big enough to supply required heat during the

---

[1] "Daily Degree-Day Data," *National Climatic Data Center*, Federal Building, Acheville, North Carolina 28801 ($1/month).

worst-case cold weather. The annual cost of heating is the amortization of the plant, plus its operation and maintenance, and the cost of energy.

MEASUREMENT OF HEAT FOR BUILDINGS

In the U.S., the energy used in heading buildings is commonly measured in British thermal units (Btu). A Btu is the amount of heat that will raise 1 lb of water 1 degree Fahrenheit. Typical energy contents of fuel used in heating buildings are:

Distillate fuel oil	138,700 Btu/gallon
Residual fuel oil	149,700 Btu/gallon
Natural gas	1,000 Btu/cubic ft
Electricity	3,412 Btu/kWh
Wood	5,000 Btu/lb, if dry
Coal, bituminous	22.4 million Btu/ton (2,000 lb)

Thermal conductivity of walls and insulation is measured in Btu/square ft/degree F. A more convenient value is "R," which is the reciprocal of conductivity. Each cubic foot of air infiltrating out of the building carries with it 0.018 Btu of heat for every degree F of temperature difference.

A degree day is the number of degrees that the average temperature falls below 65°F during the heating season. For example, if on December 15 the average outdoor temperature was 15°F, then 65 −15 = 50 degree days for that particular day. Weighted degree-data for U.S. cities is published by the American Gas Association.[2] Heating requirements vary from one year to the next. For example, at Milwaukee, Wisconsin, the 1941 to 1970 average was 7,444 degree days. In a five-year period the highest was 7,702 in 1978 and the lowest was 7,387 in 1981. The most heating, in the contiguous U.S., was needed at Duluth, with an average of 9,756 and a high of 10,329 degree days in 1982. Most northern cities were in the range of 5,000 to 7,000 degree days. Most southern cities were under 3,000. Honolulu had zero degree days.

CONSERVING HEATING ENERGY

The energy used to heat an occupied building can be conserved by reducing the winter-time temperature to the lowest tolerable value. Energy can be further reduced by limiting the heating of the building, when not occupied, to the lowest

---

[2] "Future Gas Consumption in the United States," Vol 10 (Arlington, Virginia: American Gas Association, December, 1983).

temperature that will still avoid damage from humidity, corrosion, and frozen pipes. Heating of unoccupied portions of buildings can be controlled by a computer that commands valves, switches, or dampers to close when occupants are away. Such computers are usually equipped with uninterruptible power to preserve their time schedules during electric power failures.

The energy used to heat buildings can also be conserved by insulating walls and ceilings, by installing double and triple glazing in windows, by providing air locks on doors, and by extracting heat from the air used in ventilation before it is discharged. A new building can be so designed that the cost of heating energy is trivial. Old buildings have been refurbished to make heating cost trivial as well.

The key element in evaluating conservation features is their cost relative to savings in future fuel expense. Future fuel cost must be converted into present value, as described in Chapter 9. For example, assume that the sum of the present values of future annual fuel costs for a house is $10,000 during a twenty-year period. Assume also, that the capital cost of proposed conservation measures is also $10,000. Then an alternative investment yielding 10 percent interest would pay the fuel bills and also preserve the capital funds. The actual comparison is more complicated because a rising fuel cost will skew the expense.

Factors other than economics affect the desirability of conserving energy. For example, stores, theaters, and restaurants choose not to risk losing business by dropping interior temperature during the winter. Firms that build new houses on speculation would not build windowless structures even if they consume less heating energy than do houses with windows.

HEATING STRATEGIES

The energy loss from planned buildings can be calculated, and the loss from existing buildings can be estimated from heating costs. The energy systems engineering problem is usually one of optimizing heat escape and heat generation in a manner that achieves minimum life-cycle cost. The process is made complex by the virtually infinite possible combinations of insulation, heating apparatus, and construction techniques, which must be analyzed in terms of acquisition cost, operating cost, tax benefits, and future inflation. A rigorous analysis becomes so complex that using a computer is about the only practical way to make it.

An example of a comprehensive analysis is one done by T. A. Vineyard and associates of the Department of Energy's Oak Ridge Laboratory.[3] They optimized

---

[3] T. A. Vineyard, and associates, "Analysis of Conservation and Renewable Options for New Single-Family Residences." *Proceedings of the 18th Intersociety Energy Conversion Engineering Conference,* AIChE, 1983, pp. 2050–2058.

new house designs for Phoenix, San Diego, Salt Lake City, Washington, D.C., and Boston. Their analysis considered variations of the following elements of life-cycle cost:

> Insulation in walls and ceiling, from R19 to R49.
> Two and three layers of glass in windows.
> South-facing windows and thermal shutters.
> Infiltration rates down to 0.2 changes per hour.
> Trombe walls and water storage of solar heat.
> Solar and natural gas heating.
> Tax benefits and future tax costs.
> A real discount rate of 10 percent.

The key conclusion from the study was that the lowest life-cycle cost for the Salt Lake City house resulted from the best practical insulation in the walls and ceilings, triple glazing, and low infiltration. These features reduced heating requirements to a point where the added cost of solar collectors didn't pay off, and the heat could best be supplied from a simple pulse-type gas furnace.

Even old houses can be retrofitted with generous insulation. A three-story town house in St. Louis, built in the 1840s, was re-insulated to levels of R-52 in the walls, R-65 in the ceilings, and R-32 in floors. Four air-to-air heat exchangers recovered the heat from ventilating air discharged out of the building. An air lock entry and multiple glazed windows contributed to lowering the winter heating cost from $100/month to $100/year.[4]

A strategy for heating buildings is to minimize life-cycle cost, which in general terms involves spending the right amount of capital initially to realize reasonable operating-cost savings in the future. Future benefits become marginal when the capital needed to achieve them could alternatively produce an even greater return. Unfortunately, the analysis has to use uncertain predictions of future fuel costs, inflation rates, and interest rates.

A. E. Oviatt calculated the cost and benefit of varying thicknesses of insulation.[5] Table 13–1 shows the cost and benefit from insulating 1,000 square ft of ceiling with various thicknesses of insulation for three of the seven cities analyzed by Oviatt. The term "U" represents the conductivity of the insulation in terms of

---

[4]"Superinsulation, New Energy Saving Technique for Any House," *Better Homes and Gardens,* January, 1984, pp. 23–27.

[5]A. E. Oviatt, "Optimum Insulation Thickness in Wood-Framed Homes," *Pacific Northwest Forest and Range Experiment Station,* U.S. Department of Agriculture, Portland, Oregon, p. 7–8.

TABLE 13-1. Heating and Cooling Costs Per Year Per 1,000 Square Feet of Ceiling (Various Locations)

Item	\<div>Insulation thickness (in inches)\</div>										
	0	1	2	3	4	5	6	7	8	9	10
U value construction[1]	0.292	0.147	0.099	0.074	0.060	0.050	0.042	0.037	0.033	0.030	0.028
				Btus per hour per square foot per degree Fahrenheit							
Cost per year of insulation, 40-year amortization[2]	0	7.45	8.94	10.43	11.92	13.41	14.90	16.39	17.88	19.37	20.86
					Dollars						
Miami:											
Heating	4.05	1.98	1.34	1.00	.81	.68	.57	.50	.45	.41	.38
Cooling	60.00	29.40	19.80	14.80	12.00	10.00	8.40	7.40	6.60	6.00	5.60
Total	64.05	31.38	21.14	15.80	12.81	10.68	8.97	7.90	7.05	6.41	5.98
Total insulation and operation	64.05	38.83	30.08	26.23	24.73	24.09	23.87	24.29	24.93	25.78	26.84
St. Louis:											
Heating	44.75	21.90	14.75	11.03	8.95	7.45	6.25	5.52	4.92	4.47	4.17
Cooling	39.90	19.55	13.15	9.85	7.98	6.65	5.58	4.92	4.38	3.99	3.72
Total	84.65	41.45	27.90	20.28	16.93	14.10	11.83	10.44	9.30	8.46	7.89
Total insulation and operation	84.65	48.90	36.84	31.31	28.85	27.51	26.73	26.83	27.18	27.83	28.75
Montpelier:											
Total	168.00	82.20	55.30	41.30	33.60	27.90	23.50	20.70	18.47	16.80	15.66
Total insulation and operation	168.00	89.65	64.24	51.73	45.52	41.31	38.40	37.09	36.35	36.17	36.52

[1]The R-value is the reciprocal of U.
[2]Oviatt's analysis was published in 1975. The cost of insulation has changed since then.

Btu/h/square ft of area/in of thichness. The R value, as mentioned earlier, is the reciprocal of conductivity. Note that the difference in cost per year, amortized over forty years, of installing eight-inch thick rather than five-inch thick insulation is only $4.47/1,000 square ft. In contrast, the highly insulated walls and ceilings added $7/square ft of floor area to the cost of renovating the 1840 town house in St. Louis. The $7/square ft was many times the original cost of building the house.

CALCULATING HEAT LOST FROM A BUILDING

Heat is carried from a warm object by radiation, convection, and conduction. Heat dissipated by radiation varies as the fourth power of the absolute temperature of the radiating body. A black surface at a temperature (T) facing another black surface at absolute zero temperature will radiate heat at a rate (Qr):

$$Qr = 1.797 \times 10^{-8} \times T^4 \text{ Btu/square ft}$$

If the colder object is above zero, it will radiate heat back to the warmer object at a rate corresponding to its temperature and emittance. However, the fourth-power of temperature relationship usually makes the reverse radiation trivial. The significance of radiation loss depends on temperature. A hot tungsten filament of a lamp dissipates virtually all of its power by radiation. At 2,800 K, its surface radiates 1 million Btu/square ft per hour, or 350 watts/square cm.

A black body at 70°F would radiate 140 Btu/square ft if it faced deep space. If it faces another 70°F surface, it collects as much heat as it radiates, so all objects in a room ultimately reach the temperature of the interior walls. The deep space that the outer surface of a roof sees on a clear night is at 4 kelvin. However, the heat reaching the roof surface is limited by the conductivity of the insulation underneath. The roof is also warmed by convection from the air flowing past it.

In windows radiation is even helpful because the glass is transparent to incoming visible sunlight, but opaque at the infrared wavelengths where the interior objects radiate heat. This is the principle of the greenhouse. Radiated heat affects comfort. A person occupying a space filled with cold air, but surrounded by warm objects feels comfortable.

Transfer by convection. Heat is transferred by convection when air flows past an object that is warmer or colder than the air stream. For example, air blown over the bonnet of a furnace or through the fins of a heat exchanger becomes heated. The outside surface of a wall, door, or window dissipates heat into the outside air mostly by convection, and the heat loss is greater if a wind is blowing on the building. However, changes in wind velocity have trivial effects on heat transfer through a wall that is insulated to reduce conduction loss.

Heat transfer by conduction. The third mode of heat transfer through walls, doors, and windows is conduction, which relates to the nature of heat itself. In a volume of material at absolute zero, −459.7°F, or −273.15°C, all molecules are at rest. Adding heat makes the molecules vibrate, and the temperature of the volume is a measure of how much energy has been invested into causing this vibration. The molecules vibrating in a volume of hot material will rattle the molecules of adjoining materials, thus transferring heat. This process, called "heat conduction," is expressed in heat-transfer tables as k, which has units of Btu/h/square ft of area/in thickness/degree F. The equivalent metric expression of heat conductivity is kilowatts/square m/m/degree Kelvin.

Low heat transfer through walls, windows, and doors is achieved with materials that have low thermal conductivity.

Heat loss through walls and ceilings. The R-value, a reciprocal of k, is more convenient to use in insulation calculations because the R-values of components in a wall or ceiling can be added to get the total R-value of the structure. An example calculation from Ref. 6 is reproduced in Figure 13–1. R-values for some insulating materials, from Ref. 7, are in Table 13–2. An outstanding insulation is polyurethane foam, which has an R-value ranging from 5 to 7 per inch, depending on formulation.

The example in Figure 13–1 shows a heat loss of 22,987 Btu/h through 2,118 square ft of wall with a temperature difference of 46°F between the inside and outside of the wall. A 6.73 kW electric heater could supply this heat, since 1 kW corresponds to 3,412 Btu/h. At 4.5 cents/kWh, this would cost 30 cents/h. This cost could be reduced to 10 cents/h by using an electric heat pump with a coefficient of performance of 3.0. With natural gas costing 60 cents/therm of 100,000 Btu, the heat could be supplied for 15 cents/h if the furnace were 90 percent efficient.

For heat paths in parallel, the U-value of conductivity in Btu/square ft/degree F is convenient to use. The heat flowing through wall and ceiling areas having different conductivities can be added together. Summing the product of

[6]Chuck Eberdt, "Heat Loss Calculations," (Washington Energy Office, Washington Energy Extension Service, Olympia, WA, EY0050), January 1984.

[7]"Living in a Glass House, Energy Savings Through Insulation," bulletin, *Bonneville Power Administration*, U.S. Deparrment of Energy, November 28, 1977.

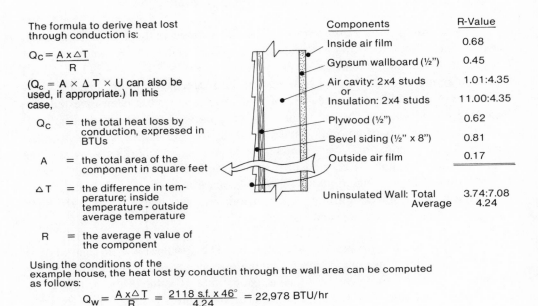

The formula to derive heat lost through conduction is:

$$Q_c = \frac{A \times \Delta T}{R}$$

($Q_c = A \times \Delta T \times U$ can also be used, if appropriate.) In this case,

Q_c = the total heat loss by conduction, expressed in BTUs

A = the total area of the component in square feet

ΔT = the difference in temperature; inside temperature - outside average temperature

R = the average R value of the component

Components	R-Value
Inside air film	0.68
Gypsum wallboard (½")	0.45
Air cavity: 2x4 studs	1.01:4.35
or	
Insulation: 2x4 studs	11.00:4.35
Plywood (½")	0.62
Bevel siding (½" x 8")	0.81
Outside air film	0.17
Uninsulated Wall: Total	3.74:7.08
Average	4.24

Using the conditions of the example house, the heat lost by conductin through the wall area can be computed as follows:

$$Q_w = \frac{A \times \Delta T}{R} = \frac{2118 \text{ s.f.} \times 46°}{4.24} = 22,978 \text{ BTU/hr}$$

Figure 13-1

degree days and heat loss through exterior walls gives part of the energy required to heat a building. Heat is also lost by air infiltration.

OPTIMIZING INSULATION LEVEL

Local building codes specify minimum insulation effectiveness. For example, at one time Oregon and Washington required R-19 insulation in ceilings, R-11 insulation in walls and R-9 insulation under floors in new houses. At the time these values were adopted they were appropriate. Rising fuel prices have made the optimum value higher. FHA standards require that doors and windows in buildings with more than 4,500 degree days in the heating season have double glazed construction. An alternative is storm doors and windows. Chicago, for example, has an average of 6,600 degree days.

TABLE 13-2. Insulation Value of Common Materials

Material	Thickness (in)	R-value	Material	Thickness (in)	R-value
Air and Film Spaces:					
Air space, bounded by ordinary materials	¾ or more	.91	Vertical tongue and groove board	¾	1.00
Air space, bounded by aluminum foil	¾ or more	2.17	Drop siding	¾	.94
Exterior surface resistance	—	.17	Asbestos board	¼	.13
Interior surface resistance	—	.68	⅜" gypsum lath and ⅜" plaster	¾	.42
Masonry:			Gypsum board (sheet rock)	⅜	.32
Sand and gravel concrete block	8	1.11	Interior plywood panel	¼	.31
	12	1.28	Building paper	—	.06
Lightweight concrete block	8	2.00	Vapor barrier	—	.00
	12	2.13	Wood shingles	—	.87
Face brick	4	.44	Asphalt shingles	—	.44
Concrete cast in place	8	.64	Linoleum	—	.08
Building Materials—General:			Carpet and fiber pad	—	2.08
Wood sheathing or subfloor	¾	1.00	Hardwood floor	—	.71
Fiber board insulating sheathing	¾	2.10	Windows and Doors:		
Plywood	⅝	.79	Single window	—	.88
	½	.63	Double window	—	1.60
	⅜	.47	Exterior door	—	approx. 2.00
Bevel-lapped siding	½ × 8	.81	Insulation Materials		
	¾ × 10	1.05	(mineral wool, glass wool, wood wool):		
			Blanket or batts	1	3.30
				3½	11.00
				6	19.00
			Loose fill	1	2.5
			Rigid insulation board (sheathing)	¾	2.10

A. E. Oviatt optimizes the thickness of insulation by first calculating an energy cost index (Ce) for the local area:[8]

$Ce = 0.36\ D\ C t + 0.146\ \Delta T\ CH\ C k$
D = degree days in the heating season
Ct = cost of natural gas in dollars/therm
ΔT = design equivalent sol-air temperature difference, degrees F
CH = summer cooling h/year
Ck = cost of electricity in dollars/kWh

His equation applies only to natural gas heating and electric cooling, and assumes operating efficiencies of 0.67 for the gas heater and 2.0 for the cooling unit. The solar heat gain increases the outside temperature considerably when the sun is shining, and its effect varies with location and the nature of the surface. The *American Society of Heating, Refrigeration, and Air Conditioning Engineers Handbook*[9] lists these design equivalent temperatures and explains their selection. Oviatt explains that his equation gives reasonably accurate insulation thickness with an error of less than 1 in, plus or minus, even in Miami where much cooling is needed.

The total cost of heating 1,000 square ft of area for a year then becomes Ce/R, where R is the value of the insulation. The cost of heating with natural gas costing Ct then becomes:

$$\text{Cost} = \frac{D \times Ct \times 24 \times 1{,}000}{R \times 100{,}000 \times Eg}$$
(Eg = efficiency of gas furnace)

For No. 2 heating oil having 140,000 Btu/gallon and costing Cg/gallon:

$$\text{Cost} = \frac{D \times Cg \times 24 \times 1{,}000}{R \times 140{,}000 \times Eo}$$
(Eo = efficiency of oil furnace)

For cooling with electricity costing Ck per kWh:

$$\text{Cost} = \frac{CH \times Ck \times \Delta T \times 1{,}000}{3412 \times Ee}$$
(Ee = efficiency of cooling unit)

[8] A. E. Oviatt, "Optimum Insulation Thickness in Wood-Framed Homes," *Pacific Northwest Forest and Range Experiment Station,*" U.S. Department of Agriculture, Portland, Oregon, pp. 7–8.

[9] "Handbook of Fundamentals," *American Society of Heating, Refrigerating, and Air Conditioning Engineers,* (New York, N.Y. 1972), Chapter 22, Table 50.

Optimizing Insulation Level

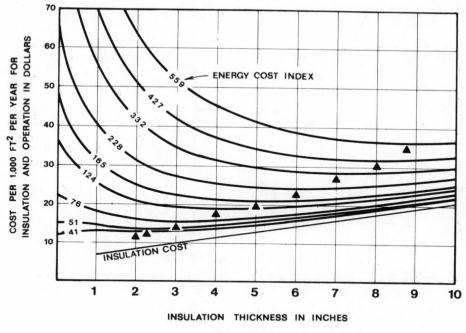

Figure 13-2

Oviatt then plots his calculated costs per 1,000 square ft as a function of insulation thickness. In Figure 13-2 are his plots of optimum roof-ceiling insulation thickness with energy cost index as a parameter. His plots on the cost of wall insulation (Figure 13-3) show optimizations for energy cost indexes of 58, 80, 123, and 490. Note the jump in cost as stud widths increase from 4 in to 6 in and from 6 in to 8 in. Note also that Oviatt's optimizations were based on R-values of fiberglass insulation available at the time. Polyurethane insulation with an R-value of 5 to 7 per in of thickness would change the results.

Heat loss through infiltration. Warm air escaping from a building carries away the energy invested in heating the air. The rate of leakage can be defined in terms of the air changes per hour (ACH). The number of changes per hour can vary between 0.1 and 5. The heat carried away by an air change (Qi) is:

$$Qi = 0.018 \times V \times \Delta T \times ACH$$

where 0.018 = the heat required to heat 1 cubic ft of air 1 degree F
V = volume of space being evaluated
ΔT = temperature difference between outside and inside of space

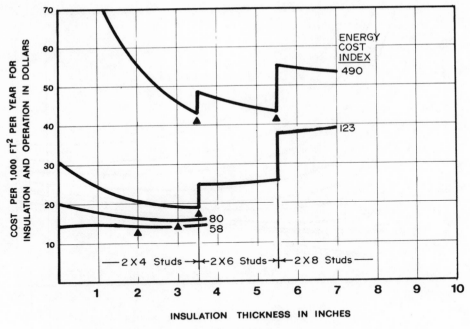

Figure 13-3

For example, consider a structure that has 20,000 cubic ft of volume, with the outside temperature 24°F, and the inside temperature 70°F, giving a temperature difference of 46°F. If there are to be 1.5 air changes per hour, then the heat loss would be:

$$Qi = 0.018 \times 20,000 \times 46 \times 1.5 = 24,840 \text{ Btu/h}$$

Burning about one quart of fuel oil per hour would replace this heat loss.

Variables affecting the number of air changes per hour include tightness of doors and windows, effectiveness of air barriers in the wall, and wind. Utility auditors, building departments, and heating and ventilating contractors use figures that are comparable to the following ranges:

Construction	*Air changes per hour*
New houses, 1 to 5 years old	
with electric baseboard heat	0.3
with central heating	0.6 to 0.8
Houses 5 to 30 years old	0.8 to 1.5

Air to air heat exchangers. Warm air escaping from a building in the winter carries with it heat energy that must be replaced. An air-to-air heat exchanger captures the heat from escaping air before it is released. Descriptions and costs of these devices are published in Ref. 10.

HEAT ENERGY SOURCES FOR BUILDINGS

A building can be designed so that an energy-consuming boiler or furnace for heating is not required. For example, in the North Kitsap High School in Washington, heat energy is delivered into the buildings by students and faculty in the form of metabolism, and by the local public utility in the form of power for lights, pumps, blowers, and heat pumps. Extra heat from daytime operations is stored in a swimming pool and water tanks for night time withdrawal by heat pumps.

A similar heat storage is used in the 880,000-square ft Transportation Building in downtown Boston. The heating was designed by Henry Eggert and Howard Mckew of Shooshanian Engineering Associates of Boston, who predict that they were able to avoid buying 740,000 gallons of oil per year. Thick walls and double-glazed windows in deep recesses reduce heat loss during winter and heat gain during summer. The building is owned by the Commonwealth of Massachusetts. A non-government owner might choose to heat the building with a fuel-burning boiler, and rent out the space needed for the three 250,000-gallon water tanks that store heat for the three consecutive coldest days expected in the winter.

Important considerations in heating buildings include:

- Gradual depletion of fossil-fuel reserves will drive upward the cost of energy during the 15-to-30-year life of a building.
- Congeneration pays if there is a continuous market for the heat and power produced. Otherwise cogeneration is competing with the local utilities that have access to efficient power plants that generate power with low cost fuel.
- Multi-fuel adaptability in a heating plant facilitates post-construction strategies.

HEAT FROM HIGH-TEMPERATURE SOURCES

High-temperature heat comes from combustion of fuels, electricity, sunlight, and nuclear reactors. When we use high-temperature heat for heating buildings, we

---

[10]David Schaub, "Air-to-Air Heat Exchangers," (Washington State Energy Office, Washington Energy Extension Service, Olympia, Wa.), May, 1983.

discard the temperature difference between the source and the heated space. This temperature difference could have been used to generate power.

Available high-temperature electric, gas, and oil heaters do not offer much opportunity for efficiency improvements. For example, all of the electric power delivered to an electric heater in the middle of a room is released as heat in the room. Rocks heated by off-peak electric power may be a practical mechanism for heating buildings with energy from nuclear power plants. An EPRI study showed that 850°F is the best temperature.

The most efficient natural gas furnaces for houses have heat exchangers that condense the water vapor in the stack gases, thus recovering even the latent heat of vaporization of the product water. The condensate can be readily discarded, but the remaining nitrogen and carbon dioxide aren't warm enough to drift up the chimney. The solution has been to pulse the gas combustion to generate the force that drives out the products of combustion. Where an exact efficiency of a gas furnace is not known, a 90 to 95 percent approximation will be close.

The stack gases of an oil-burning furnace cannot be cooled below 212°F without condensing water that forms sulfuric acid with the sulfur-containing gases in the stack. With new burners, efficiencies as high as 85 percent are claimed. Older furnaces were rated in gallons per hour of fuel consumption and Btu output. For example, a furnace with a 1 gallon/h nozzle and a 100,000 btu/h output would have this efficiency:

$$\text{Efficiency} = \frac{\text{output}}{\text{input}} = \frac{h \times 100{,}000 \text{ Btu} \times \text{gallon}}{1 \text{ gallon} \times h \times 137{,}750 \text{ Btu}} = 72.6 \text{ percent}$$

FIREWOOD

When dried in air, wood has an energy content of 5,000 to 6,000 Btu/lb. Yellow pine, an exception, has up to 9,000 Btu/lb. However, most wood that is burned is not dry. The water in wood that is not dry has to be boiled off during combustion, consuming over 1,100 Btu/lb of water.

The efficiency of wood-burning apparatus varies from 0 to 80 percent. Fireplaces are often net consumers of heat, the heat energy in the air carried up the chimney being greater than the heat energy released to the room by the burning wood. Ordinary stoves are around 40 percent efficient, with much of the loss attributed to combustible tar and other partly burned fuels discharged into the chimney.

A wood stove heating a given space in a building must have a rated output that corresponds to the heat loss occurring during the coldest outside-air temperature that the building will experience. At other times, the stove will generate excess heat and consume wood at its worst-case rate unless its heat-generation rate is reduced by restricting air flow with dampers. With less than sufficient air for stoichiometric combustion, unburned gases escape into the chimney, making the combustion inefficient.

Research at the University of Maine has produced the high-efficiency wood-burning stove in which high-temperature combustion is achieved by injecting secondary air into the final combustion zone. The wood is burned rapidly, but only part of the generated heat is immediately released into the building, the rest being stored in a water tank for release after the fire has died.

Air pollution released by simple wood-burning stoves in some communities has used up pollution allowances that could be better used by employment-generating industries. This problem has inspired various types of regulation. For example, the state of Oregon has limited the selling of wood stoves to models that have passed a certification test. Missoula County, in Montana, fines owners of chimneys that produce smoke that's too thick.

Comparison of fuel costs. The relative costs of electricity, natural gas, oil, and wood for heating buildings can be easily compared. Electricity is simple. Every kilowatt-hour releases 3,412 Btu of heat, with 100 percent efficiency, in the space where it is dissipated. On the power bill the electricity cost is printed.

Natural gas is sold by the producer to a pipeline company at a negotiated price in units of 1,000 cubic ft at sea-level atmospheric pressure. The gas is tailored with methane to give it an energy content of 1,000 Btu/cubic ft. In Seattle, for example, natural gas was sold for around 60 cent/therm in 1984. Washington Water Power Company's average residential revenue in 1984 was 58.44 cents/therm. A therm is a unit of 100,000 Btu.

Two kinds of oil are used in heating buildings. One is called "residual fuel," because it is left in the refinery process after all more-valuable products have been extracted. Residual fuel is so viscous that it has to be heated before it can be pumped. It is sold by the barrel to industries equipped to handle it. A 1984 price was around $30/barrel. The more common fuel oil is essentially diesel fuel. It is sold by the gallon, which has an energy content of around 137,750 Btu. The energy content varies, depending on the source of the crude oil.

In comparing heating costs, the efficiency of the furnace must be considered. In the example in Table 13-3, wood having an energy content of 5,000 Btu/lb was used.

LOW-TEMPERATURE HEAT SOURCES FOR BUILDINGS

Buildings can be heated with sources that are at a temperature lower than that of burning fuels or boiling water. Common to low-temperature heat sources is the need for big ducts or pipes to transport the heat. For example, to heat a 70°F space

Table 13-3. Examples of the Cost of Heating Fuels

Heat source	Cost	Btu per unit	Cost per 100,000 Btu	Efficiency	Heat cost per 100,000 Btu
Electricity	6 cents/kWh	3412/kWh	$1.76	1.0	$1.76
Natural gas	$0.60/therm	100,000/therm	$0.60	0.95	$0.63
Fuel oil	$1/gallon	137,750/gallon	$0.73	0.85	$0.85
Wood	$80/cord	17.9×10^6*	$0.44	0.4	$1.12

*5000 Btu/lb, 50 lb/cu ft, 56 percent solid wood.

Low-Temperature Heat Sources for Buildings

with hot air, twice as much ducted air must flow when 105°F solar heat is substituted for 140°F furnace heat. The air flow would have to be tripled if stored heat is recovered from 94°F rocks or water.

Useful sources of low-temperature heat are:

Power plant losses.
Heat pumps that extract heat from the air, water, or ground.
Solar heat stored in water, rocks, earth, and phase-change salts.
Geothermal water.

Geothermal heat sources have been described in Chapter 4 and solar heating has been discussed in Chapter 6.

Power plant losses. Some of the energy in the fuel consumed by a heat engine ends up in the power output. The rest of the energy is dissipated as waste heat. For example, a 1,100 MW nuclear power plant operating at 30 percent efficiency delivers 2,567 MW or 8.7 billion Btu/h of heat, mostly into its cooling tower. To generate that much heat you would have to burn 1,500 barrels of oil per hour. Steam power-plant waste heat is conveniently available at around 100°F, or at higher temperature in steam bled from the turbine.

In Europe, where power plants are located within cities, the waste heat is often used for heating buildings. In America, coal-burning and nuclear power plants are generally built outside of cities. Nuclear plants, in particular, are built on reservations where billion-Btu-per-hour heat customers are hard to find.

District heating of buildings. District heating means supplying heat to many buildings from a central source. In downtown zones of cities are boilers that deliver heat to nearby buildings through underground steam or hot water pipes. This avoids having a boiler and licensed personnel to watch it in each building. Most American district heating plants burn fuel to produce heat, but generate no power. By installing steam turbines, generators, and condensers, the owner of a district heating plant could generate power as well as steam. Up to 80 percent of the fuel energy can be used in this manner.

American cities in the Southwest provide a less favorable opportunity for heating with power plant losses. For example, assume that a 33 percent efficient power plant costs $2,000/kW. For every kW generated it consumes 3 kW of fuel energy and releases 2 kW or 6,824 Btu of heat. A 20 percent finance charge represents a $400 per year or $33 per month cost that persists whether the plant is running or not. This corresponds to $17 per month per kW of heat-producing capacity. If the heating season lasts only two months, then the owner has to choose between shutting the plant down for ten months and absorbing the losses, or generating electric power. The utility to which he would sell his power is being pressed to hold the power rate low by generating power in coal plants or buying surplus hydro from Pacific Northwest.

On the other hand, the owner of the district heating plant could buy a simple natural gas boiler that costs $20 per kW of heat produced. At 20 percent finance cost, he pays $4 per year or 33 cents per month for his idle capacity during summer months.

A district heating plant that also produces power is physically the same as a cogeneration plant. The word "cogeneration" commonly refers to a plant that provides heat and power to a single user.

Cogeneration in high-rise office building. The usefulness of cogeneration depends on the need for the product heat. On the low end of the scale of usefulness was a modern high-rise office building analyzed by J. H. Eto of Berkeley Laboratories.[11] Eto's thorough analysis included consideration of environmental effects, regulatory factors, taxes and tax benefits, and alternative fuels and power plants. His optimizations were mechanized with a computer program that would be useful to anyone evaluating cogeneration for buildings.

Eto's approach was to calculate the value of future benefits from (a) the electricity that didn't have to be bought from the utility, (b) the electricity sold to the utility, and (c) the recovered waste heat. For all values he used a range to establish sensitivity. For example, he calculated the first cost of a power plant at $500/kW, but he perturbed his calculations with a range of $400 to $1,000 a kW. He priced natural gas fuel at 52 cents/therm, and owning and maintenance costs at 0.5 cents/kWh, and again determined sensitivity with excursions. He used 5 cents/kWh for power that he generated, and a range of 5.5 to 6.0 cents for power sold to the utility. He estimated price escalation to be 7 percent each year.

Eto's economic analysis technique was to compute the present value of future benefits, using 10, 25, and 35 percent discount rates. The future benefits were reduced by a 50 percent income tax, but enriched by tax benefits from a five-year depreciation write-off and a 10 percent first-year tax credit. A 10 percent discount rate means that today's value of a $10,000 after-tax profit five years from now is

$$\text{Present value} = \frac{\$10,000}{(1.10)^5} = \$6,209$$

Eto concluded that no strategy that he could devise produced a positive present value with his inputs or excursions of them. He observed that a well-designed office building doesn't need a lot of heat, especially in San Francisco,

---

[11]Joseph H. Eto, "Optimal Cogeneration Systems for High Rise Office Buildings," *Proceedings of the 18th Intersociety Energy Conversion Engineering Conference,* AIChE, 1983, pp. 1825–1831.

Low-Temperature Heat Sources for Buildings

and it is occupied only during business hours. The rest of the time the cogeneration plant is either standing idle, consuming interest on the investment, or burning natural gas to generate power for Pacific Gas and Electric.

It is worth noting that Eto's clients were builders of office buildings who were interested in rates of return of 10, 25, and 35 percent, and were not interested in any returns that come more than 10 years after the building is completed. They apparently had to consider the option of investing money in additional office buildings, instead of cogeneration plants.

Cogeneration with diesel engines. The Public Utilities Regulatory Policies Act (PURPA) could make diesel heating plants that would otherwise be marginal practical. PURPA requires power-supplying utilities to buy energy generated by customers. The price paid has to be related to the avoided cost to the utility for building a new generating plant. This pricing policy is being interpreted in various ways. For example, Washington Water power offers to buy non-firm power from plants producing less than 100 kW for 2.7 cents/kWh. Prices for firm power are negotiated, presumably at higher rates.

A diesel engine releases its waste heat into the cooling water, into exhaust gases, and into the space surrounding it. Virtually all of the heat in the cooling water, which is around 200°F, can be extracted with a water-to-water or water-to-air heat exchanger. A turbo charged engine discharges its exhaust at around 700°F. Recovery of this heat requires a high-temperature heat exchanger that is made of corrosion-resistant materials. An engine also heats its surroundings with convection and radiation, mostly from exhaust manifolds, silencers, and stacks. This heat is generally carried away by the engine-room ventilation.

A Cummins 1,000 kW turbo-charged and aftercooled generator set, when running at full load, has an efficiency of 34 percent. It consumes 73 gallons of 137,750 Btu/gallon fuel each hour.

Input = 73 gallons/h × 137,750 Btu = 10.05 million Btu/h

The 1,000 kW generated during this hour carries away:

Output = 1,000 kWh × 3412 Btu/kWh = 3.4 million Btu

The difference, 6.6 million Btu is released in these forms if the engine is equipped with water cooled manifolds:

Engine heat dissipation	Btu per hour	Percent
Engine cooling water	2.95 million	44.7
Exhaust gases	3.07 million	46.5
Engine-room ventilation	0.58 million	8.8

The exhaust heat exchanger should not significantly increase back-pressure. Otherwise engine efficiency would drop.

A diesel engine is not likely to be the best alternative source of heat when:

- The heat output is not needed continuously. When heat is not needed the plant is generating power with costly diesel fuel. For example, the cost of fuel alone for a 15 kW engine-generator using 1.3 lb/hour of $1 a gallon diesel fuel is 8.6 cents/kWh. Utah Power's coal costs less than 2 cents/kWh generated.
- The plant is small, say 25 kW. Maintenance and interest cost on a per-kWh basis are higher for small plants.

Packaged natural gas engine for hospitals. Martin Cogeneration Systems, in a development sponsored by the Gas Research Institute, is building a packaged 300 kW plant for a 150-bed hospital, and a 450-kW plant for a 250-bed hospital. The plant includes a natural gas engine generator, controls, cooling towers, and heat exchangers for capturing heat from both the exhaust and the cooling water of the engine (Figure 13–4). The target cost is $900/kW. Assuming

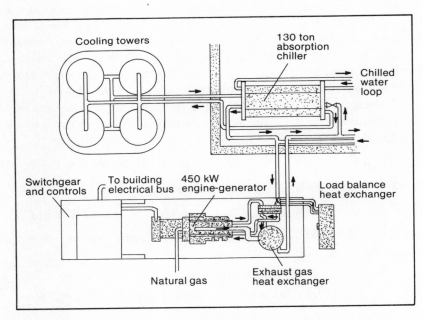

Figure 13-4

8,000 h/year of operation, $5/1,000 cubic ft of gas, and 8 cents/kWh for delivered power, the sponsors expect to see a three-year payback period.

A hospital is a good application for cogeneration. It needs hot water twenty-four hours every day. Furthermore, many codes require hospitals to have standby power plants, which might as well be generating power and heat.

GAS TURBINE COGENERATION

Gas turbines are being used in cogeneration to produce both power and heat. The gas turbine's 20 to 35 percent efficiency is about the same as the 35 percent for an alternative diesel engine. However, the gas turbine has two important advantages:

- The gas turbine runs well on natural gas fuel, which generally costs less than diesel fuel.
- A prosperous air transportation industry has motivated the development of long-life propulsion gas turbines. Industrial derivatives of an aircraft turbine will run for up to 30,000 hours or over three years without overhaul.

The cost of a simple-cycle gas turbine is around $200/kW of electrical output. To this must be added the cost of an unfired boiler for extracting heat out from the exhaust. For example, Vulcan Materials planned a $48 million cogeneration plant running on natural gas to produce 100 MW of electric power plus steam. Cogeneration gas turbines have been installed in industries where there is a continuous use for the by-product heat, and the costly gas turbine produces power and heat 24 hours a day, 365 days a year. Examples of cogeneration gas turbines include:

- Procter & Gamble, in 1982, installed a 20,000 kW General Electric LM2500 unit that supplies 900°F gas into a Vogt 38,000 lb/hour boiler. The plant uses the heat for drying paper, and exports about 10,000 kW to Southern California Edison.
- Dow Chemical has four IM 5000 gas turbines at its Stade chemical processing plant in West Germany. Heat recovery boilers generate steam for processing chemicals. Dow expects to save $30,000/day in fuel with these units.

PUMPS THAT MOVE HEAT

A heat pump extracts heat from one medium, lowering its temperature, and delivers the heat at a higher temperature to another medium. The heat-pump losses can be added to the delivered heat. Forms of heat pumps are piston compressors, turbo compressors, thermoelectric junctions, and chemical reactions. The performance of a heat pump is measured in terms of coefficient of performance (COP), which is the ratio of quantity of heat delivered to the heat equivalent of the energy that drives the pump.

Energy features of a heat pump include:

- All heat pumps are Carnot-cycle limited in efficiency. If it were not so, a heat pump could generate heat that would drive a perpetual-motion engine.
- The COP of a heat pump falls as the temperature difference between the input and delivery ports becomes greater. A good heat pump achieves a COP of around 3 when pumping heat from 45°F outside air into a 70°F room.
- Ambient air and water are the common heat sources for heat pumps.
- Simple control components will convert a heat pump into an air-conditioning cooler.

Practical limits to the use of heat pumps include:

- Heat pumps are not practical in arctic climate.
- Natural gas fired, electric-power plants have fuel rates of 9,000 to 10,000 Btu/kWh. A heat pump with a coefficient of performance of 3 will deliver 10,242 Btu for each kWh of energy consumed. Without even considering transmission and distribution losses, the customer would be better off burning the natural gas himself, rather than buying natural gas-generated electricity for running a heat pump. The same logic applies to power from oil-burning plants.
- When power is generated in nuclear, hydro, or coal plants, the customer does not have convenient access to the original fuels. With such power the value of a heat pump can be determined by comparing life-cost of the heat pump and its alternatives.
- A heat pump may require more maintenance than a furnace requires.

The most common heat pump is arranged around a power-driven compressor that pressurizes a refrigerant to the point where it will condense if allowed to dissipate heat. The heat is dissipated by a heat exchanger into the air circulated by a fan through the building being heated. The condensed refrigerant is then piped to another heat exchanger where it evaporates, absorbing heat from the outside air or water (Figure 13–5). The soil is also a usable heat source. A new development is a Ditch Witch trencher that carves deep vertical slots in the ground. The shape of the buried heat extracting pipe then becomes like a snake in the vertical plane. Thus the heat pump can extract heat from large volumes of earth in a limited area.

Heat pumps for heating houses come in three forms. One has a compressor, fan, and evaporator coil mounted on a pad out of doors. Copper tubing carries the compressed gas to an indoor condenser, and the condensed refrigerant back to the

Pumps That Move Heat

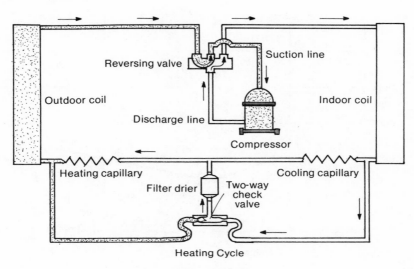

Figure 13-5

evaporator. Ducts and blowers circulate the building air though the condenser. The second form is an add-on heat pump that has the condenser heat-exchanger mounted in the exit duct of an existing furnace. A third version of the heat pump has all components in one package that mounts into a window or hole cut in the wall of a building. Most heat pumps are equipped with valves that reverse refrigerant flow to achieve cooling in the summer.

The quantity of heat per cubic foot of exit air flow is an important characteristic of a heat pump. The heat carried from the condensing coil to the room in a cubic foot of air is its specific heat, about 0.24, times its mass, about 0.074 lb, times the temperature difference between the intake and output of the heat exchanger. The output of a furnace can be 150°F or greater; the output of a heat pump is below 100°F if high COP is desired. As a result, delivering heat at a given rate from a heat pump requires more circulating-air volume than would be required from a furnace. This could mean bigger ducts.

A Friedrich YS 09E33 heat pump illustrates the point. It delivers 9,300 Btu/h when the outside air temperature is 62°F. Inside of the building room, air is sucked into the unit and delivered out with a 23.1–degree temperature rise. When the outside air drops to 42°F, the heat delivery drops to 7,000 Btu/h, and the temperature rise falls to 17.1 degrees. A furnace that heats the air by 70°F would need only one-fourth the air flow to deliver the same amount of heat.

The add-on heat pump neatly avoids this duct problem. The heat pump is operated in moderate weather when the designed air flow is adequate to heat the

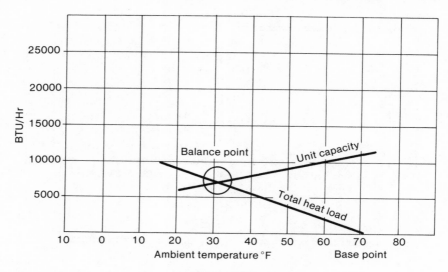

Figure 13-6

house. During the coldest days, the heat pump is turned off and the furnace supplies heat at the appropriate temperature for the existing ductwork.

As the source temperature drops, the heat pump becomes less efficient. A balance point is reached when the COP becomes 1 (Figure 13–6). Most heat pumps are provided with electric resistance heaters that supply the heat when the source temperature falls below the balance point.

Evaporator coils in heat pumps extracting heat from outside air will become coated with ice when the refrigerant is at less than 32°F and the air contains moisture. Controls recognize this freezing condition and periodically reverse the heat pump to melt off the ice.

Performance of a heat pump. A measure of the efficiency of a heat pump is its coefficient of performance, which is the delivered heat divided by the consumed power over a given time period. Two refrigeration words sometimes used in heat-pump literature are:

- Energy-efficient ratio (E.E.R.) is the number of Btus transferred with the expenditure of 1 watthour of electrical energy. One watthour is 3.412 Btu. If 6.828 Btu are transferred with 1 watthour, the ratio is 6.8.
- A ton of refrigeration is the cooling obtained by melting a ton of ice in 24 hours. It corresponds the 12,000 Btu/h.

Pumps That Move Heat

Heat pump example. A Friedrich "Twin Temp" Model YS 09E33 heat pump is a single unit. It heats in the winter and cools in the summer. It runs on 208-230 volts 60 Hz ac power, and provides 8,800 Btu/h of cooling or 7,600 Btu/h of heating. Outside air is the heat source for heating and the heat sink for cooling. The refrigerant flow is shown in Figure 13–5.

Performance of the unit as a heat pump is specified as 8.2 E.E.R. That means that at an outside temperature of 40°F, it delivers 8.2 Btu/watt-hour of electricity consumed. The corresponding coefficient of performance (COP) is:

$$COP = \frac{\text{Heat Out}}{\text{Energy In}} = \frac{8.2 \text{ Btu} \times \text{watt-hour}}{1 \text{ watt} \times 3.412 \text{ Btu}} = 2.4$$

At 52 °F outside air temperature the COP is 2.8.

Friedrich provides a simple chart for calculating the heating required from a room unit (Table 13–3). The calculated heat required at 40°F establishes the rating of the heat-pump unit. Below 40°F electric resistors supply adequate heat.

TABLE 13-4. Heating Load Form
Friedrich Room Unit-Heat-Pumps

		Btu/h per °F Temp. Difference
Walls: (Linear Feet)		
2" Insulation	Lin. Ft × 1.6	
Average	Lin. Ft × 2.6	
Windows & Doors: (Area, Square Ft)		
Single Glass:	Sq. Ft × 1.13	
Double Glass:	Sq. Ft × 0.61	
Infiltration—Windows & Doors: Avg.	Lin. Ft × 1.0	
Loose	Lin. Ft × 2.0	
Ceiling: (Area, Square Ft)		
Insulated (6")	Sq. Ft × 0.07	
Insulated (2")	Sq. Ft × 0.10	
Built-up Roof (2" insulated)	Sq. Ft × 0.10	
Built-up Roof (½" insulated)	Sq. Ft × 0.20	
No Insulation	Sq. Ft × 0.33	
Floor: (Area, Square Ft)		
Above Vented Crawl Space		
Insulated (1")	Sq. Ft × 0.20	
Uninsulated	Sq. Ft × 0.50	
Slab on Ground		
1" Perimeter Insulation	Lin. Ft × 1.70	
Based on Linear Feet of Outside Wall	Lin. Ft × 1.00	
	Total Heat Loss per °F	Btu/h/°F

Multiply total Btu/h/°F × 30 and plot on Figure 13-6 at 40°F. Draw straight line from 70 base point thru point plotted at 40°F. Intersection of this heat loss line with unit capacity line represents the winter design temperature in which the unit will heat the calculated space.

14

Non-Fossil Fuels: Hydrogen, Alcohol, and Wood

Petroleum, coal, and natural gas are fossil fuels that supply most of the world's energy. Fuels such as gasoline and diesel fuel that are derived from petroleum can be replaced with liquid fuels such as methanol, which can be derived from coal. However, the burning of fossil fuels adds an extra 5 billion tons of carbon dioxide each year to the atmosphere. Since the beginning of the industrial revolution, the carbon dioxide content of the atmosphere has grown from around 270 to 340 parts per million. The sulfur and nitrogen oxide emissions from burning fossil fuels can be controlled, but not the carbon dioxide. Hydrogen is a fuel that could be generated with nuclear power without contributing carbon dioxide to the atmosphere.

Burning of alcohol contributes to the carbon dioxide load in the atmosphere. However, as an additive to gasoline it makes engines more efficient and reduces the total amount of fuel that must be consumed. The lowest cost alcohol is made from natural gas, petroleum, or coal.

Features of hydrogen, alcohol, and wood are described in the sections that follow.

HYDROGEN: A GOOD FUEL BUT HARD TO STORE

Hydrogen is a gas that burns in air and produces steam as a product of combustion.

Energy characteristics of hydrogen include:

- Hydrogen, a byproduct of petroleum refining, is sometimes burned in a flare because there is no market for it. Hydrogen can also be produced from natural gas.
- Hydrogen is used in the U.S. to make ammonia, which is a fertilizer that is applied directly as anhydrous ammonia or indirectly in ammonia compounds.
- About four lb/cubic ft is the most compact stored volume of hydrogen, obtainable with metal hydrides or in cryogenic tanks.
- Hydrogen can be used as a fuel, displacing petroleum products for propulsion of vehicles. However, many of the non-petroleum sources of hydrogen are also useful for fuels.

Convincing scenarios suggest that petroleum should be reserved for lubricants, airplane fuel, and chemical feedstocks, and hydrogen should be used as an automobile fuel as soon as a distribution network can be established.

Characteristics of hydrogen are summarized in Table 14–1. Hydrogen is hard to store. Optimized containers for storing hydrogen gas are heavy, for example, 100 lb of calcium-mischmetal hydride/lb of hydrogen stored. To make hydrogen into a liquid takes about one-fourth the energy of the product hydrogen, and a high-quality 1,000 gallon cryogenic hydrogen tank for liquid hydrogen would have a boiloff rate of 0.5 percent/day. An aluminum liquid hydrogen tank designed at Beech for a hydrogen bus would be 48 inches in diameter, and would weigh 600 lb empty and 795 lb filled.[1] The hydrogen would be at a pressure of 75 psia and boiloff would be 1 percent/day.

For vehicles, metal hydride seems to be the best way of storing hydrogen. The hydrogen is absorbed, forming a metal hydride at one temperature, and released at a higher temperature. For example, a mischmetal-nickel-aluminum alloy can be charged to 90 percent of its hydrogen capacity in 5 minutes at 112 psi (0.8 Mpa) at 77°F (25°C). At a lower pressure it releases hydrogen only when supplied heat. The performance of metal hydrides for storage of hydrogen is described in Ref. 2.

Mercedes, in Germany, built a small fleet of hydrogen-powered buses, with hydrogen stored in iron-titanium hydride. Heat required to release the hydrogen is recovered from the engine exhaust. When the hydride bed is recharged at night this heat is released in the garage where the buses are stored. Thus the heat,

---

[1]David N. Ferrence, "Hydrogen Transit Bus Program," *Proceedings of the 19th Intersociety Energy Conversion Engineering Conference,* ANS, 1984, pp. 1433–1442.

[2]G. D. Sandrock, "A New Family of Hydrogen Storage Alloys Based on the System Nickel-Mischmetal-Calcium," *Proceedings of the 12th Intersociety Energy Conversion Engineering Conference,* 1977 pp. 951–958.

Hydrogen: A Good Fuel But Hard to Store

Energy released, $H_2 + O_2 \rightarrow H_2O$

Energy measured in--	Water in the form of--	
	Liquid	Steam
Btu per pound of hydrogen	62,032	51,593 (100°C)
Btu per ft³ of hydrogen	325	275
Joules per gram of hydrogen		120,900
Calories per gram of hydrogen		28,876
Calories per cm³ of hydrogen		2.414
kWh per gram of hydrogen		0.03358
MJ per ft³ of hydrogen		0.286287 (21°C)
kWh per pound of hydrogen		15.233
Joules per cm³ of hydrogen		10.11 (21°C)
kCal per mole (18 grams of water) 25°C		57.7976

Equivalent voltage $\dfrac{285{,}800}{2 \times 96{,}500} = 1.48$ V

Fuel-cell maximum: $\dfrac{237{,}200}{2 \times 96{,}500} = 1.23$ V

Compressed-hydrogen content of a 2640 cu in. tank (8.5 OD x 51 in).

[Graph: Hydrogen in tank, cu ft. vs Gauge pressure, psig, at Temperature, °F: 0, 40, 70, 100, 130]

---

originating in the engine losses, is recovered for heating the building, in effect achieving cogeneration.

Hydrogen production. The cost of hydrogen varies, depending on where it is and how much it is needed. For example, Dow Chemical in its Stade, West Germany, chemical plant, was burning unneeded hydrogen in a flare until 1985 when it installed a pair of IM-5000 gas turbines made by General Electric to generate electric power plus steam. The turbines burn a nominal mixture of 50 percent hydrogen and 50 percent natural gas. The natural gas costs $6/million Btu.

Electrolysis of hydrogen with commercial power is around 50 percent efficient, considering rectifier, electrolyzer, and transformer losses. Assuming that the power costs 8 cents/kWh, the cost of electrolyzed hydrogen (Ch) would be:

$$Ch = \frac{\$0.08 \times kWh \times 120{,}000}{kWh \times 0.5 \times 3412 \; Btu \times 120{,}000} = \$5.63/120{,}000 \; Btu$$

A gallon of gasoline, costing around $1.00, also contains 120,000 Btu.

General Electric has a catalyzed solid-polymer electrolyte that could be adapted to produce hydrogen with 80 percent efficiency in the electrolyzer cell.[3] They estimate that with their technology, hydrogen would cost $13/million Btu if electricity costs 4 cents/kWh. However, at $0.04/kWh, the cost of a million Btu (Cm) of energy is:

$$Cm = \frac{\$0.04 \times kWh \times 1{,}000{,}000}{kWh \times 3412 \; Btu \times million} = \$11.72/million \; Btu$$

Thus the $13/million Btu must be based on a 90 percent efficiency with no allowance for other costs of buying and running the plant. As a reference, natural gas at retail costs around $6/million Btu.

In another estimate, Gerald A. Crawford described development of electrolysis having 80 to 90 percent efficiency and costing $400 to $800 per kW.[4] He observed that if surplus hydro power could be had for 2 cents/kWh, his plant could compete with natural gas by producing hydrogen for $5/million Btu.

Water can also be thermally broken down to hydrogen and oxygen, but at a temperature so high (4,500°R or 2,500°K) that it is not feasible to separate the gases before they recombine during cooling. An alternate way of making hydrogen from water is with heat and chemical processes, thus avoiding the some 60 percent heat-to-electricity conversion loss. For this process to be efficient, the heat flows must be cascaded and balanced so that the right amount of heat is available at the right temperature for each stage, and all rejected heat by one part of the process must be used by other lower-temperature parts. A plant with such features must be large to produce hydrogen at reasonable cost. H. H. Otsuki and his associates describe such a hydrogen-producing plant that is powered by a nuclear reactor which can deliver process heat at 1160°K (1628°F).[5] Both the plant and reactor could be developed by the year 2000.

---

[3]M. Bonner and associates, "Current Research in Advanced Water Electrolysis in the U.S. and Abroad," *Proceedings of the 17th Intersociety Energy Conversion Engineering Conference*, IEEE, 1982, 1197–1201.

[4]Gerald A. Crawford, "The Greening of Large Industrial Electrolytic Hydrogen (EH2)," *Proceedings of the 19th Intersociety Energy Conversion Engineering Conference*, ANS 1984, pp. 1435–1440.

[5]H. H. Otsuki and others, "Recent Developments in the Engineering and Chemistry of the Zn Se Thermochemical Hydrogen Cycle," Proceedings of the 12th Intersociety Energy Conversion Engineering Conference, ANS, 1977 pp. 939–946.

Hydrogen: A Good Fuel But Hard to Store

A study done for NASA provides these cost estimates for fuels derived from coal[6]:

Product	Cost updates to 1980 dollars[7]	
	Per GJ	Per million Btu
Synfuel (equivalent of aviation grade kerosene)	8.04	8.47
Liquid methane	8.39	8.89
Liquid hydrogen	10.45	11.01

Hydrogen engines. Hydrogen can be converted into mechanical power with engines or with fuel cells and electric motors. The fuel cell, described in Chapter 8, contains a costly plantinum-type catalyst for which an interest charge persists every day, whether or not the cell is used. The fuel cells will become practical for 8-to-16-hour-per-day service before they are practical for vehicles. A laundry, which requires heat and electricity, is a good example of an early application for fuel cells.

Gasoline engines have been adapted for hydrogen use, and engines have been designed specifically for hydrogen fuel. Hydrogen is not a good fuel for diesel engines because it must be compressed before being injected into the engine. This consumes power. Devices for injecting gaseous hydrogen into diesel engines have not been developed.

The successful hydrogen engines have used spark-ignition and the Otto cycle. Efficiencies as high as 32 percent have been measured. For example, ninteen engine configurations were tested at the University of Miami with variations in the combustion chamber configuration, timing, mixture formation method, exhaust gas recirculation, and water and air injection.[8] Compression ratio was 9:1, and the measured thermal efficiencies were not significantly different from those obtained with gasoline as a fuel.

In search of a way to reduce emissions within a mine, Eimco Mining

---

[6]Robert D. Witcofski, "Comparison of Alternate Fuels for Aircraft," NASA TM 80155, September, 1979.

[7]D. Daniel Bruner, "Is Liquid Hydrogen the High Cost Option for Aircraft Fuel?" *Proceedings of the 17th Intersociety Energy Conversion Engineering Conference,* IEEE, 1982, 1191–1195.

[8]M. R. Swain and associates, "Hydrogen Engine Design Data Base Summary," *Proceedings of the 18th Intersociety Energy Conversion Engineering Conference,* AIChE, 1983, pp. 536–542.

Machinery International and Hydrogen Consultants modified a Caterpillar 3304 turbocharged and aftercooled diesel engine to operate with hydrogen fuel.[9] The compression ratio was reduced from 17.5:1 down to 10.5:1 by substituting Caterpillar gas-engine pistons. Spark-plug adapters were likewise available for gas conversions. The engine was equipped with parallel induction in which hydrogen and air flow through separate passages into the manifold. The intake valve was arranged to delay hydrogen flow until air flow had started. Hydrogen was supplied from a metal hydride fuel source. The turbocharged engine gave virtually the same performance with hydrogen as it would have given with diesel fuel. Emission of NOx was less than 1 gram/kWh, compared with 7 to 10 g/kWh for diesels. The hydrogen engine, of course, generated no carbon monoxide or carbon dioxide.

METHANOL AND ETHANOL FOR FUELS

Denatured alcohol, which is ethanol with a poison added, is used in alcohol blowtorches and stoves. Wood alcohol, also called "methanol," is used in stoves for heating. These alcohols, which can be derived from biomass, are often suggested as a fuel for automobiles and power plants.

Energy features of thes alcohols include:

- Methanol is economically and efficiently produced in a petroleum refinery with a catalyst from methane (CH_4) and hydrogen. Methanol can also be made from wood.
- Ethanol can also be manufactured economically in a refinery.
- Methanol, when burned in air, releases 9,603 Btu/lb compared with 18,000 to 20,000 Btu/lb for gasoline.
- An Otto-cycle spark-ignition engine will burn methanol more efficiently than it burns gasoline.

Fuels that are alternatives to gasoline and diesel fuel are not, in general, economically practical when they are made from petroleum. They do become useful when they can be produced with non-petroleum energy at lower cost than equivalent petroleum products.

Methanol engine for recovering exhaust heat. Once when testing a Volkswagen with a 1,350 cc air-cooled engine with gasoline-methanol fuel mixtrues, we discovered that the car delivered more miles per gallon than it did with

---

[9]N. Baker and associates, "A Hydrogen Engine for Underground Mining Vehicles," *Proceedings of the 18th Intersociety Energy Conversion Engineering Conference,* AIChE 1983, 569–574.

pure gasoline. This was surprising because methanol has less energy than gasoline:

Fuel	Energy in Btu	
	Per lb	Per gallon
Gasoline	18,320	125,000
Methanol	9,603	64,914

The reason turned out to be that methanol requires more heat for evaporation than does gasoline. Specific values are 353 Btu/lb for gasoline and 512 Btu/lb for methanol. This cools the intake of the engine, improving efficiency. Professor L. C. Lichty notes that as a result, the heating value of air-fuel mixtures is 94 Btu/cubic ft for gasoline, and 104 Btu/cubic ft for methanol.[10] Thus the power output of the engine as well as its efficiency are improved by using methanol.

Pincas Jawetz observed that the nation's oil imports could be reduced by making alcohol from petroleum and selling a motor fuel that is 10 percent alcohol and 90 percent gasoline.[11] He calculated that even if it took 2 energy units of petroleum to make 1 energy unit of alcohol, the net result would be about 3 miles/gallon average improvement in the nation's vehicle fleet. He noted that tests with fleets showed a 5.3 to 6.1 percent improvement in miles/gallon when gasahol was substituted for gasoline as a fuel. This improved efficiency comes from the higher compression ratios that become possible because alcohol inhibits knocking in engines.

ALCOHOL FROM AGRICULTURE PRODUCTS

Farms produce such products as food, fibers, flowers, and tobacco. Farms also produce crops that can be converted alcohol. Ethanol or grain alcohol has a heating value of 12,000 Btu/lb, or 83,650 Btu/gallon, so many have suggested its use for automobile fuel.

Use of ethanol is encouraged in grain-growing states by taxation formulas. The real cost of gasoline derived from Mid-East oil is probably less than 15

---

[10] Lester C. Lichty, "Internal Combustion Engines," 6th ed. (New York: McGraw Hill, 1951), p. 146.

[11] Pincas Jawetz, "Improving Octane Values of Unleaded Gasoline via Gasahol," *Proceedings of the 14th Intersociety Energy Conversion Engineering Conference,* ACS, 1979, 301–302.

cents/gallon, including well drilling, pumping, transporting, and refining costs. Most of the selling price is taxes, levied by governments in various guises. Thus the use of alternatives to gasoline can be easily encouraged by manipulating tax formulas.

Making ethyl alcohol requires more energy than the alcohol contains. For example, the typical American farm is not a positive supplier of energy because the energy in fuels and fertilizers that it consumes can exceed the energy content of its products. The energy in eggs, beef, and milk, is as little as one-tenth of the energy used in producing them. Intensive corn growing consumes, in petroleum products about one-half the energy that appears in the corn. Furthermore, not all of the energy in the farm-produced grain can be recovered in the alcohol. Some is rejected in the heat and byproducts of the fermentation process.

Finally, the conventional distillation that produces high-purity alcohol consumes more heat energy than the product alcohol contains. Thus, the logical use of grain for energy is to stoke it into boilers like pulverized coal, and use the displaced oil for vehicles. This then becomes a prohibited logic because it inspires visions of Americans burning wheat while people in other countries starve.

An alternative to distillation for separating ethanol from water is extraction with a solvent that is completely miscible with ethanol at one temperature, and partly miscible at another. The highest temperature in the process analyzed by Malcolm D. Fraser is 115°C.[12] Heat exchangers reuse as much of the heat as possible in the process. White paraffin oil was the best solvent of those analyzed, and the final product would be 94.7 percent ethanol, 4.5 percent oil, and 0.8 percent water by weight. The thermal energy required would be somewhere between 4 and 34 percent of the fuel value of the product.

Methanol manufacture. Methanol is a byproduct of petroleum refineries. It is also made from the methane in natural gas. Low-cost natural gas is available in remote oil fields, where it is otherwise burned in flares to avoid fire hazard. The Signal Companies are designing methane-to-methanol converters for use in such remote areas.

Saudi National Methanol Co. converts natural-gas feedstock into 700,000 tons of methanol per year at Jubail.

Prohibited approaches. Certain feasible energy-use alternatives are not acceptable because they arouse emotions, or because they are not compatible with the promoter's interests.

One example is emotional opposition to garbage burning. European firms have developed boilers that efficiently burn garbage, extracting some 5,000

[12]Malcolm D. Fraser and Gurmukh D. Mehta, "The Feasibility of a Novel Extraction Process for Separating Ethanol and Water, *Proceedings of the 18th Intersociety Energy Conversion Engineering Conference,* AIChE, 1983, 592–597.

Btu/lb for steam turbines or district heating. The combustion in the boiler and the filters for the stack gases are designed to limit emissions to legal limits. Such garbage burning units are operating in Saugus, Massachusetts, and Baltimore, Maryland. An installation was proposed for Oregon City, Oregon, where garbage is buried in a landfill, and where paper mills could use the generated steam. The opposition effectively organized a campaign to prohibit construction of the plant on the basis that neighboring communities would furnish much of the garbage that would be burned.

BURNING WOOD FOR POWER GENERATION

Continuous need for process steam made cogeneration a good replacement for boilers heated with natural gas and oil at Dow Corning's Midland, Michigan plant that manufactures chemicals. The cogeneration plant, developed by Bernard A. Bartos, manager of the firm's Power and Utilities department, has 1250 psi, 900°F steam driving a 22.4 MW turbo generator. Steam is bled from the turbine at 165 lb/in^2 for chemical processes and 30 lb/in^2 for heating buildings.

Unique to the plant is wood fuel. From 390 to 500 dry tons of wood is needed each day to generate the 275,000 lb/h of steam. This requires forty truckloads to be delivered each day, five days a week, to a four-acre concrete-paved storage area. Around one-half of this wood is recovered from sawmill wastes, power line trimming, tree service activities, and industrial scrap such as broken pallets.

The other half of the fuel supply is harvested from Michigan forests by contractors directed by a Dow seven-person team of foresters and wild-life biologists. The team contracts with landowners for thinning or clearing ten to fifty acre cuts as appropriate. For example, Dow contracted with the U.S. Forest Service to salvage dead wood from a 3,000 acre burn site. Around 4,000 acres of forest each year must be harvested to keep the plant going. This is 0.1 percent of Michigan's 4 million acres of forest. The wood is within seventy-five miles of the plant. Hydraulic feller-bunchers, grapple skidders, and chippers are used in harvesting.

Electrostatic precipitators were sufficient to clean the emissions discharged out of a 200-ft stack. An air emissions permit was required, even though the cogeneration plant replaced oil and gas-fired boilers. Fire protection was required for the woodpile, and explosion-proof electrical equipment was required in wood-dust areas.

FUEL FROM FOREST WASTE

Waste wood in forests can supply heat energy at a cost lower than natural gas or oil. An objective of forest management is to maximize the yield of high-value tree

350   Non-Fossil Fuels: Hydrogen, Alcohol, and Wood   Chap. 14

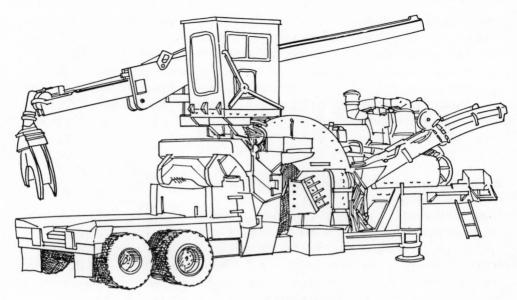

Figure 14-1

species. This is achieved in part by removing dead and diseased trees, and excess trees that limit the growth of the desired product. The waste wood can be recovered and delivered to a boiler for around $1.50/million Btu, or around one-third the cost of natural gas.

The keys to successful waste wood fuel are these:

- Providing owners of small woodlots the forest appraisal and management techniques that give them the best yield in forest products.
- Using equipment that can extract designated trees from a forest without disturbing those that are to remain.
- Converting the wood into chips that can be blown into trucks and unloaded into a hopper.

An example of the use of wood chips is a study by Progressive Engineering Consultants for the Central Michigan University on ways of reducing the cost of heating the campus.[13] Paul T. Spelman and his associates found, in a life cycle

---

[13]Paul T. Spelman and associates, "Central Michigan University Technical Assistance Report, Boiler Plant," Progressive Engineering Consultants, Grand Rapids, Michigan, April 6, 1982.

Fuel from Forest Waste

cost analysis, that a $2.7 million investment would save $20 million during the subsequent ten years by substituting wood fuel for 346 million cubic feet of gas being burned each year. The wood would be burned in a Lamb wet cell burner, with the product hot gas ducted into one of the existing Wickes boilers. Other existing natural gas boilers would be operated only during heating peaks in January and February each year.

Forest resources. In supporting the Central Michigan University study, Satellite Sauchip System assessed the woodlots within a 9.5-mile radius of the campus. They found 37,937 wooded acres in the 218,880 acre area. They estimated that the average forest there contained 100 tons of wood/acre, of which 40 tons would be harvested during initial thinning. After that the regrowth would be 6 tons/acre, or 227,622 tons/year for the 37,937 acres of forest.

In an example analysis of a 15-acre woodlot, the forestry firm showed how the owner could sell timber worth $10,600 during ten years, yet end the period with a forest having essentially the same value as it had at the start of the harvesting. The key is selectively harvesting wood that is suitable only for fuel and paper mill chips, worth $0.75 to $2.25 per ton. The growth of trees that yield high-grade logs worth $7 to $15 per ton is enhanced.

Waste wood recovery in the forest. Morbark Industries, Inc. in Winn, Michigan, a manufacturer of forestry processing equipment, suggests this basic combination for producing wood chips:

Quantity	Type
2	Mark IV Feller Buncher
1	20-inch Super Buncher
3	Grapple Skidder (Handles 150 to 250 tons per shift)
1	Model 27 whole tree chipper (800 to 900 tons per shift)

The feller buncher has hydraulic shears that cut trees at the ground surface. Its grapple is arranged to hold several previously cut trees while cutting another one. The grapple skidder hauls trees from the forest to the nearby chipper. These units ride through the forest on low-pressure tires that minimize soil disturbance.

The Model 27 chipper is 33 ft long, 10 ft high, and 13.5 ft wide (Figure 14–1). It weighs 62,000 lb, but can be hauled over roads from site to site like a truck trailer. The operator feeds trees into the chipper with a grapple, and the chips are blown into a truck that hauls them to the customer's plant. The highest rate of feed of 27-inch logs into the chipper is 80 ft of log/minute, and it then produces 3.4 tons of 1 inch chips/minute.

Cost of wood chip fuel. Jerry Morey of Morbark Industries provides this cost estimate for recovering wood fuel from a hardwood or pine stand in the eastern U.S.

Whole Tree Chipper: One Morbark Model 27 Total Chiparvestor, including spare parts kit, extra knives, delivered	$257,875
Feller Bunchers: Two Mark IV Feller Bunchers with spare parts kit, delivered	118,350
One Super Buncher, delivered	88,000
Skidders: Three Grapple Skidders	210,000
Yard tractor with blade	17,000
Maintenance and fuel truck	15,000
Total Equipment Cost	$706,225

Morey then calculated the cost of operation for five years, disregarding the salvage value of the equipment at the end of the period:

Original equipment cost	$706,225
Interest, 8 % add-on (approximately 16 % simple), 5 years	282,490
Insurance, based on 3 % of the original cost per year	105,934
Maintenance, including labor, engine oil, lubrication and hydraulic oil, chipper knives, counter knives, and miscellaneous supplies (grows from $65,600 the first year to $120,500 the last year)	465,325
Total, equipment and its operation	$1,559,974

While the equipment can produce 500 tons/day of chips, Morey assumed a conservative 350 tons/day and 200 days/year, for a cost of produced fuel (Fc):

$$Fc = \frac{\$1,559,974}{350 \text{ tons/day} \times 200 \text{ days/year} \times 5 \text{ years}} = \$4.46/\text{ton}$$

To this cost Morey adds fuel, labor, and miscellaneous elements:

	Cost per ton of chips
Fuel consumption, using an average of 300 gallons per day of fuel costing $1.04 per gallon	$0.89
Labor, using $450 per week per person of the 8-person crew for direct and indirect cost for 40 working weeks plus 2 weeks vacation	2.16
Administration, taxes, office rental	0.27
Total, delivered into a van	7.78

This total cost is sensitive to annual production tonnage. The variable costs such as fuel, have only a slight effect on the cost per ton, so Morey ignored them in this sensitivity analysis:

Fuel from Forest Waste

Annual tonnage production	Cost per ton
60,000	9.08
70,000	7.78
80,000	6.81
90,000	6.05

Morey derives the delivered cost of wood fuel as follows:

Harvesting cost	$7.78
Trucking (25 tons per load, $1 per mile, 80 miles round trip)	3.20
Stumpage (compensation given to the land owner, runs from $0.50 to $1.00 per ton)	1.00
Profit	3.00
Total, delivered chips	14.98

Alternatives to Wood Fuel. Wood is generally burned as received without drying, so it has around 5,000 Btu/lb. The cost of the heat (Ch) then is:

$$Ch = \frac{\$14.98 \times \text{ton} \times \text{lb} \times \text{million}}{\text{ton} \times 2,000 \text{ lb} \times 5000 \text{ Btu} \times \text{million}}$$
$$= \$1.50/\text{million Btu}$$

This compares with natural gas at $3.00 to $4.00/million Btu. Residual fuel oil containing 5,850,000 Btu in each $30 barrel would cost $5.12/million Btu. Coal at $40/ton would cost $2 per million Btu. Coal, when burned, releases sulfur compounds, which must be captured by cleaning up the stack gases. Wood has no significant sulfur content, and when properly burned produces no visible smoke or fly ash.

In his analysis of supplying heat for the Central Michigan University, Spelman noted that the wood burning unit can operate automatically. The incoming truck drivers operate the weigh scales, position their trucks on the dumper, and unload their own loads. Spelman also observed that the money spent for the wood fuel remained in the community.

The analysis also indicates that there is a limit to the availability of wood. The sustained yield was 6 tons per acre per year, and 17 percent of the land surveyed was wooded. We could assume that Consumers' Power chose to replace the 808 MW Midland 2 nuclear power plant with a 30 percent efficient wood burning power plant. The required forest are a (Af) would be:

$$Af = \frac{808 \text{ MW} \times 1{,}000 \text{ kW} \times 3412 \text{ Btu} \times 8766 \text{ h} \times \text{acre} \times \text{ton} \times \text{lb}}{0.30 \times \text{MW} \times \text{kWh} \times \text{year} \times 6 \text{ tons} \times 2{,}000 \text{ lb} \times 5000 \text{ Btu}}$$
$$= 1{,}342{,}608 \text{ acres of forest, or}$$
$$= 1{,}342{,}608 \text{ acres}/0.17 = 7{,}897{,}700 \text{ acres of total land}$$

This corresponds to 12,340 square miles, or a circle with a radius of 62.6 miles. Thus such wood-burning power plants, if operating in an area which is 17 percent forest, would have to be spaced 125 miles apart.

Sawmill waste fires a 42.5 megawatt power plant. A sawmill cuts logs into boards, and the byproducts are bark, sawdust, and long slivers. The slivers could be used for making paper and particle board. Sawdust and bark, called "hog fuel," are a suitable fuel for boilers. Pollution control laws in the late 1970s prohibited open burning, so mill waste had to be burned in a controlled manner, hauled to land fills, or hauled to users. Mills producing seasoned lumber burn the waste in boilers that generate steam for the drying kilns.

The cost to the customer of the waste is mostly the cost of transporting low density material. A firm evaluating the construction of a wood-waste compacting plant in Tacoma, Washington, was offered delivered hog fuel for around $17/ton.

Washington Water Power Co. completed in 1983 a 42.5 megawatt wood-burning power plant at Kettle Falls, Idaho at a cost of $87.1 million, or $2,050/kW. At about the same time Snohomish County Public Utility District No. 1 in Washington completed its Sultan River 111.8 megawatt hydro project for $180 million, plus financing costs, or $1,610/kW.

The Kettle Falls wood-waste fired generating station uses private contractors to truck in 500,000 tons a year of sawdust, bark, and planer shavings from sawmills within an 85-mile radius from the plant. The moisture content of the wood ranges from 40 percent to 55 percent and the heating value at 50 percent moisture content is 4,300 Btu/lb. Wood waste is stored around a 1,600-ft conveyor so that the generating station can be scheduled to generate power during the 7,000 h/year when wood-waste power fits the needs of the utility. The 42.5 MW from this plant supplies only a small part of the 1,200 MW peak generation available to the utility.

This plant illustrates the limits of the wood-waste as a power plant fuel. It takes fifty truckloads a day for the 1,714 tons of fuel needed to keep the plant operating at full power. Increasing the power to 85 MW would require 100 truckloads a day, brought in from a 120-mile radius containing 45,238 square miles. This assumes that waste-producing sawmills are uniformly distributed. The state of Washington contains only 681,192 square miles, and much of it has no sawmills.

Energy from garbage. Garbage burning, a practical source of energy, is being adopted in many places in the world and even in the U.S. The key component is the boiler, which accepts raw unsorted garbage, and burns the combustible components completely with controlled temperature and secondary air added to combustion zones. The plant produces steam and electric power in various combinations. It accepts garbage directly from trucks, charging a fee that is competitive with alternative methods of disposal.

Signal Refuse Energy Systems Company (RESCO) builds plants that burn 2,250 tons of garbage per day and generate 55 MW of electric power. For example, the plant built for Pinellas County, Florida, began operation in 1983, selling 50 MW of power to Florida Power Co. RESCO builds and operates the facilities, collecting revenue from disposal fees and sale of power and steam.

In Albany, New York, solid waste is shredded. Iron is then extracted from the product, leaving "refuse-derived fuel" (RDF). A. R. Nollet and R. H. Greeley report that the heating value of moisture-free RDF ranged from 5,563 to 6,011 Btu/lb (12,940 to 13,980 kJ/kg).[14] Ash ranged from 35 to 40 percent. They estimate that the average RDF, as produced, will contain 23 percent water and have a higher heating value of 4,620 Btu/lb (10,750 kJ/kg). With a 65 percent boiler efficiency they calculate that 35 tons of RDF per hour will produce 200,000 lb of steam/h.

---

[14] A. R. Nollet and R. H. Greeley, "Startup and Shakedown of Albany, New York Solid Waste Energy Recovery System," Transactions of ASME, Journal of Energy Resource Technology, vol 105, September 1983, p. 401.

Index

Adiabatic engine, 42
Aerodynamic drag, 294
Airplane economics, 279–81
Airplanes, 277–86
  comparison with ships, 285
  cost of operating, 282–84
  future, 284
Alaska ferries, 270
Alaska hydro power, 106–107
Alcohol (*see also* Methanol)
  ethyl alcohol, 347
  from agriculture products, 347–48
Alternatives:
  achieving equivalence, 9
  batteries, storage, 204
  electric cars, 298
  energy storage, 213–14
  geothermal power, 128–29
  postulating, 6
  ranking, 12
  solar heat, 178
  solar power, 174–76
  solar vs Lurgi coal gasification, 183
  wind turbines, 190–91
  wood fuel, 353

ARCO solar power history, 171
Automobile:
  energy destination, 294
  Stirling engine, 27
  cost of engines, 27
  efficient engines, 292
  electric, 295–98
  energy efficiency, 289–95
  friction losses, 294
  methane powered, 298
  route to energy conservation, 290, 293–95
  solar powered, 289
  testing fuel efficiency, 292–93

Batteries (*see also* Primary batteries; Storage batteries)
  alternatives, 204
  charge-discharge cycling, 196
  comparisons, 209–10
  electric vehicle, 199–200
  gas-metal, 208
  lead acid, 196–201
    cost, 199–201

Batteries (cont.)
    GNB 20-year-life Absolyte, 201
    life, 197, 199–201
    temperature effects, 200
    load leveling, 2
    low-maintenance, 197–98
    miner's cap lamp, 203
    nickel-cadmium, 201–208
        cost, 207–208
        life vs depth of discharge, 206–207
        temperature effects, 202, 204–205
    nickel-hydrogen, 208
    redox, 208
    reliability, 16, 207
    sodium sulfur, 209–210
    starting-lighting-ignition, 199
    30-year life, 198
Bicycles, 286–288
    energy consumption, 287
Binary fluid, Rankine cycle, 123–26
Biomass methane production, 93–94
Boiler feed pump:
    optimization, 6
    performance, 33
Boilers, steam, 34
Brayton cycle (see Gas turbines)
Building heat loss, calculation, 320–22
Buildings that don't need fuel, 327
Buses, passenger carrying, 275–77

Carbon dioxide in atmosphere, 130, 341–55
Cargo ship:
    characteristic hull speed, 303
    cost of operation, 306–307
    design, 303
    fuel used, 301–303
Carnot cycle efficiency, 21
Cars (see Automobiles)
Cheng/DFC combined cycle, 63
Coal, 67–74
    cost of fuels from coal, 345
    cost of mining, 72

future cost, 74
gasification, solar, 181–83
heating values, 70
mining, 70
quality, 68
reserves, 71
synthetic fuels from coal, 95
Cogeneration:
    diesel engine, 333–34
    high-rise office building, 332–33
Combined cycles, 60–65
Comparison:
    airplane and ship, 285
    airplane and train, 283
    auto weight and fuel consumed, 295
    cogeneration for office building, 332
    energy storage mechanisms, 213–14
    freight hauling, 285
    hydrofoils vs airplanes, 274
    truck and freight train, 313
    batteries, 209–210
    fuel costs for heating buildings, 329
Compressed-air energy storage, 59
Conserving energy:
    automobiles, 299
    heating, 316–17
    heating old house, 318
Construction time, Lugo 1-MW solar plant, 172
Cool Water generating station, 95
Copper-indium-selenide solar cells, 167
Cost:
    air freight vs ship, 285
    airplane flying, 282–84
    auto transportation, 290–91
    balance of system, solar, 174
    coal, future, 74
    combined cycle plants, 64
    diesel engines, 37
    freight train operation, 310
    fuels from coal, 345
    gas turbine for cogeneration, 335
    heating of buildings, 329
    hydro power, 106–11

Index

hydrogen, 344
lead-acid batteries, 199–201
natural gas, 89, 94–95
natural gas pipelines, 259
nickel-cadmium batteries, 207–208
nuclear power, 141, 145, 146
oil from Prudhoe Bay, 86–87
oil, world prices, 234
passenger buses, 276–77
passenger train, 262
petroleum, future, 76
power plant personnel, 143
pricing methods, 231
pumped hydro storage, 114–16
ship operation, 305–307
shipping oil in pipelines, 255–56
shipping oil in tankers, 257
solar energy storage, 158
solar gasification of coal, 182
solar power plant, 173–76
sources of data, 231
steam plants, 33–34
synfuel, 78
train crew and fuel, 261
transmission lines, 245
truck operation, 312–13
underground pumped hydro, 119
uranium enrichment, 134
Volkswagen cylinders, 44
water tunnel diameter, 108
wind turbines, 188–90
wood fuel, 350–51
wood fuel harvesting, 352–53
world oil prices, 234
zinc-air battery power, 219

Dams, hydro, 102–103
Demand charge, 2
Diesel engines, 36–44
  cogeneration for hospitals, 334–35
  efficiency, 42, 44
  fuel consumption, 37
  marine, 41
  ship propulsion, 268, 303
  superchargers, 39
Diesel plant, preliminary design, 8
Direct-current power transmission, 250
District heating, 331–32
Duke Power, Belews station, 33

Economic trends, predicting, 230
Economics, 229–41
  airplanes, 279–81
  discounted future expense, 237–38
  elasticity, 235–37
  law of supply, 233
  learning curves, 239–40
  power grid opportunities, 240–41
  present value in heating, 317
  ship operation, 305–307
  supply and demand, 232
Efficiency:
  airplane and ship, 285
  airplanes, 278
  automobiles, 289–95
  Carnot cycle, 21
  diesel engine, 42, 44
  energy storage, superconducting, 211
  gas and oil furnaces, 328
  geothermal power, 125
  improvement with methanol, 347
  Otto cycle engines, 47
  Rankine cycle steam plant, 30
  ship propulsion, 305
  sodium-sulfur battery, 211
  solar cells, 159, 167
  solar coal gasification, 183
Elasticity, types, 235–37
Electric automobiles, 295–98
  alternatives, 298
  nickel-cadmium battery, 208
Electric power transfers, 240
Electric power transmission, 242–53
  direct current, 250
  example network, 247–49
  grids, 240–41

Electric power transmission (*cont.*)
   identifying voltage, 250
   optimization, 245
Electrochemical energy conversion, 215–28
Energy consumption:
   bicycling, 287
   freight trains, 308–10
   highway trucks, 311–12
Energy content:
   coal, oil, gas, 67
   methanol, 347
Energy efficiency testing, auto, 292–93
Energy from garbage, 355
Energy in falling water, 98
Energy measurement, units, 17
Energy movement, pipelines and ships, 252–59
Energy storage (see also Battery), 192–214
   comparisons, 213–14
   flywheels, 212–13
   limits, 193
   measurements, definitions, 193
   pumped hydro, 111–116
   solar, 158
   superconducting, 211
Energy systems engineering, 1–16
Engines:
   diesel
   heat, 19–65
   hydrogen fueled, 345–46
   Otto cycle, 44–47
   performance limits, 20, 23
   performance measurement, 21
   Rankine cycle, 28–36
   routes to fuel efficiency, 291
   steam, 22
   Stirling, 23–27
Ethyl alcohol, production, 347–48
Evaluation:
   cogeneration for office building, 332
   power sources, 14

Failure rate:
   data sources, 7
   fuel cells, 11
   power transmission components, 251
Feedwater pumps, optimization, 6
Ferries, 267–71
Firewood (see also Wood Fuel), 328–29
Flywheel energy storage, 212–13
Fossil fuels, 66–90
Free piston Stirling engine, 26
Freight movement (see also Cargo Ships), 300–13
Freight trains, 307–10
   cost of track, 310
   energy consumed, 308
Freight trucks (see Trucks)
Friedrich heat pump, 337–40
Fuel cells, 223–28
   alternatives, 228
   efficiency, 225–26, 227–28
   failure rate, 11, 12
   4.8 MWe, 225–26
   40 kW natural gas, 226–27
   fuels, 224
   phosphoric-acid, 25
   spacecraft, 225
Fuel consumption:
   airplanes, 279–81
   buses, 275–76
   cargo ships, 301
   freight trains, 309–10
   hydrofoils, 273
   passenger ferries, 269
   passenger trains, 261
   steamship, 303
Fuel cost, in airplane ticket, 282

Gallium arsenide solar cells, 167
Garbage, methane recovery, 91–93
Gas furnace efficiency, 328
Gas turbines, 47–60
   cogeneration, 335

configurations, 48
derating, 57
developments, 51
fuel consumption, 54
performance limits, 48
performance maps, 52–56
regenerators, 50
starting, 58
Geothermal power, 119–29
alternatives, 128–29
binary cycle plant, 123–26
efficiency, 125
fluid temperature, 122–23
hot dry rocks, 128
Raft River plant, 123–26
rotary separator turbine, 126–28
Greenhouse effect, 154, 320

Hauling energy, pipes and ships, 252–59
Heat conduction, 321
Heat convection, 320–21
Heat sources for buildings, 327–40
Heat engines, 19–65
Heat exchangers:
air-to-air, 327
high-temperature, 50
Heat loss from buildings:
calculation, 322
infiltration, 325–27
through walls and ceiling, 321–22
Heat pumps, 335–40
add-on, 337–38
natural gas, 89
performance, 338–39
Heat radiation, 320
Heat rate, Belews station, 33
Heat recovery, marine engine, 307
Heating buildings, 314–40
fuel costs, 329
high-temperature heat sources, 327–28

limits, 315
low-temperature sources, 330
with power plant losses, 331–34
with solar energy, 177
Heating strategies, 315, 317
Heating value:
coal, 70
gasoline, 45
wood, 2
Heber geothermal plant, 125–26
Helms pumped hydro project, 111–17
Human powered vehicles, energy consumption, 287
Hydro power, (see also pumped hydro) 97–119
cost, 106–107, 109–11
dams, 102–103
energy in water, 98
estimating power resource, 101
losses, 102
measurements, 98
permits, 112
river flow, 99–100
small plants, 99
Sultan River project, 107–11
turbines, 103–105
Hydrofoils, 272–75
Hydrogen, 341–46
characteristics, 342–43
cost, 344
engines, 345
production, 343–45
storage, 342
storage heat recovery, 342–43
vehicles, 342, 345–46
Inflation rate, estimating, 12
Insulation, heat:
optimization, 318–20
value of materials, 323
Interest rate, estimating, 12
Intermodal freight transport, 313
Intersociety Energy Conversion Engineering Conference, 7

Kettle Falls wood-burning power plant, 354
K value, heat conduction, 321

Landfill methane, 92
Learning curve, 239–40
Life cycle cost, computation, 9
Life, spacecraft batteries, 206–207
Limits:
   airplanes, 278
   electric automobiles, 296
   electric power transmission, 243, 245
   energy used in heating buildings, 315
   ferries, 268
   freight trains, 308
   heat pumps, 336
   heating buildings, 315
   hydrofoil ships, 272–75
   moving people, 260
   oil tankers, 256
   passenger ships, 265
   sailplanes, 288
   storage batteries, 194
   train speed, 262
Load leveling evaluation, 2
Lugo 1-MW solar power plant, 171

MAN B & W diesel engine, 304
Measurements:
   energy storage, 193
   engine performance, 21
   heating of buildings, 316
   hydro power, 98
   nuclear power, 131
   oil, gas, and coal energy, 67
   petroleum, 75–76
   solar, 152, 155
   solar-cell performance, 161
   storage batteries, 194
   wind power, 184
Methane (see also natural gas):
   automobile fuel, 94, 298
   from cattle manure, 92
   from garbage, 91–93
   from plants, 93–94
   separation from impurities, 91
Methanol fuel, 346–49
   from wasted methane, 348–49
   production from coal, 96
   purification, 348
Moving energy, 242–59
Moving freight, 300–313
Moving people, 260–99

Napier grass for methane, 93
Natural gas, 87–90
   cost, 89
   economics, 94–95
   energy content, 89
   formation, 89
   fuel cell, 227–28
   pipelines, 258–59
   shipping as liquid, 259
   shipping from Alaska, 90, 258–59
   uses, 90
Nickel-cadmium batteries, 201–208
Non-fossil fuels, 341–55
Nuclear energy for autos, 299
Nuclear power, 130–150
   boiling water reactors, 137
   breeder reactors, 149
   Candu reactor, 140
   cost, 141
   cost, foreign plants, 146
   fission energy, 132
   fuel cost, 143
   fusion, 131, 150
   high-temperature gas-cooled reactor, 138
   limits, 130
   measurements, 131
   plant availability, maintenance, 145
   pressurized water reactors, 135
   safety, 139
   space use, 168

Index

Trojan plant, 136
uranium enrichment, 133–34
waste reprocessing, 147–49
nuclear ship, Savanah, 230

Ocean thermal gradient power, 117–19
Oil:
  furnace efficiency, 328
  pipelines, 253–56
  reserves, 88
  tankers, 256–57
Oil wells (see petroleum)
Optimization:
  electric power transmission, 245
  heat insulation, 318–20, 322–25
  heating a house, 317–18
  solar concentrators, 179
Otto cycle engines, 44–47
  efficiency, 47

Pacific Gas and Electric:
  generation options, 230
  Geysers geothermal plant, 128
  Helms pumped hydro project, 111–17
  power source scenarios, 14
Passenger buses, 275–77
Passenger ships, 264–75
  ferries, 267–71
  France, 266
  fuel consumption, 266
  hydrofoil, 272–75
  Queen Mary, 265
  speed, 265
  unmanned engine room, 270
Peaking power plants, 58
Permits and approvals, hydro power, 112
Petroleum, 75–88
  artificial lift, 82
  cost of Prudhoe Bay oil, 86–87
  drilling oil wells, 80
  enhanced recovery, 84

exploratory drilling, 79
finding, 78
future cost, 76
limits, 75
measurements, 75–76
refining, 85
reserves, 88
secondary recovery, 83
Photovoltaic energy converters (see also Solar cells), 160
Piggyback freight train, 313
Pipelines:
  natural gas, 258–59
  oil, 252–56
Porsche, Dr. Ferdinand, 290
Postulating alternatives, 6
Power grids, 240
Power plant:
  availability, nuclear, 145
  coal consumption, 68
  cost, nuclear, 142, 145
Preliminary design, extent of, 7
Present value calculation, 237–39
Pressurized water reactor, 135
Pricing methods, 231
Primary batteries, 216–22
  aluminum, 220
  cost, zinc-air, 219
  definitions, 216
  gas-metal battery, 222
  high-performance, 220
  limits, 216
  lithium, 221
  zinc-air cell, 217–219
  zinc-bromine, 222
  zinc chlorine, 222
Progressive generation, 65
Prohibited approaches, 348–49
Propulsion power:
  freight trains, 309–10
  ships, 301
Prudhoe Bay oil, cost, 86–87
Pumped hydro energy storage, 111–19
  alternatives, batteries, 117

Pumped hydro energy storage (*cont.*)
  alternatives, gas turbines, 116–17
  cost, 114–17
  underground, 117–19
Pyrheliometer, 156

Railroad train (see Freight trains, Passenger trains)
Rankine cycle engines, 28–36
  binary fluid, 123–26
Ranking of alternatives:
  example, 13
  Pacific Gas and Electric power sources, 14
  non-quantifiable factors, 13
Reliability:
  airplanes, 283–84
  battery, 16
  diesel engines, 37
  electric power transmission, 247–49
  failure rates, 5
  MTBF and MTTR, 5, 15
  nickel-cadmium batteries, 207
  prediction, 15
  probability of success, 15
  required redundancy, 9
  sources of failure-rate data, 7
  transmission line components, 251
Requirements:
  perturbing, 4
  real, derived, assumed, 4, 6
  reliability, 5
Reserves, coal, 71
Rolls Royce RB211-535 gas turbine, 52–56
Roosevelt Hot Springs geothermal plant, 128

Sailing ships, 307
Sailplanes, 288
Sasol synthetic fuel plant, 96
Sawmill waste, 354

Sensitivity, wood fuel cost to production, 353
Ships, passenger carrying, 264–75
Ships, sailing, 307
Solar arrays:
  cost, 171
  design, spacecraft, 169
  design, terrestrial, 170
  output variation, 172
  sun-following benefit, 172
Solar cells, 159–169
  connecting and installing, 165
  copper-indium-selenide, 167
  effect of temperature, 164, 170
  efficiency improvement, 167
  gallium arsenide, 167
  performance, measurement, 161
  physics, 161
  power output, 163
  silicon, manufacture, 161
  spacecraft, 167
  spectral response, 154, 162
  volt-ampere curve, 164, 165
Solar coal gasification, 181
Solar concentrators, optimization, 179
Solar energy:
  cost of storing, 158
  future, 180
Solar heat collection, 176
  buildings, 177
  transporting heat, 179
Solar heated engines, 178–79
Solar heating, alternatives, 178
Solar hot water, Virginia buildings, 177
Solar intensity, 154
Solar measurements, 152, 155
Solar power, 161–83
  comparison with coal, 174
  limits, 152
  Lugo substation, 171
  plant cost, 172–74
  Ranch Seco 1.2 MW plant, 172
  seasonal variation, 172
Solar power satellite, 169

Index

Solar powered airplanes, 289
Solar powered cars, 289
Solar radiation:
  greenhouse effect, 154
  map, U.S., 157
  spectrum, 153
Space nuclear power, 168
Spacecraft:
  batteries, life vs depth of discharge, 206–207
  energy storage, flywheels, 213–14
  fuel cells, 225
  solar cells, 167
St. Lucie nuclear plant, 144
Standby power, 41
Steam power plant:
  improvements, 36
  efficiency, 30, 32
  temperatures and pressures, 31
Stirling engines, 23–27
  cost, 27
  development problems, 25, 26
  heater temperatures, 27
  rice hull fuel, 25
Storage batteries (see also Batteries) 193–211
  applications, 195
  cycling life, 196
  measurements, 194
Subsidy:
  passenger buses, 276
  passenger trains, 264
Sultan River hydro project, 107–11
Synthetic fuel from coal, 95–96

Tax benefits, solar power, 175
Tidal power, 117–19
Trains, passenger, 261–64
  high-speed, 264
  propulsion power, 263
Trans Alaska pipeline, 253–54
Transmission line voltage, 243
  recognizing voltage, 250

Transportation, passengers, 260–99
Trolley buses, 277
Trucks:
  energy consumption, 311
  freight hauling, 310–13
Tunnel boring machine, 108
Turbines, hydro, 103–105
Turbines, rotary separator, 126–28

Units:
  cargo ships, 301
  electric power transmission, 244
  energy measurement, 17
  energy movement, 253
  heat pumps, 337
  heating buildings, 316
  transportation of people, 261
Uranium enrichment, 133

Ward Leonard drive for ferries, 268
Washington Public Power Supply System, 1, 229
Washington Water Power Co., wood burning plant, 354
Wind power, 183–91
  alternatives, 190–191
  cost, 188–90
  farm windmills, 184
  limits for extraction, 184
  measurements, 184
  small, 188
Wind velocity distribution, 186
Wood fuel:
  cost, 350, 352–53
  for power generation, 349–54
  forest area required, 354
  forest management, 349, 351
  from forest waste, 350–51
  recovery equipment, 351
Wood, heating value, 2